T0317608

MOLECULAR MODELING OF CORROSION PROCESSES

THE ELECTROCHEMICAL SOCIETY SERIES

ECS-The Electrochemical Society
65 South Main Street
Pennington, NJ 08534-2839
http://www.electrochem.org

A complete list of the titles in this series appears at the end of this volume.

MOLECULAR MODELING OF CORROSION PROCESSES

Scientific Development and Engineering Applications

CHRISTOPHER D. TAYLOR

Strategic Research & Innovation, DNV GL, Dublin, OH, USA

Fontana Corrosion Center, Department of Materials Science and Engineering, The Ohio State University, Columbus, OH, USA

PHILIPPE MARCUS

Institut de Recherche de Chimie Paris/Physical Chemistry of Surfaces, Chimie ParisTech-CNRS, Ecole Nationale Supérieure de Chimie de Paris, Paris, France

The Electrochemical Society Series

Library of Congress Cataloging-in-Publication Data:

Molecular modeling of corrosion processes : scientific development and engineering applications / edited by Christopher D. Taylor, Philippe Marcus.
  pages   cm
Includes index.
ISBN 978-1-118-26615-1 (cloth)
1. Corrosion and anti-corrosives.   2. Corrosion and anti-corrosives–Mathematical models.
  I. Taylor, Christopher D., 1978– editor.   II. Marcus, P. (Philippe), 1953– editor.
  TA418.74.M656 2015
  620.1′1223–dc23

                              2015000031

Set in 9.5/11.5pt Times by SPi Publisher Services, Pondicherry, India

Printed in the United States of America

10 9 8 7 6 5 4 3 2 1

1   2015

CONTENTS

LIST OF CONTRIBUTORS

Dominique Costa, Institut de Recherche de Chimie Paris/Physical Chemistry of Surfaces, Chimie ParisTech-CNRS, Ecole Nationale Supérieure de Chimie de Paris, Paris, France

Michael F. Francis, École Polytechnique Fédérale de Lausanne (EPFL), STI IGM LAMMM, Lausanne, Switzerland
Brown University, School of Engineering, Providence, Rhode Island, USA

Julian D. Gale, Department of Chemistry, Nanochemistry Research Institute, Curtin University, Perth, WA, Australia

Edward F. Holby, Materials Science and Technology Division, MST-6, Los Alamos National Laboratory, Los Alamos, NM, USA

Gang Lu, Department of Physics and Astronomy, California State University Northridge, Northridge, CA, USA

Noelia B. Luque, Institute of Theoretical Chemistry, Ulm University, Ulm, Germany

Philippe Marcus, Institut de Recherche de Chimie Paris/Physical Chemistry of Surfaces, Chimie ParisTech-CNRS, Ecole Nationale Supérieure de Chimie de Paris, Paris, France

Alexis Markovits, Laboratoire de Chimie Théorique, Université Pierre et Marie Curie—CNRS, Paris, France

Vincent Maurice, Institut de Recherche de Chimie Paris/Physical Chemistry of Surfaces, Chimie ParisTech-CNRS, Ecole Nationale Supérieure de Chimie de Paris, Paris, France

Christian Minot, Laboratoire de Chimie Théorique, Université Pierre et Marie Curie—CNRS, Paris, France

Steve Policastro, Center for Corrosion Science and Engineering, Chemistry Division, Naval Research Laboratory, Washington, DC, USA

Paola Quaino, PRELINE, Universidad Nacional del Litoral, Santa Fé, Argentina

Elizabeth Santos, Institute of Theoretical Chemistry, Ulm University, Ulm, Germany
Faculdad de Matemática, Astronomía y Física, IFEG-CONICET, Universidad Nacional de Córdoba, Córdoba, Argentina

Wolfgang Schmickler, Institute of Theoretical Chemistry, Ulm University, Ulm, Germany

Hans-Henning Strehblow, Institute of Physical Chemistry, Heinrich-Heine-Universität, Düsseldorf, Germany

Christopher D. Taylor, Strategic Research & Innovation, DNV GL, Dublin, OH, USA Fontana Corrosion Center, Department of Materials Science and Engineering, The Ohio State University, Columbus, OH, USA

Xu Zhang, Department of Physics and Astronomy, California State University Northridge, Northridge, CA, USA

FOREWORD

There has been an explosion in the application of atomistic and molecular modeling to corrosion and electrochemistry in the past decade. The continued increasing computational power has allowed the development and implementation of atomistic and molecular modeling frameworks that would have been impractical even a short time ago. These frameworks allow the application of fundamental physics at the appropriate scale on assemblies of atoms of a size that provides a more realistic basis than ever before. In some cases, that level is the determination of the electronic structure based on quantum mechanics. Such is the case when determining the energetics of surface structures and reactions. In other cases, the appropriate scale requires the forces between atoms or ions to be calculated, and the effects those forces have on the configuration of atoms and how it changes with time. Surface and solution diffusion are prime examples.

This first edition of *Molecular Modeling of Corrosion Processes* will be a foundational piece of work, representing the state of the art at this point in time. The publication of this book occurs at an important time in the science of corrosion as the recent results and promise of computational methods have caught the imagination of many in the field. The authors of the chapters in this book have demonstrated the powerful information and understanding that well-designed calculations can provide on intrinsically complicated processes, such as those involved in corrosion. The pitfall of such excitement can be expectations that get too far ahead of reality, leading to disappointment and eventual setting back of the field. The promise needs to be balanced by an understanding and appreciation of the current ability of the methods as well as the likelihood of near-term advancements. This book provides the right balance in its descriptions of the successes, opportunities, and challenges of molecular modeling in the context of corrosion science. In particular, the emphasis on the need for well-designed experiments to provide both input data that cannot be calculated and validation data to assess the accuracy of models is a recurring theme. A marriage of experiment and modeling is necessary.

Part of the power of molecular modeling lies in its ability to isolate aspects of a phenomenon in ways that are simply not possible by experiment. The effects of bond energy on the dissolution or surface diffusion allows one to "turn off" surface diffusion, for example, to quantitatively determine what effect it has on the dissolution rate and the resulting nanostructure. This example is one of many in which the dependence of critical parameters on the atomistic and molecular composition, as well as the local structure, including defects, can be determined. The insights that such calculations can provide into the overall thermodynamic, mechanical, and kinetic properties of a system are substantial.

This edition of *Molecular Modeling of Corrosion Processes* provides an overview of some of the critical questions that are being addressed. These questions include those surrounding surface reactivity and the complexities inherent in the interactions of surfaces with an aqueous environment as described in Chapter 2. Although most corrosion works focus on dissolution, no corrosion can occur without cathodic reactions consuming the electrons so liberated. Chapter 3 focuses on understanding one of the most important cathodic reactions, hydrogen evolution, using methods that are extendable to other cathodic reactions, including oxygen reduction. As mentioned earlier, not all questions involving corrosion are best addressed at the quantum level. The factors controlling the rate of dealloying and the resulting surface morphology are best probed at the atomistic level, as described in Chapter 4. Chapter 5 shows that the range of mechanisms by which organic molecules can inhibit corrosion can be addressed with quantum chemistry. In many instances, bare metal surfaces are exposed to the environment, and the understanding of what occurs under those conditions is important in localized corrosion and environment-assisted cracking. Chapter 6 addresses the utility of electronic structure techniques to calculate surface phase diagrams. Nanometer-thick oxide films are often the only barrier preventing high rates of metal dissolution; without them, many metallic structures could not survive. Chapter 7 reviews work on the well-recognized, but poorly understood, effects of halide ions in attacking these oxides locally. The last chapter, Chapter 8, presents a compelling argument that the use of *ab initio* multiscale modeling approaches can provide quantitative insights into proposed models for hydrogen embrittlement.

Molecular Modeling of Corrosion Processes has the ambitious goal of describing the state of the art in a field in its infancy. There is some danger in such an endeavor, but the contributors do an outstanding job in covering the wide range of corrosion issues that exist, the wide array of molecular modeling techniques that have been developed, and then demonstrating where the application of the latter to the former provide heretofore unobtainable insights into corrosion processes.

PREFACE

Molecular modeling and computational materials science have the potential to transform the way chemicals and materials are synthesized and built by providing a "virtual laboratory" for the testing of materials properties, including their response to the environment, before engaging in any of the financial, safety, or environmental risks that accompany the bench-top synthesis of a new material. Furthermore, a sophisticated virtual laboratory can be used to predict the behavior of materials currently in service in new environments, or modifications of old environments, providing scientists and engineers with an additional tool for risk assessment prior to making changes in operating procedures or conditions. Finally, molecular modeling and computational materials science provide a "third eye" into the mechanisms of materials transformation that occur in the nanoscale world, the fundamental landscape in which many materials/environment effects take place, but only few contemporary measurement techniques can access, and even then, only under well-controlled conditions. For these reasons, the modern virtual laboratory is bringing about a scientific revolution, and each day new publications appear in the literature, illustrating how molecular modeling uncovers unprecedented insights into the inner workings of processes in both chemistry and materials science or predict, for the first time, the properties of new compounds.

Some of these papers are tackling the difficult problem of materials corrosion. Corrosion, at its core, is the exchange and reaction of electrons, atoms, or molecules between a material and its environment; and hence, it is highly predisposed to a molecular modeling approach. Yet, corrosion, despite its ubiquity and long history of study, remains a fiercely challenging problem. It bridges topics of metallurgy; solid-state physics; and physical, inorganic, and organic chemistry. It is controlled by processes that occur not only at the atomic length scale, but at the nanoscale, mesoscale, microscale, and beyond. Corrosion may be a fleeting, transient event, but lifetimes of materials must often be assessed for years, decades, or millennia. But it is also an urgent problem—corrosion costs industrial and agrarian economies hundreds of billions of dollars each year, some have estimated trillions, creating along with it significant human and environmental costs—and, therefore, demands the best of our scientific theoretical and experimental efforts to be applied. This book highlights some of the areas in which molecular modeling and computational materials science have been applied to this endeavor, and tries to shine a light ahead toward future fields of application in corrosion science and to the techniques that still need to be developed to take us there.

Two of the biggest challenges facing the molecular modeler in any field of inquiry are "which technique do I use?", and "how do I simplify my complex materials system into the key thousand or so atoms that represent the active site of interest?" This book, therefore, provides a number of case studies that demonstrate how scientists over the past few years have found answers to these questions, and at what success they arrived. Some of the time- and length-scales surrounding corrosion phenomena are too big for a simplistic atomistic approach; but by judicious choice of representation or the use of multiscale modeling, reasonable approximations can be obtained and an important complementary viewpoint to the field of experimental corrosion science provided.

As a consequence, this is a book that should be read by students interested in beginning a career in corrosion modeling, by corrosion scientists eager to add another tool to their research group, by researchers already engaged in modeling but not yet aware of the opportunities that exist in the world of corrosion, or by project leaders in industry who may be curious about investing in molecular modeling research but uncertain as to what value such a project could provide. Through interacting over the years with professionals in each of these categories, it became apparent that a resource of this nature was in demand, and could help encourage the growth of modeling efforts in the field of corrosion science.

Two recent workshop series have been initiated that have also greatly helped to promote the use of molecular modeling to critical corrosion problems facing industry today: the Quantitative Micro-Nano workshops in stress corrosion cracking, principally organized by Dr. Roger W. Staehle, and the International Winter School on Corrosion Modeling, co-organized by one of the editors of this book, Dr. Philippe Marcus. Both of these events have worked to foster a sense of community amongst those scientists who commonly think about corrosion science problems from an atoms and bonds viewpoint, experimentally and theoretically.

The first chapter of this book was inspired by the presentations given by Professors Julian Gale, Philippe Marcus, and Christopher (Chris) Taylor at the Winter School, which was held in Saclay, France, in December 2012.

The remainder of the book was assembled through soliciting contributions from colleagues who we have come to know through workshops, conferences, and academic collaborations over the years. Their hard work on these chapters is gratefully acknowledged. Thanks to all the contributors the present volume spans topics ranging from the initial stages of oxidation of metal surfaces through to the kinetic simulation of metal dissolution, and the mechanisms of inhibition through to hydrogen embrittlement. Their collective breadth of expertise speaks to the necessity of engaging experts across the fields of chemistry, physics, and materials science and engineering for the challenges corrosion presents.

The editors would like to personally thank colleagues who have provided encouragement along the road to completing this volume: Rob Kelly and Matthew Neurock at the University of Virginia, Scott Lillard at the University of Akron; Rudy Buchheit and Jerry Frankel at the Ohio State University; and Mark Paffett at Los Alamos National Laboratory.

Los Alamos, NM CHRISTOPHER D. TAYLOR, PH.D.
August 21, 2013

Paris France PHILIPPE MARCUS
September 21, 2013

1

AN INTRODUCTION TO CORROSION MECHANISMS AND MODELS

CHRISTOPHER D. TAYLOR[1,2], JULIAN D. GALE[3], HANS-HENNING STREHBLOW[4] AND PHILIPPE MARCUS[5]

[1]*Strategic Research & Innovation, DNV GL, Dublin, OH, USA*
[2]*Fontana Corrosion Center, Department of Materials Science and Engineering, The Ohio State University, Columbus, OH, USA*
[3]*Department of Chemistry, Nanochemistry Research Institute, Curtin University Perth, WA, Australia*
[4]*Institute of Physical Chemistry, Heinrich-Heine-Universität, Düsseldorf, Germany*
[5]*Institut de Recherche de Chimie Paris/Physical Chemistry of Surfaces, Chimie ParisTech-CNRS, Ecole Nationale Supérieure de Chimie de Paris, Paris, France*

1.1 INTRODUCTION

The history of mankind is distinguished by the pronounced effort to understand the processes of nature and to manipulate these processes for the improvement of the human condition and survivability. One remarkable instance of this effort can be found in the realm of materials engineering; here, advances have been so significant that historical eras are frequently named by the materials that characterized them: the Stone Age, Bronze Age, and Iron Age [1]. Materials engineers in our present age have a significant number of alloy components and fabrication techniques at their disposal. Consequently, there is a vast range of properties for which materials can be tailor-made, based on considerations such as materials lifetime, strength, ductility, and temperature range. In the world of metals, corrosion is an ever-present concern, and there are an often bewildering number of modes via which materials failure by corrosion may occur, such as localized corrosion (pitting), stress corrosion

Molecular Modeling of Corrosion Processes: Scientific Development and Engineering Applications, First Edition.
Edited by Christopher D. Taylor and Philippe Marcus.
© 2015 John Wiley & Sons, Inc. Published 2015 by John Wiley & Sons, Inc.

cracking, galvanic corrosion, crevice corrosion, uniform corrosion, or hydrogen embrittlement [2]. As in past ages, corrosion scientists and engineers of today must confront such problems by conceptualizing the modes *via* which these effects may occur, leading to direct testing of these hypotheses through a combination of modeling and experiment. Ultimately, the objective is to provide solutions, such as ever more durable materials and/or processes for current and future applications.

In the precomputation era, materials modeling was restricted to mechanical (i.e., physical, tangible) models, hypothetical *Gedanken* experiments, or analytic calculations based on the continuous phenomena known to materials physics. However, it is known that physicochemical processes such as corrosion ultimately occur as discrete, atomistic events that involve the making, breaking, or rearrangement of bonds between atoms, such as dissolution, substitution, and diffusion. To date, kinetic models for these processes have been fitted to experimental data in an effort to understand their relative importance and hence develop predictive models for corrosion rates. However, the ability to extrapolate such models to as yet unstudied conditions, in which case the mechanisms of corrosion may be subtly (or drastically) changed, is hindered by the lack of a first-principles justification for the values obtained from such fitting experiments. Empirical *rules of thumb* may be useful in, for example, identifying the relative importance of such alloying ingredients as chromium, molybdenum, or nitrogen to the corrosion resistance of a series of iron-based alloys under a given set of conditions [3]. However, such rules do not provide the fundamental insights as to the mechanism *via* which these effects occur, as would be necessary to significantly advance materials development. In an age where the ability to control materials design at the nanoscale is beginning to appear feasible, it is important to develop the theoretical tools to guide such low-level design effects and to reach levels of corrosion resistance optimization previously considered unthinkable.

Molecular modeling provides the ability to simulate and analyze hypothetical processes at the atomic level [4]. Such models have been used extensively in the realm of chemistry [4, 5], as well as in solid-state physics and materials science [6, 7]. Now, we have progressed to the point where the conjunction between chemistry and solid-state physics, namely, interfacial science, is also being tackled by such methods [8]. While heterogeneous catalysis initially dominated this particular application of molecular modeling to interfacial systems, the field of corrosion has recently been attracting more and more attention. This increase in interest can be seen from the analysis in Figure 1.1, which shows that the current interest in applying molecular modeling techniques (highlighted here as *molecular dynamics* and *density functional theory (DFT)*) to problems in corrosion science is roughly equivalent to the surge of interest that spurred the application of these techniques to problems in catalysis at the beginning of the twenty-first century.

Molecular modeling of corrosion, by definition, requires the construction of a model system, which consists of atoms and molecules, that is as faithful as possible a representation of the corroding system of interest. Since corroding systems contain an extraordinary degree of complexity—encompassing features such as: the microstructure of the material; the elemental composition; multiple phases including metals, alloys, and oxides; surface–solute–solvent interactions; and electrochemical interfaces—the construction of an appropriate model therefore requires an advanced understanding of corrosion mechanisms. As shown in Figure 1.2, interfacial processes such as corrosion lie at the intersection of chemistry and materials science. A particular challenge for molecular modeling is that, at least for present computers, only a finite number of atoms and molecules (somewhere between dozens and, at best, billions) can be simulated [9–11]. Thus, one must be especially judicious in choosing the

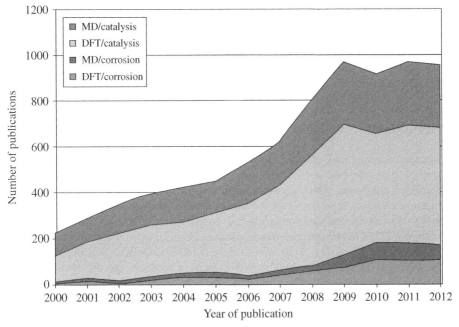

FIGURE 1.1 Number of peer-reviewed publications with topics *molecular dynamics* or *density functional theory* combined with either *catalysis* or *corrosion*. Data obtained from the Web of Knowledge database by Thomson Reuters, 2013. https://access.webofknowledge.com/.

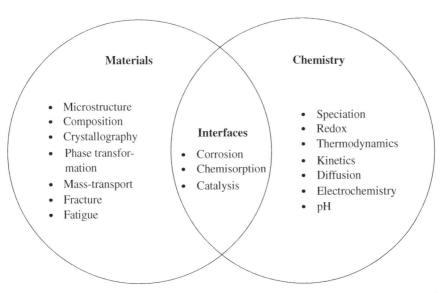

FIGURE 1.2 Interfacial problems, such as corrosion, occur at the intersection between materials science and chemistry.

representation of the corroding system (recall that even a billion atoms of Fe weigh $<1 \times 10^{-10}$ mg). The challenge is, therefore, significant; how does one model a macroscopic and inherently multiscale process such as corrosion using molecular modeling techniques? Approaches to tackle this challenge form the scope of this book. We begin in this chapter by introducing a series of corrosion mechanisms and then discuss the principal modeling techniques used to represent and simulate these systems from the molecular perspective.

1.2 MECHANISMS IN CORROSION SCIENCE

The conversion of metals back into their native oxide forms, or some similarly preferred thermodynamic state, can take many forms and can be exacerbated by various chemical or environmental conditions. Thus, corrosion itself has no single mechanism of action [12]. Various mechanisms have been proposed to account for effects such as atmospheric corrosion [13–15], chloride-induced pitting [16–25], sulfidation [26–40], acidic dissolution, crevice corrosion [13, 41–44], stress corrosion cracking [45–49], ammoniacal attack [18, 19, 23–25, 29–31, 33, 50], and so on. In order to make the following discussion as general as possible and to avoid going into detail regarding some of the more specific corrosion instances that will be addressed in later chapters, we here take the liberty of dividing corrosion processes into a number of sequential (although they often occur in parallel) steps, each of which may be subject to chemical and/or environmental modification. By categorizing the leading causes and mechanistic pathways of corrosion in this way, we will then show how, in each case, molecular modeling could feasibly be applied to provide greater insights into the fundamental processes and hence guide next-generation corrosion mitigation strategies and the design of corrosion-resistant materials.

The Collins English Dictionary [51] defines corrosion as:

> a process in which a solid, especially a metal, is eaten away and changed by a chemical action, as in the oxidation of iron in the presence of water by an electrolytic process.

Corrosion is a chemical process that occurs at the solid/environment interface, that is, at the external or internal surface of a material where it meets a fluid (gaseous or solvent) phase. The sciences appropriate to the study of this process include: surface science; electrochemistry; physical, inorganic, and analytical chemistry; physics; materials science; metallurgy; and, finally, theory, which encompasses both modeling and simulation. The chief reactions that contribute to aqueous corrosion are electrochemical in nature and, for this reason, can be broken down into anodic metal dissolution and the corresponding cathodic reactions of hydrogen evolution and oxygen reduction [3]. Factors that may affect these reactions, and thus the overall process of corrosion, include the structure and composition of the bulk solid phase, as well as the composition of the electrolyte and the changes that these subsystems undergo during the process of corrosion.

This breakdown is given further elaboration in the list of topics that accompanies the Venn diagram in Figure 1.2. Small changes in materials microstructure, alloy composition, and mass-transport pathways (such as the *short-circuit* diffusion routes in passive oxide films) [52–57] can be particularly significant for the overall corrosion properties of the material. Similarly, phase transformations that occur as a function of temperature, or the impurity content at either grain boundaries or in the bulk, may again affect the resilience of a material to corrosion, especially when the material is also placed under mechanical strain (such as in instances of stress corrosion cracking). Even factors such as the materials texture [58], grain-boundary engineering [55, 59, 60], and the crystallographic orientations that are exposed at the surface can affect its reaction with the environment.

From the chemistry side, the environmental conditions can be highly important. The pH may affect speciation of organic or mineral acid moieties, thus changing their propensity for surface reaction. Variations in the electrode potential (that can be induced galvanically, from the materials side, or chemically, via ionic concentration gradients in the electrolyte) can play a large role in biasing a system toward or away from passivation. The myriad of physical and chemical interactions that go into surface adsorption, bond breaking and bond formation, and solvation by the solution play a continuous role in mediating the overall thermodynamics and kinetics of corrosion reactions [61, 62]. Chemical species, such as inhibitors, can be introduced to bias the surface chemistry such that corrosion kinetics are subdued, whereas other chemical species (such as halides) can accelerate them [25b, c, 32, 38, 63, 64].

These two features of corrosion—materials control and environmental control—meet in the very processes that occur at the interface, which is the critical region where corrosion reactions take place, but is also the least understood [65–68]. Hence, there is a strong motivation for developing the theoretical framework via which these reactions can be simulated, predicted, and, ultimately, controlled.

1.2.1 Thermodynamics and Pourbaix Diagrams

The starting point for the scientific understanding of chemical processes is thermodynamics. For a metal subject to aqueous, that is, electrochemical, corrosion, the scheme introduced by Marcel Pourbaix provides the framework for a first assessment of materials stability [69]. Pourbaix considered that the Gibbs free energy associated with an electrochemical reaction is a function of two key variables: the electrode potential and the pH. These two variables can be used to define the Gibbs free energy of the dissolution, oxidation, hydroxylation, and hydride reactions. Since these are the primary products of corrosion under aqueous conditions, a broad range of corrosion phenomena can be surveyed visually with Pourbaix's approach. This visual aspect is conveyed via the *Pourbaix potential–pH diagram*—a phase diagram that outlines the most stable phase (as defined by the phase with the lowest Gibbs free energy) in the two-dimensional space limited by reasonable values that can be expected for the pH and potential (typically between −2 and +2 V vs. normal hydrogen electrode (NHE) and pH 0–14). As an example, the Pourbaix diagram for iron is provided in Figure 1.3.

The electrochemical model for corrosion breaks the process down into anodic and cathodic *half-cell* reactions, each having their own equations, which may include:

For dissolution:

$$M \rightarrow M^{z+} + 2e^-$$

For oxygen reduction:

$$O_2 + 4H^+ + 4e^- \rightarrow 2H_2O$$

For hydrogen reduction:

$$2H^+ + 2e^- \rightarrow H_2$$

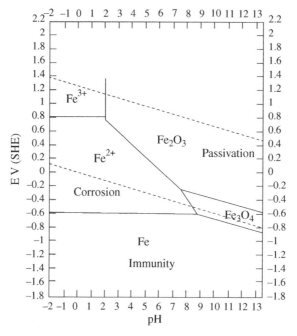

FIGURE 1.3   The Pourbaix diagram for iron. Adapted from Strehblow and Marcus [70].

Reactions that involve a number of protons or hydroxide ions being exchanged are dependent upon pH, via the equation [71]

$$\Delta G \approx \Delta G^0 \pm 2.303 mRT\, \mathrm{pH}$$

where m is the number of protons exchanged in the reaction and the $\pm$ depends on whether protons are produced or consumed, respectively. T is the temperature, R is the ideal gas constant, and ΔG^0 is the standard Gibbs free energy (activity $a = 1$, pressures of gases $p = 1$ atm $= 1.013$ bar).

Reactions that involve the loss or gain of electrons (oxidation or reduction, respectively) are dependent upon the potential. The reaction free energy varies with potential according to the equation [70]

$$\Delta G = -nFE$$

where n is the number of electrons consumed for the cathodic direction of the reaction, that is, the reduction process, and ΔG the related change of the Gibbs free energy. The counter reaction for the production of the electrons is by convention the oxidation of hydrogen gas to hydrogen ions for the standard hydrogen electrode with a hydrogen pressure of $p(H_2) = 1$ atm $= 1.013$ bar and a hydrogen ion activity of $a(H^+) = 1$. The following two equations give an example for the reduction of metal ions Me^{n+} and the compensating hydrogen oxidation:

$$M^{n+} + ne^- \rightarrow M$$

$$\frac{n}{2}H_2 \rightarrow nH^+ + ne^-$$

For a reactive metal like iron, the discussed electrodeposition associated to hydrogen oxidation is not possible with a strongly positive ΔG. The reverse process, that is, iron dissolution associated to hydrogen evolution, is the spontaneous reaction with a very negative ΔG. E is the electrode potential measured against the standard hydrogen electrode. With $a = 1$ for all activities and $p = 1$ atm for gas pressures, the electrode potential equals its standard value $E = E^0$ for the electrochemical reaction of interest. For these conditions the relation $\Delta G^0 = -nFE^0$ holds with the standard Gibbs free energy change ΔG^0 holds. The standard electrode potentials E^0 are listed for all metal–metal-ion and redox electrodes [71b] similar to the standard Gibbs free energy ΔG^0 of electrochemical reactions. For Fe dissolution to Fe^{2+}, the electrode potential equals $E^0 = -0.44$ V, a relatively negative value for a reactive metal.

Reactions that are only dependent upon pH are represented in the Pourbaix diagrams by vertical lines. Reactions that are only dependent upon potential are marked as horizontal lines. Reactions that involve both will have some slope that depends jointly upon the ratio between proton/hydroxide activities and electron transfer coefficients, in which case both of the above equations have to be applied.

As seen in Figure 1.3, the typical Pourbaix diagram consists of three regions, corresponding to immunity (the metallic phase has the lowest free energy), passivation (the oxide phase has the lowest free energy), and corrosion (either the metal cation phase is most stable or the soluble hydroxides/oxyanions). The electrochemical boundaries associated with the oxidation and reduction of water are also typically shown as boundary points within these diagrams via the use of dotted or dashed lines.

As described in more detail in subsequent chapters of this book, the Pourbaix principle has been extended to include adsorbed surface phases and other variables, including the presence of sulfides or chlorides [62, 73–76]. The analog for high-temperature corrosion by gaseous agents is the Ellingham diagram, in which the free energy associated with oxidation reactions is plotted against temperature [77]. However, as with any thermodynamic analysis, one should keep in mind that the kinetics often play a crucial role in the observed behavior. For instance, Anderko and Shuler's study of iron sulfide phases indicated that many metastable phases are significant and should be displayed in the practical Pourbaix diagram due to the slow kinetics associated with the formation of the most stable thermodynamic phase [78].

1.2.2 Electrode Kinetics

While thermodynamics provides a starting point, kinetics is essential for providing any corrosion model of practical utility. The term *electrode kinetics* is often used as, in the electrochemical paradigm, the oxidation and reduction occur at independent sites, which can be considered as separate electrodes marking a solid/electrolyte interface at which the half-cell reactions take place. In the case of chemical corrosion, these half-cell reactions can take place at the same location, in which case there is no external current flow between the half-reaction centers, but instead, direct charge transfer between the reactants via electronic contact at the same metal site.

The electrode potential exerts a powerful control over corrosion kinetics, just as the chemical potential or the electrochemical potential does in thermodynamics. The deviation of the electrode potential E from its equilibrium value E_{eq} given by the Nernst equation, the so-called overvoltage $\eta = E - E_{eq}$, is the driving force for the kinetics of the electrode process. This kinetic control can be broken down into the following categories:

1. Transport of reactants and products, controlled by the diffusion overvoltage
2. Chemical reaction rates, controlled by the reaction overvoltage
3. Adsorption and desorption on the electrode surface with charge transfer, controlled by the charge transfer overvoltage

The charge transfer overvoltage is fairly well understood in terms of the Butler–Volmer equation: [71a]

$$i = i_0 \left(e^{\alpha z F \eta / RT} - e^{-(1-\alpha)zF\eta/RT} \right)$$

where z is the number of elementary charges of the species passing the electrode–electrolyte interface ($z = -1$ for an electron, $z \geq +1$ for a metal cation), α is the transfer coefficient, and η is the overpotential, which marks the deviation of the electrode potential from the equilibrium potential for the reaction being considered. F is Faraday's constant. Inspection of this equation shows that, for large positive overpotentials, the first exponential dominates, whereas, for large negative overpotentials, the second exponential is more critical. In these two cases, one can show that the overpotential should be linear with respect to the log of the current density. This relation is known as the Tafel equation.

In the limit of small overpotentials ($\eta \sim 0$), the Butler–Volmer equation can be linearized to yield the charge transfer resistance (R_{CT}):

$$i = \frac{i_0 z F \eta_{CT}}{RT}$$

$$\eta_{CT} = i R_{CT}$$

$$R_{CT} = \frac{RT}{i_0 z F}$$

Observed deviations from this behavior can be attributed to the interference of diffusion control.

Mass-transport (i.e., diffusion or electromigration) effects are particularly acute in the cases of cracking, pitting, and crevice corrosion, whereby occlusion effects can create highly concentrated solutions that move an otherwise stable system into regions of thermodynamic instability at the local level [13, 14, 41–43, 79–82]. When porous films or particular solution flow conditions exist, mass-transport effects should also be taken into account [83, 84]. Molecular dynamics and Monte Carlo simulations of interfaces over the past few decades have provided some insight into the concentration gradients that occur close to the electrochemical interface [85–91], and these, coupled with computational fluid dynamics simulations, can indicate the extent to which mass-transport effects can dominate an overall corrosion scenario [92].

At the same time, surface effects can be strongly dependent upon local species that adsorb on the exposed surface and thereby modify in a direct way the rates of those chemical reactions that contribute to a given corrosion pathway. When comparing mass-transport effects to surface phenomena, therefore, such scenarios should also be taken into account. Researchers in the field of heterogeneous catalysis, for example, have applied molecular modeling to understand the effects by which environmental species and/or reaction by-products can poison certain reaction pathways and favor others [93]. In the world of corrosion, certain species are suspected to promote hydrogen uptake by metals [45, 94–98] and, in other cases, to inhibit surface passivation [99–102]. The tools that have been successfully applied to assess such surface processes in the field of catalysis can and have been applied to similar problems in corrosion science [65, 93, 103–108].

1.2.3 Metal Dissolution

The most quintessential element of corrosion is the loss of mass from a structural component, which may be embodied by the generic metal dissolution reaction:

$$M \rightarrow M^{z+} + ze^-$$

The metal may be directly passing from a bare metal surface into an aqueous solution [109–111] or passing through an oxide (alternatively, sulfide, chloride, etc.) film via point defect-mediated mass transport [112], among other mechanisms. This reaction may be accelerated (or, potentially, retarded) by the complexation of the surface metal atoms by species in the environment. For instance, sulfur, hydroxide, or chloride could form bonds with the surface adatoms on a bare metal surface, thereby weakening the metal/surface bonds and facilitating the corrosion reaction [25b, c, 113]. Some atomistically resolved investigations of this process suggest that the surface structure provides a high degree of control, leading to such features as surface faceting and crystallographic pitting [25b, c, 113b, 114].

In the case of iron dissolution, the following reactions are believed to play a role according to K.E. Heusler:

$$H_2O_{(l)} \rightarrow OH^-_{(ads)} + H^+_{(aq)} \quad \text{(fast)}$$

$$Fe_{(s)} + OH^-_{(ad)} \rightarrow FeOH_{(ads)} + e^- \quad \text{(fast)}$$

$$FeOH_{(ads)} + OH^-_{(ads)} + Fe_{(s)} \rightarrow FeOH^+_{(aq)} + FeOH_{(ads)} + 2e^- \quad \text{(slow)}$$

$$FeOH^+_{(aq)} + H^+_{(aq)} \rightarrow Fe^{2+}_{(aq)} + H_2O_{(l)} \quad \text{(fast)}$$

Thus, surface adsorption of water and subsequent dissociation and reaction with surface iron adatoms form a complete mechanism for dissolution [114b]. In addition to experimental investigations of reaction kinetics and surface characterization, molecular modeling can play a key role in assessing the various aspects associated with these mechanisms. The Butler–Volmer framework also provides a means for assessing the role of overpotential in affecting the overall rate of reaction [71a].

Thermodynamically, dissolution should occur whenever the free energy of the solvated metal ion at a given concentration is lower than the free energy of the atom in the metal, plus the thermodynamic potential of the electrons exchanged during the reaction via the standard hydrogen electrode. Such thermodynamic conditions are summarized in the series of potential–pH phase diagrams that have been extensively collated by Pourbaix [69]. Even in cases where a metal is covered by a thin protective oxide, deleterious corrosion effects can

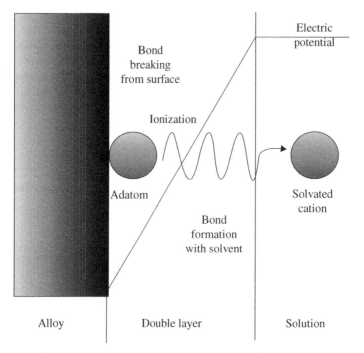

FIGURE 1.4   Schematic of the transport of a metal atom across the electrochemical double layer (from left to right).

arise from the dissolution of metal atoms directly exposed by virtue of cracks or pores in the protective film. The dissolution of metal atoms at the bare metal/solution interface, and also at the oxide/solution interface, requires the metal atoms to traverse the so-called electrochemical double layer [71a]. This concept, arising from electrochemistry, is based on a potential difference between the metal and the solution, due to aligned water dipoles, the ionic strength of the solution, and any bias existing on the metal [89, 115–121]. Dissolution of metal atoms, therefore, involves the breaking of cohesive forces between the metal atom and its neighbors in the solid state, electron transfer resulting in an ion formation (it is presently unclear exactly when this step occurs) [109–111], the formation of a solvation sphere around the nascent metal ion, and the movement of the ion through the potential gradient existing at this *double layer* (Fig. 1.4) [110]. Clearly, such a process involves a number of atomistic events that are sensitive to several highly localized environmental parameters that are amenable to a molecular modeling analysis.

1.2.4   Hydrogen Evolution and Oxygen Reduction

Since aqueous corrosion is often electrochemical in nature, the reactions that lead to loss of metal via dissolution require a cathodic counterpart, which is often either hydrogen evolution or oxygen reduction [3]. Hydrogen evolution involves the reduction of a proton (or a hydronium ion, i.e., a proton within a water molecule) to form molecular H_2:

$$2H^+ + 2e^- \rightarrow H_2 \quad \text{(acid solution)}$$

$$2H_2O + 2e^- \rightarrow 2OH^- + H_2 \quad \text{(alkaline solution)}$$

The hydrogen evolution reaction in acid solution has been broken down into the following steps:

$$H^+ + e^- \rightarrow H_{(ads)} \quad \text{(Volmer reaction)}$$

$$H_{(ads)} + H_{(ads)} \rightarrow H_{2(ads)} \quad \text{(Tafel reaction)}$$

$$H_{(ads)} + H^+ + e^- \rightarrow H_{2(ads)} \quad \text{(Heyrovsky reaction)}$$

$$H_{2(ads)} \rightarrow H_2 \quad \text{(molecular dissolution followed by gas bubble formation)}$$

In experimental reality, the presence of oxygen is found to greatly accelerate the rate of corrosion, which is due to the action of oxygen as an oxidant via the oxygen reduction reaction. The overall reaction is expressed as

$$2O_2 + 4H^+ + 4e^- \rightarrow 2H_2O$$

The reaction steps differ in acidic and basic reactions, namely,

$$O_2 + 2H^+ + 2e^- \rightarrow H_2O_2 \quad \text{(acidic)}$$

$$H_2O_2 + 2H^+ + 2e^- \rightarrow 2H_2O$$

$$O_2 + 2H_2O + 2e^- \rightarrow H_2O_2 + 2OH^- \quad \text{(basic)}$$

$$H_2O_2 + 2e^- \rightarrow 2OH^-$$

The oxygen reduction reaction has received a large amount of attention recently, from both modeling and experimental investigations, due to its importance in the development of fuel cell systems [74, 103, 122–125]. The rate of corrosion reactions may be controlled by either the anodic or cathodic process, depending upon which of the two half-cell reactions is rate limiting. This combination is elaborated further in the following text.

1.2.5 The Mixed Potential Model for Corrosion

In the electrochemical mechanism of corrosion, the metal dissolution—which involves the loss of electrons vis-à-vis oxidation—must be accompanied by a cathodic reaction that consumes electrons, which is typically oxygen or proton or water reduction. According to the Butler–Volmer equation, the current density for the anodic reaction varies according to

$$i_a = i_{a,0} e^{zF(E-E_{0,a})/RT}$$

A similar expression exists for the cathodic reaction:

$$i_c = i_{c,0} e^{-zF(E-E_{0,c})/RT}$$

Here, $E_{0,a}$ and $E_{0,c}$ are the equilibrium potentials for the anodic and cathodic process, respectively, and $i_{a,0}$ and $i_{c,0}$ are the respective current densities at these potentials. In a corroding system, where both reactions are occurring simultaneously, the total current is the difference of these terms:

$$i = i_a - i_c = i_{a,0} e^{zF(E-E_{0,a})/RT} - i_{c,0} e^{-zF(E-E_{0,c})/RT}$$

Under open-circuit conditions (i.e., when the metal surface is not connected to an external potentiostat), the net current $i_a - i_c$ will be zero, and the electrode potential E will adjust according to the corrosion potential, E_{corr}. In this case, the corrosion current density (i.e., the current density associated with the loss of metal) is $i_a = i_c = i_{corr}$.

In the limit of large anodic polarization, π, relative to the open-circuit potential E_{corr}, the variation of current with potential will be log-linear and vice versa for large cathodic potentials (relative to open-circuit potential). This behavior can be modeled by the semilogarithmic Tafel equations:

$$\log i = \log i_{corr} + \frac{\pi}{b_a} \quad \text{(large anodic potential, } \pi)$$

$$\log |i| = \log i_{corr} - \frac{\pi}{b_c} \quad \text{(large cathodic potential, } \pi)$$

The corrosion potential can then be found by extrapolating the linear portions of these two equations in the plot of log current density versus potential.

The variation in current with potential implies that there exists an electrochemical resistance that results from the coupling of anodic and cathodic reactions. Following from the expressions for the corrosion current density, as given earlier, this *polarization resistance* R_p can be obtained from the equations:

$$i = i_{corr} \left(\exp \frac{\pi \ln 10}{b_a} - \exp \frac{-\pi \ln 10}{b_c} \right)$$

$$\frac{1}{R_P} = \left(\frac{di}{d\pi} \right)_{\pi \to 0} = i_{corr} \left(\frac{\ln 10}{b_a} + \frac{\ln 10}{b_c} \right)$$

$$R_P = \frac{b_a b_c}{\ln 10 (b_a + b_c) i_{corr}}$$

These equations, therefore, allow the experimentalist to extract fundamental quantities from a mixed potential system in order to understand aspects of the decoupled anodic and cathodic reactions.

1.2.6 Selective Dissolution of Alloys

Since most structural metals are, in fact, alloys, it is necessary to consider how metals corrode when found in combination. A classic system here, amenable to detailed investigation due to the nobility of the components, is the AuCu alloy system [126]. These alloys are characterized by a very low dissolution current at potentials that are positive with respect to the Cu/Cu^{2+} equilibrium potential, but negative with respect to the Au/Au^{3+} electrode. There exists a critical potential E_{cr}, however, above which the dissolution current increases sharply. This critical potential increases with the Au content of the alloy [127]. The general mechanism proposed for this phenomenon is that above the Cu/Cu^{2+} potential surface, copper atoms may dissolve, but as they do, they leave Au enriched at the metal surface. Therefore, continued Cu dissolution is blocked by Au atoms that cap the surface. When the critical potential is reached, the steep increase in current is due to the fast Cu corrosion current density corresponding to the high overpotential for Cu dissolution. The stress corrosion cracking of brass in ammonia-rich environments presents another example of this failure mode, caused by the loss of Zn from the alloy as a result of selective dealloying [128, 129].

1.2.7 Passivity of Metals and Alloys

In addition to dissolution, reaction of solution or atmospheric constituents with the metal very often results in the formation of a surface film, most commonly a metal oxide, that can be as thin as a nanometer (i.e., a few atomic layers) [112, 129b, c, 130–136]. The rate of formation of this passive film and the rate at which it dissolves into solution ultimately determine the stability of the material in its environment. In this way, iron, which is thermodynamically stable only in a small region of the Pourbaix diagram (see Fig. 1.3), can be usefully employed over a significantly wider range of conditions due to the extremely slow dissolution kinetics of the passivating oxide film. Passive film stability has been described mechanistically via several models, including the *high-field* and *point defect* models, which attribute the overall stability to kinetic factors such as the migration of vacancies or other oxide defects through the thickness of the film, as a function of the electric potential across the film [112, 129b, c, 137, 138]. Other important mechanistic details include the transfer of metal ions from the underlying metal substrate to the oxide film and the interfacial chemistry occurring between the oxide film and the solution itself. Localized corrosion may also occur when particular defects in the oxide are subject to a preferential attack from species such as chloride in the environment [25b, c, 137, 138]. Such pathways and their associated kinetic parameters could be directly assessed from first-principles and molecular modeling techniques, although a complete analysis of the many pathways and interfacial transitions remains a significant challenge. Metal surfaces exposed to aqueous solution or water vapor may have a compound passive film, which consists of an outermost hydroxide or oxyhydroxide and an inner layer oxide. Further complexities arise due to composition gradients in the material and the relative oxidation strengths of different alloy components [129b, c].

The electrochemical behavior of passivation is such that as the potential increases from the cathodic direction, one first encounters the *active region* in which typical current densities may be up to $100 \ mA/cm^2$ and more. This corresponds to free corrosion of the metal, before a protective oxide or oxyhydroxide film forms. As the potential becomes increasingly anodic, the current density diminishes, down to $1 \ \mu A/cm^2$, as the corrosion product film lowers the rate at which corrosion can occur. This is the region of passivity. It can extend over several hundred mV. Beyond this region lies the transpassive film: oxygen evolution and transpassive dissolution can occur on semiconducting passive films. For more inert materials (Al and the valve metals, Zr, Hf, etc.), the passive region can extend to a larger range of anodic potentials due to the insulating nature of the oxide films that severely retard ion mass transport and electron transfer across the passive layer.

Localized corrosion can occur in the otherwise passive region of potentials. In the case of localized corrosion, the passive film is locally breached, due to mechanical or chemical compromise of the oxide film or structural defects, such as grain boundaries, triple points, surfacing dislocations, or intermetallic phases such as carbide and/or sulfides. Localized corrosion manifests itself in the plot of corrosion current versus electrode potential (the polarization curve) as *noisy* peaks in the passivation region, due to the stochastic nature of this mode of corrosion.

1.2.8 Inhibition of Corrosion

The corrosion rate of metals may be reduced by the addition of inorganic or organic compounds, called inhibitors, to their environment [139]. The inhibitor efficiency is defined as the relative reduction of the corrosion current density, where $i_{c,0}$ and i_c are the corrosion current densities without and with addition of the inhibitor, respectively:

$$n_i = \frac{(i_{c,0} - i_c)}{i_{c,0}}$$

The mode of inhibition depends upon the material, the environment, and the mechanism of corrosion that is being inhibited. For instance, the inhibitor may adsorb at the metal surface and thus reduce its anodic dissolution [140]. Alternatively, the inhibitor may act to repress the cathodic reactions such as hydrogen evolution or oxygen reduction. In keeping with this concept of adsorption-based inhibitors, inhibitors typically are molecular species with alkyl chains and an active head group such as $-CN^-$ (cyanides), $-SH$ (thiols), $-NH_2$ (amines), $-COO^-$ (carboxylates), and $-OPO_3^{-3}$ (phosphates). Numerous approaches to optimize the design and synthesis of inhibitors have been developed based on quantitative structure–property relationship analysis, a type of molecular modeling that is coupled with statistical analysis and correlations with empirical data for inhibition efficiency [141, 142]. Such approaches correlate molecular properties, for example, the electron density at the binding head group, with the inhibitor efficiency. Only recently have molecular approaches been employed to study the mechanisms of inhibition [143].

Since inhibitors can influence either the cathodic or anodic reaction, their effect on the open-circuit (i.e., corrosion) potential will differ according to this mode of action. This relationship can be understood by comparing the Tafel diagrams in Figure 1.5. Suppression of the reduction current lowers the Tafel semilog line, thus moving the point of intersection of the anodic and cathodic regions to a lower open-circuit potential (more cathodic). On the other hand, suppressing the anodic reaction will increase the open-circuit potential into more anodic regions. Finally, the inhibitor may be mixed mode, in which case there may be little or no change in the open-circuit potential of the system. In all cases, the corrosion current is lowered by the presence of the inhibitor on the surface.

1.2.9 Environmentally Assisted Cracking and Embrittlement

Corrosion failure can also occur when environmental species absorb into the metallic lattice itself, modifying the materials properties to the point that it can no longer maintain its specified loading [46]. Most commonly, this is attributed to the absorption of hydrogen, yet embrittlement and stress corrosion cracking can also occur via other species, including sulfide, iodide, and chloride [58, 144–148]. In such cases, the chemically aggressive species

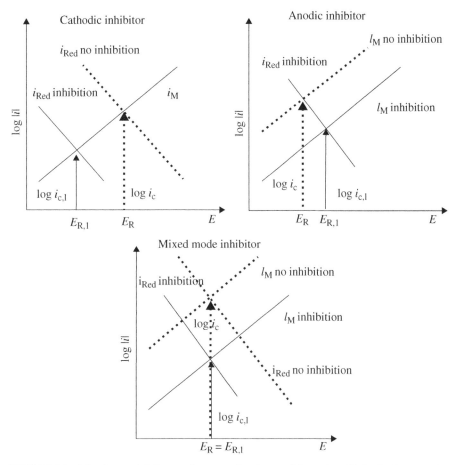

FIGURE 1.5 Mixed potential diagram demonstrating the possible modes of behavior for corrosion inhibitors. Adapted from Strehblow and Marcus [70].

clearly plays a role in altering the nature of the metal–metal bonding and the response of the metal's deformation modes, such as dislocations, elastic response, or grain-boundary motion. Therefore, the ability to interrogate how interposed environmental species modify these properties from a first-principles approach, and to simulate various hypothetical scenarios, could lead to new materials design approaches, such as alloying, surface modification, or nanostructuring to improve the longevity of materials under combined stress corrosion conditions.

1.2.10 Crystallographic Pitting

Postmortem inspection of corroded specimens often indicates a predilection for corrosion and/or stress corrosion cracking to proceed along grain boundaries in the metallic structure [149–156]. Somewhat related to this observation is the fact that under certain conditions, corrosion pits may adopt a *crystallographic* appearance, conforming to the underlying symmetry elements of the metal [25b, c, 113b, 157–159]. These observations suggest that certain features of the metal, and in particular the directionality and local sensitivity of the cohesive forces binding the metal atoms to one another, can play a significant role in directing the overall corrosion response of the metal. Therefore, powerful modeling techniques that directly incorporate the best physical models for describing interatomic interactions could be used to help interpret such observations and, through the modeling of various hypothetical scenarios, suggest physically justifiable approaches for their mitigation.

1.2.11   Summary of Corrosion Mechanisms

This extremely cursory overview suffices to show that there are several molecular and atomistic level interactions that are involved in any single corrosion event and that there is a large number of possible mechanisms that still require a rigorous theoretical treatment. In the following section, we introduce a number of molecular-level modeling techniques that have been, or could be, applied to the further detailed study of these mechanisms for materials presently used in industry and materials yet to be discovered and/or applied.

1.3   MOLECULAR MODELING

The chief aim of atomistic and molecular modeling techniques is to compute the thermodynamic, mechanical, and/or kinetic properties of a system composed of isolated or periodic, static or dynamic, arrangements of atoms and molecules [4, 7, 160–162]. Common to all these modeling techniques are physics-based representations of the interactions between the atomic and molecular constituents. These interactions may be computed based on a broad spectrum of techniques ranging from approximate pair potentials at one extreme through to accurate determination of the electronic structure based on quantum mechanics [6, 7]. Forces and charges on atoms, as well as system and subsystem energies, are the primary quantities obtainable from these methods, though quantum mechanical methods may also yield further properties that depend on the behavior of the electrons. By exploring the dependence of these quantities on the atomistic, molecular, and materials structure and composition, one can build an elaborate picture of the overall thermodynamic, mechanical, and kinetic properties of the system one is trying to represent.

At its root, molecular modeling consists of the solution of physics-based equations for the interactions between the basic constituents of matter—atoms and molecules—and any externally applied forces, such as electric forces due to potential [163]. One may use minimization techniques to find low-energy (and therefore physically reasonable) configurations of atoms and molecules [164], in order to rank different configurations, or understand bonding arrangements, or integrate the equations through time to simulate dynamical trajectories for these systems [165]. The underlying physical equations are often based upon the fundamental Coulombic interactions between nuclei and electrons that, in conjunction with the kinetic energy, ultimately determine bonding in solids and molecules and thereby all other properties. While a deeper understanding of molecular modeling techniques can be gained only through a more extensive study of the literature [4, 5, 7], we take some time here to introduce some of key concepts for the more pertinent approaches.

1.3.1 Electronic Structure Methods

Electronic structure methods comprise the most exact techniques for interrogating the internal structure and properties of molecules and materials [5, 7]. At best, they require no arbitrary parameters and can be systematically improved to obtain very nearly exact solutions [166]. At the same time, these methods are the most computationally expensive, requiring one to determine the fine structure of the electronic orbitals that define bonding in materials, some of which information may not be required to explore all problems. Schrödinger's time-independent wave equation usually forms the starting point for these methods, in that the internal energy E of a configuration of atoms and electrons is determined as the eigenvalue of the Hamiltonian operator, acting on the electronic wave function Ψ:

$$H\Psi = E\Psi$$

Here, it is usual to make the Born–Oppenheimer approximation that allows a classical treatment of the nuclei to be separated from a quantum mechanical description of the electrons. In this case, the wave function becomes just that of the electrons, and the nuclear–nuclear interaction is added to the energy as a sum over point particles. Consequently, the Hamiltonian operator H includes the kinetic energy of the electrons, the electron–electron interactions, and the electron–nuclei interactions. The wave function determined by solving this eigenproblem consists of a Slater determinant of the molecular orbitals for a molecule or, alternatively, the band structure of a solid. Unfortunately, direct solution of this equation is complicated by the electron–electron interactions. Often, it is necessary to introduce a mean-field approximation that neglects the individual dynamical electron–electron correlations but instead treats the electrons as moving in the average field created by the other electrons. Various corrections have been developed to improve upon this approximation [160, 167, 168].

In the world of molecular modeling, quantum mechanical approximations typically start from one of two approaches. The first of these is Hartree–Fock theory, which has the advantage of being systematically improvable through methods such as perturbation theory (e.g., Møller–Plesset theory, starting from second order, MP2) and configuration interaction (CI) [166]. In the world of materials, and rather commonly now in the molecular sciences, a second approach known as density functional theory (DFT) has become more widespread. As the name suggests, here, the focus is more on the electron density rather than the wave function, and Schrödinger's wave equation is replaced in most practical applications with the analogous Kohn–Sham equations [169, 170]. While DFT suffers from the disadvantage of not being able to arbitrarily converge on the exact solution, it is typically able to yield superior results to Hartree–Fock theory for a reduced computational cost relative to MP2, CI, or other such so-called post-Hartree–Fock methods.

The solution to both Hartree–Fock and DFT approaches requires a self-consistent approach, in which trial solutions are iteratively refined until convergence to within a given precision is achieved. Solutions are typically expressed as a matrix of coefficients describing the contribution of various basis functions to the electronic structure of the system. These basis sets are a numerically convenient way of representing the wave functions and can be localized atomic-like orbitals, or plane waves in the 3D space of the simulation cell, or many other forms [164]. The improvement of these techniques is an ongoing endeavor, and yet, despite this enormous challenge, the approximate functionals to date remain among the most accurate tools for evaluating structure and energy relationships in materials and molecules [161, 162, 171–175]. Functionals and semiempirical approaches have been recently developed to better describe the weak electron correlations associated with dispersion forces, for example, whereas hybrid functionals and the *Hubbard U* method have been developed to take into account strong electron correlations [7, 167, 176–179]. Because of the fundamental nature of modeling techniques based on electronic structure, they are often called first principles or *ab initio*. As in all models, much care must be taken in setting up such a calculation to ensure that the results produced are both valid and relevant to the materials system of interest [180].

More approximate solutions to the electronic structure problem come from tight-binding and semiempirical molecular orbital theory [181, 182]. These methods simplify and efficiently approximate the expressions within the quantum mechanical problem in order to reduce the computational cost. To compensate for this, empirical parameters are often introduced that are obtained from fitting to experimental or high-level quantum mechanical data. However, because of this fitting process, the parameter sets so developed can rarely be confidently extended beyond the range of systems for which they were originally designed. Since corrosion systems typically bridge length scales that incorporate bulk materials, the interfacial region and an environmental (such as aqueous) solution phase, obtaining a parameter set suitable for spanning all these behaviors presents a formidable challenge [183, 184].

1.3.2 Interatomic Potentials (Force Fields)

Whereas ab initio techniques can be applied to the various modeling scenarios described in this and the following subsections, typically a more efficient approach is required in order for systems of a more realistic size or timescale to be simulated. Based on potential energy surfaces constructed from ab initio techniques or implied from various experimental quantities, such as the bulk modulus, one can derive approximate relationships for the interactions between atoms, including the electronic effects only in an implicit way. Such relationships, called interatomic potentials [185], can be rigorously derived by expanding the energy as a series of terms that depend on increasing numbers of different atoms: [186]

$$U = \sum_i U_i + \frac{1}{2}\sum_{i,j} U_{ij} + \frac{1}{6}\sum_{ijk} U_{ijk} + \cdots$$

Despite the exact relationship expressed in this equation, the accurate forms for the interaction terms and how they depend on relative positions are not exactly known [187]. Thus, using reasonable physical arguments, a number of approximations have been derived. Furthermore, if the expansion is continued to the point where it includes all atoms, and therefore is exact, this becomes exponentially expensive to solve as the size of the system increases. Fortunately, it turns out that the magnitude of the contribution also rapidly diminishes as the number of atoms involved in the interaction term increases. Hence, the series can be truncated quite rapidly while still giving reasonable results. For ionic materials, it is often sufficient to stop at the second term, which corresponds to pairwise potentials. In the case of covalent organic materials, it is usually necessary to include three- and four-body terms too. There now exists a variety of different expressions for approximate energy calculations, each of which have their own appellation as a family of interatomic potentials (for instance, embedded atom [188–194], Stillinger–Weber [195], Buckingham potentials [177], etc.).

Through experience, it is now generally well understood which potential model is appropriate for a given type of material. For purely ionic systems, for example, the one-body term, $E_a(i)$, may be neglected, and the potential determined simply by the sum of Coulombic

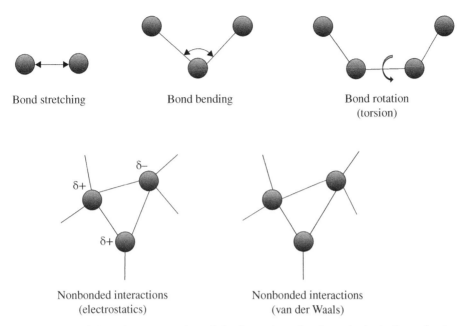

Bond stretching   Bond bending   Bond rotation
                                               (torsion)

Nonbonded interactions   Nonbonded interactions
(electrostatics)   (van der Waals)

FIGURE 1.6 Schematic representation of the interactions that form the basis for molecular mechanics simulations of bonded and nonbonded systems.

interactions between the ion and its neighbors, plus a short-range repulsive term. This description may be augmented by a *shell* model for the ion polarizability [196, 197] or self-charging terms that allow the charge on an ion to fluctuate based on the relative electronegativity and hardness of its neighbors (thus introducing a one-body term, $E_a(i)$, corresponding to the energy associated with local charging) [198, 199].

In molecular systems, three-body (angular) and four-body (torsional or out-of-plane) terms can also be added to the energy function, thus providing greater flexibility and therefore accuracy in the overall potential scheme [4]. Such a breakdown is illustrated in the schematic of Figure 1.6. This technique, known generally as molecular mechanics, is widely used in the simulation of biological molecules and polymer systems, due to its relatively rapid computation for systems of many atoms; its intuitive appeal, in terms of a ball-and-stick concept; and its ability to incorporate additional interactions, such as nonbonded dispersion forces. Here, Coulombic and other nonbonded interactions are typically excluded between nearest and next-nearest neighbors to avoid confusing the physical interpretation of the potential functions. Similarly, so-called 1–4 nonbonded interactions between atoms that are separated by three bonds are often reduced by a scale factor, so that the torsional potential dominates the energy.

Embedded atom interatomic potentials can be formally derived from DFT [188–194]. This class of interatomic potentials generally works well for bulk metallic systems (i.e., accurately reproduces mechanical and structural properties), due to its relationship to the band model of electronic structure, which is related to the overlap of local electron densities to form the *free electron* gas of the metallic system. By parameterizing the form of the electron density about an atomic center, one can build a model for the electron overlap in these systems and, hence, the *many-body* features of alloy systems. The resultant expression for the energy is the sum over the local atomic embedding energy, which is in turn a function of the local electron density—the sum of the overlaps from neighboring atoms—as well as pairwise, short-range repulsions. The functional forms utilized in the embedded atom (and modified embedded atom) approaches are derived from fundamental forms for equations of state (such as the Rose equation) [200].

A limitation of many classes of interatomic potential is that the *identity* of an atom is typically fixed, such that if it moves from a metallic to an ionic environment—as happens in corrosion, for example—the potential is unable to capture the local change in either oxidation state or bonding that occurs. New classes of potentials have been developed to overcome this restriction. Reactive force fields, in particular, are attracting increasing interest for interfacial problems, due to their inbuilt ability to accommodate changes in coordination and oxidation state [11, 201–203].

The criteria for assessing the usability of a particular interatomic potential include:

1. Transferability: A potential derived from the properties of one particular material should remain accurate when applied to the same atom type in a different environment. The extent to which transferability is desired may depend upon the range of applications the *potential engineer* wishes to explore: potentials derived for transition metals as complexes dissolved in aqueous solutions are unlikely to be useful for exploring the properties of those same metals in the solid state.

2. Computational efficiency: The amount of computational effort required to evaluate the energy of an atom in a given context should be small relative to the effort required to do so from first principles and sufficiently fast to simulate the time and length scales desired for a given application. General rules of thumb (which will likely change in the near future due to the development of new algorithms and computational architectures) are:
   - Quantum mechanics: <1000 atoms; <100 ps for molecular dynamics
   - Interatomic potentials: <1,000,000 atoms; <1 μs for molecular dynamics

3. Extensibility: This last requisite is often overlooked in the fashioning of a given potential scheme but refers to the ability for *potential engineers* to extend the accuracy or transferability of the interatomic potential in rigorous and systematic ways. This may include the addition of further terms in a Taylor expansion or the inclusion of higher multibody terms to the potential.

Once confidence has been established in the fidelity of a given interatomic potential, that potential may then be used to simulate collective properties of an ensemble of configurations representing the corrosion scenario of interest by techniques such as energy minimization, thermodynamic analysis (lattice dynamics and free energy calculation), transition state theory, molecular dynamics, or Monte Carlo simulation. For studying diffusion properties in liquids, interatomic potentials are generally preferred to ab initio methods, due to the large timescales that must be simulated. For solid-state diffusion, either quantum mechanical or interatomic potentials have been used, since the typically larger barriers are suited to the use of transition state theory, rather than molecular dynamics [56, 204–206]. Corrosion problems, due to their complexity and the fundamental transformations that take place in a material, have chiefly been studied with quantum mechanics or reactive force fields [16, 143, 207, 208], although there exist many phenomena relevant to corrosion that can be studied with pair-potential techniques, such as mass transport in fluids or the accumulation of defects in a material.

1.3.3 Energy Minimization

Statistical mechanics teaches us that the most probabilistic states of a system are those that possess the lowest Gibbs free energy [209]. At low temperatures, these states are among those with the lowest potential energy—although kinetic barriers can lead one to be *trapped* in a local minimum state, rather than the true global minimum. This is demonstrated by the coexistence of both diamond and graphite at ambient conditions [210]. At higher temperatures, entropic effects begin to dominate and so must also be accounted for.

The technique of energy minimization is used to vary the atomic coordinates of the system (i.e., relative positions in Cartesian coordinates or, alternatively, the bond lengths and bond angles of a molecular system) until a local minimum in energy is reached [211]. In the case of a solid, this can also be supplemented by varying the lattice parameters until the material is no longer under stress. For a reactive system, several local minima will exist, corresponding to reactant and product states, as well as many alternative configurations that may not significantly contribute to the reaction of interest, but may nevertheless be important to understand as potential side reactions or intermediates that link the reactant and product states. For very *soft* systems, such as in solution, the number of local minima may be exceedingly large, as water molecules can arrange into multiple configurations of similar thermodynamic stability. Furthermore, in such systems, temperature imparts a high degree of thermal motion to the solvent molecules, thus making transit across the barriers separating these minima facile. In solid systems, this is less often the case: barriers between local minima tend to be higher relative to the thermal energy, kT. The kinds of interfaces found in a corroding system, therefore, will require a mix of approaches to the determination of the most stable states of the system.

Energy minimization techniques include, most commonly, the methods of steepest descents and conjugate gradients [4]. Here, steepest descents follow the lowest energy pathway, as determined by the gradient (i.e., first derivative of the energy) at each point, while conjugate gradient methods begin similarly but include a *history* component, which accelerates the optimization process. While these two methods are widely used, due to the need to only compute the gradients, there are many more sophisticated optimization algorithms that also employ the Hessian (i.e., the matrix of second derivatives of the energy). Such Hessian-based methods offer convergence in fewer steps, at the expense of increasing the computational cost per step. There are also many intermediate methods that use an approximate initial Hessian (e.g., a unit matrix) and then apply an updating formula that improves this guess as the optimization progresses.

The molecular modeler typically begins with an educated guess for the atomic configuration of the system, based on crystallographic and/or molecular symmetry constraints, as well as the typically bond lengths and angles for such systems, and then refines possible structures based on these computational techniques. Frequently, multiple hypothetical structures can be fine-tuned in this way, and their energies compared to see which states are more stable than others.

Energy minimization produces a structure that is thermodynamically stable at 0 K (not necessarily the lowest energy one, however). In order to produce results that are meaningful at higher temperatures, one must consider the ways in which the system will vary as the temperature is increased. The first approximation utilized in this context (from the solid-state perspective) is lattice dynamics [6]. Equivalently, for molecular approaches, one considers the entropic and enthalpic contributions arising from the rotational, translational, and vibrational degrees of freedom [5]. In solid-state systems, the effect of temperature manifests itself through the excitation of phonons, and so the key quantity of interest is the phonon density of states. The quasiharmonic approximation makes the assumption that the main effect of phonons is through the thermal expansion of the unit cell. Here, the lattice parameters are varied so as to minimize the sum of the potential energy and the vibrational free energy at a given temperature, while the atomic positions within the unit cell relax according to the internal energy. Utilization of this approximation is usually valid to half the melting point. Above this temperature, anharmonic effects become significant, and so molecular dynamics becomes the preferred means to simulate thermal expansion.

Sometimes, the actual structure of a system is hard to anticipate simply based on chemical intuition. In this case, techniques such as simulated annealing or machine-learning approaches can be used to randomly search the configuration space spanned by the suspected reactants and/or products and identify a series of local minima. Such techniques are known as *global search algorithms*, although, in reality, a global search can rarely be guaranteed unless the configuration space is narrow enough that a systematic search can be made [4, 5].

The outputs of an energy minimization can be highly instructive. Starting with a minimized structure, one can use the arrangement of atoms in the system (such as bond lengths and angles) to compare with diffraction or NMR techniques, for example. One can also compute the vibrational spectrum to compare with optical spectroscopy techniques. The relative energy of different states can be an indicator of their contribution to a given reaction scheme. Once energy minimization has been performed for several hypothetical states that are assumed to be pertinent for a given reaction mechanism or pathway, statistical mechanics interpretations can then be applied to make some specific thermodynamic assertions. For instance, the enthalpy of reaction can be estimated, or the probabilistic distribution of particles among two distinct states that differ in energy can be determined via the Boltzmann distribution [209]:

$$\frac{n_1}{n_2} = e^{-\Delta E/kT}$$

Other postprocessing techniques allow the computation of the density of states/molecular orbital structure, local charges on atoms or fragments of the system, dipole and multipoles, magnetic properties, and the electrostatic potential. Energy minimization can also be performed in the presence of perturbations, such as external fields or imposed electrode potentials [212].

1.3.4 Transition State Theory

While thermodynamic analysis provides guidelines as to the equilibrium state of a system, for truly predictive modeling—especially at relatively low temperatures where activation barriers may exceed the average thermal energy and level of excitation—kinetics can be equally as important. Whereas estimates of rates can be made from thermodynamic reaction energies, using paradigms such as the Evans–Polanyi relationship through proper configurational sampling activation energies for reactions can be directly determined via molecular modeling [104, 213]. The Arrhenius equation can then be applied to provide predictions of rates, provided a suitable prefactor can be assumed [214]:

$$k = Ae^{-\Delta G^*/RT}$$

In the case of diffusion, this expression takes the form of the Eyring equation:

$$k = \kappa \left(\frac{kT}{h}\right) e^{-\Delta G^* / RT}$$

where κ is the transmission coefficient. These rates can then be used to directly assist in the interpretation of experimental data, or to compare alternative reaction pathways with one another, or combined to run stochastic or integrodifferential simulations of overall reaction sequences. For instance, if an experimentally determined Arrhenius barrier is available, it could be compared to the Arrhenius barriers obtained computationally for several different hypothetical pathways, and the correct pathway distinguished if one is found to be in agreement. A chief difficulty with Arrhenius-based modeling is in the selection of the proper prefactor, though methods do exist to compute the transmission coefficient, such as the reactive flux technique.

The activation energy can be determined once an atomic configuration, corresponding to the transition state, has been located. Transition states are local maxima along a reaction coordinate that connect two local minima on the potential energy (the reactants and products). Additionally, they are also saddle points in the local potential energy space, as they should be local minima for all directions other than that connecting the reactants and products. Thus, transition state searches rely upon computing the derivatives of the potential energy with respect to the atomic coordinates. Most search methods for transition states involve the first derivatives only, such as the synchronous transit or nudged elastic band methods, though, again, algorithms exist that can exploit the Hessian matrix, where available [215–219]. The popular nudged elastic band method begins with a series of geometries (images) supposed to lie upon a hypothetical reaction pathway and then uses a constrained minimization to find the optimal path connecting reactants and products while minimizing the forces between the images (hence, the term nudged elastic band).

1.3.5   Molecular Dynamics

Molecular dynamics simulation is perhaps the most directly accessible of molecular modeling tools conceptually speaking. However, its use as an interpretive modeling tool is perhaps one of the most limited due to the necessity of modeling relatively long stretches of time that do not necessarily include the events of interest to the modeler. Molecular dynamics simulation involves the time propagation of some initial atomic configuration within some suitable thermodynamic ensemble. A common ensemble used is the NVT ensemble, which indicates constant number of chemical species, constant volume, and constant temperature. Temperature control is typically maintained in a molecular dynamics simulation by coupling the kinetic energy of the particles in the simulation with a virtual thermal reservoir/thermostat [220, 221]. Time propagation is most commonly achieved by applying the Born–Oppenheimer approximation (i.e., the behavior of nuclei is classical and slow compared to the motion of electrons) and classical Newtonian dynamics. Molecular dynamics simulation can be achieved by using ab initio forces on the particles or by using suitable interatomic potentials. The former method has appeal, in that ab initio methods can implicitly model changes in chemical bonding that may occur during the simulation—typically interatomic potentials do not allow changes in bond types between atoms, although recent potentials have been devised to try to overcome this [202]. Molecular dynamics simulations of interfaces, such as appear in corrosion scenarios, are challenged by the difficulty of adopting a potential set that applies equally well to the solid state, as to the solution state. Molecular dynamics approaches have been applied to understanding the role of hydrogen embrittlement of metals [222–224]; to the structure, dynamics, and transport processes in ionic solids [225, 226]; and for understanding solvation of ions in dilute and concentrated solutions [85, 88, 90, 227].

In addition to providing a microscopic tool for observing the outcomes of physicochemical processes in extraordinary detail, molecular dynamics simulations can, in principle, provide a valuable technique for obtaining thermodynamic variables and rate constants via integration over selected portions of the molecular dynamics trajectory. Several techniques have been recently employed that allow this kind of analysis, even with the present limitations regarding length and timescales, such as time-accelerated molecular dynamics [228, 229].

1.3.6   Monte Carlo Simulation

Monte Carlo simulations take advantage of probabilistic sampling methods to investigate the equilibrium behavior of systems as a function of temperature (Metropolis Monte Carlo) or to advance the state of a reactive system through time (kinetic Monte Carlo). In Metropolis Monte Carlo simulation, the energy required to effect a discrete change to the system (i.e., in bonding configuration, addition or subtraction of species, atomic diffusion) is evaluated via either first principles, interatomic potentials, or a lookup table [230, 231]. If the discrete change is exothermic (i.e., negative energy change), then the event is allowed to proceed within the simulation. Otherwise, the Boltzmann probability for this event is compared to a randomly generated number between 0 and 1. If this probability exceeds the random number, then the change is effected within the simulation. In this way, an equilibrium state will ultimately emerge that corresponds to the predicted state of the system at that given temperature. The accuracy of such a prediction will depend upon the accuracy of the underlying potential energy surface and the extent to which the environment is adequately representative of the situation to which the results will be applied.

Kinetic Monte Carlo simulation proceeds in a somewhat different fashion, in that a list of all possible discrete reaction steps is compiled, or sometimes determined *on the fly*, for events such as atomic diffusion, dissolution, dissociation, or chemical combination, and a probabilistic method is used to select a reaction based on the relative likelihood it will occur in a collective portion of time [232]. Reactions with a high rate will be more likely selected than reactions with a slower rate. Reaction rates can be derived ab initio, from interatomic pair potentials, approximate functional relations, or a lookup table. Once a reaction is selected, the time step is advanced based on all possible rates for reactions involving the current configuration, and the system state is updated. A new reaction list is compiled, and the probabilistic selection process continues. In this way, a kinetic Monte Carlo simulation somewhat resembles a molecular dynamics trajectory, although the events are discrete rather than finite time step *continuous*, and typically, much greater timescales can be considered. Rather than waiting throughout the span of a molecular dynamics simulation for some significant event to occur, one can simply *step* ahead to the expected time it would take for that event to occur. In addition, the ability provided by kinetic Monte Carlo to consider competing reaction mechanisms and devise simulations with complex topological dependencies can also provide significant insights.

1.4 BRIDGING THE REALITY GAP

Just as an experiment must be well designed to provide an answer that is relevant to its practical motivation, so too must a model be well designed to mimic the physical phenomenon under consideration. For the purposes of corrosion, multiple phenomena should be given consideration beyond the immediate reactive molecular environment. These include the estimation of thermodynamic variables at realistic temperatures and pressures, the effects of solvation, control of chemical potential, electric fields and variance from charge neutrality, as well as point, line, and planar defects in the material, materials inhomogeneities, and long-range elastic effects. Although state-of-the-art efforts to incorporate these phenomena within molecular simulations will be discussed within the following chapters, a brief overview of contemporary approaches is presented in the following subsections.

1.4.1 First-Principles Thermodynamics

The techniques of energy minimization and the calculation of the internal energy of a configuration of atoms can provide much information regarding hypothetical corrosion mechanisms, yet, in order to make more quantified statements about system behavior, it is pertinent to extrapolate from the static, 0 K configurations, to provide an estimate of the system thermodynamics at the temperatures of interest. By coupling the tools of molecular modeling with statistical mechanics, it is possible to begin constructing phase diagrams for molecular and materials systems in which competing states can coexist, depending on their position in phase space. Thermodynamic variables relevant to corrosion may include the pH, electrode potential, concentrations of ions, or gas pressures. In some cases, extrapolation of first-principles internal energies can be made using tabulated data for the variation of free energy with the variables of temperature or pressure. In other cases, it may be necessary to derive all of these terms ab initio by first calculating the vibrational, torsional, and rotational degrees of freedom for a molecule or materials system (or subsystem) and then determining the zero-point energy and the partition function as a means for arriving at the overall free energy. Free energies can then be adjusted for the chemical potential by using the standard Gibbs expression. This technique has been applied to both surface chemistry and bulk alloy materials [233, 234].

1.4.2 Solvation Models

Since the properties of molecules and interfaces can be significantly modified when they are solvated, it seems critical for models, particularly in aqueous corrosion, to consider the influence on the properties computed by molecular modeling techniques. The challenge posed by solvation is that the arrangement of solvent molecules about a molecule or interface is frequently dynamic, and as such, either multiple configurations should be sampled—as in a molecular dynamics or Monte Carlo approach—or a continuum approach used to mimic the solvent properties. One popular approach that has most often been employed is the polarizable continuum model for solvent [235]. The solvent is treated as a polarizable medium with a given dielectric constant, and a method is used to project the influence of this solvent on the electronic structure calculations used to determine the molecular properties. This approach provides a correction to the free energy of the system. Typically, this correction is made using parameterized data that has been shown to provide accurate solvation energies for molecules under ambient conditions. There therefore remains some uncertainty when applying such techniques to nonambient pressures and temperatures. The representation of the solvent by explicit molecules can also be performed and may provide some particular advantages, including being able to directly model the bonded and nonbonded interactions that may develop between the molecule and material of interest and the solvent itself [184, 236]. However, one must then take proper care to sample an appropriate number of solvent configurations so that the study does not become biased toward one particular configuration that may be statistically insignificant in a thermalized system.

1.4.3   Control of Electrode Potential and the Presence of Electric Fields

Since many corrosion processes occur at the electrochemical double layer, the influence of variations in the electrode potential and the electric field should also be given consideration in advanced modeling approaches. In slab models or 2D periodic models, it is possible to apply an electric field, typically a constant electric field, across the supercell that represents the materials system, such that there exists a bias that favors the separation of charge states and the alignment of dipoles in a given direction [89, 116]. This procedure can be applied to both first-principles electronic structure codes and also to techniques that use pair potentials and charges on atom or molecular centers. An alternative approach for modeling electrode systems has been to vary the net charge on the system and make appropriate corrections to determine the role of the induced electrochemical potential in modifying the system free energy [237]. In both cases, serious consideration should be given to the relationship between the submicroscopic system, composed of molecules, atoms, and ions, and the continuum concepts of electrochemical potential.

1.4.4   Materials Defects and Inhomogeneities

Corrosion is a process that is commonly initiated at weak points at the materials/environment interface, presented by microstructural inhomogeneities such as grain boundaries, triple points, or secondary phases [238]. In many cases, the atomic-level structure of defect sites may not be well characterized or can manifest in multiple variations, such that deciding on one representation for a molecular modeling study can be challenging. At present, there are no trivial ways to simplify this problem; however, as experimental characterization techniques continue to improve the knowledge base regarding the atomic-level structure of materials defects and as computational methods expand to provide the ability to model more complex atomic configurations, this seems to be an area in which significant growth and materials insights can be expected in the near future.

1.5   MOLECULAR MODELING AND CORROSION

As indicated in Figure 1.1, the application of molecular modeling techniques to corrosion science is still in its infancy. In this first edition of *Molecular Modeling and Corrosion*, we introduce several areas in corrosion science to which molecular modeling has been successfully applied. In Chapter 2, Taylor discusses the application of molecular modeling techniques to study the reactivity of species introduced from the environment on the surface of metals and the current challenges to developing models that properly account for the defect structure of the material, as well as the configurational complexity of an aqueous phase environment. In Chapter 3, Schmickler et al. discuss the development of models to take into account surface reactivity and to describe cathodic phenomena such as the oxygen reduction and hydrogen evolution reactions. In Chapter 4, Policastro describes the current state of the art in modeling dealloying via kinetic Monte Carlo simulation techniques. Costa and Marcus describe in Chapter 5 the application of molecular modeling to investigate the adsorption properties of organic molecules on metal surfaces—a topic relevant to the development of corrosion inhibitors, as well as understanding the response of metals to biological media. In Chapter 6, Holby and Francis discuss the application of electronic structure techniques to predict the oxidation properties of fresh metal surfaces, the precursor phases to passivity of those metals. Maurice et al. present, in Chapter 7, a review of investigations of the reactivity of metal-oxide films to environmental species that induce localized corrosion and pitting. Finally, Lu and coworkers present, in Chapter 8 of this edition, a description of the application of multiscale methods to study the challenging problem of hydrogen embrittlement, which bridges mechanical failure modes in a material with the chemical interactions between H and materials defects.

As the brief survey we have made of the aforementioned corrosion mechanisms reveals, there remain many more scientific and engineering problems that can be tackled via molecular modeling approaches. The combination of modern developments in computational architectures, algorithms, and molecular modeling approaches with the age-old problems presented by corrosion science (as well as emerging problems that develop as both existing and new materials are pushed to greater extremes) will no doubt continue to increase the application of these techniques to the field of corrosion science. As a deeper understanding of corrosion mechanisms grows through this analysis, advanced strategies will be developed for the mitigation of corrosion of materials, and the science of interfacial reactions will itself be greatly developed with gains in areas not only in the field of corrosion but in catalysis, electrochemistry, and other disciplines in which interfacial properties are of key significance.

REFERENCES

1. R. E. Hummel, *Understanding Materials Science: History, Properties, Applications*. (Springer, New York, NY, 2004).
2. *Corrosion Mechanisms in Theory and Practice*, 3rd edition, Edited by P. Marcus (CRC Press, 2012).
3. E. McCafferty, *Introduction to Corrosion Science*. (Springer, New York, NY, 2010).
4. A. Leach, *Molecular Modeling: Principles and Applications*. (Prentice Hall, Upper Saddle River, NY, 2001).
5. I. N. Levine, *Quantum Chemistry*, 5 edn. (Prentice Hall, Upper Saddle River, NY, 2000).
6. E. Kaxiras, *Atomic and Electronic Structure of Solids*. (Cambridge University Press, Cambridge, UK, 2003).
7. R. M. Martin, *Electronic Structure: Basic Theory and Practical Methods*. (Cambridge University Press, Cambridge, UK, 2004).
8. C. D. Taylor and M. Neurock, *Curr. Opin. Solid State Mater. Sci.* 9, 49–65 (2005).
9. A. B. Anderson and N. C. Debnath, *J. Am. Chem. Soc.* 105, 18–22 (1983).
10. A. B. Anderson and N. K. Ray, *J. Phys. Chem.* 86, 488–494 (1982).
11. A. Nakano, R. K. Kalia, K. Nomura, A. Sharma, P. Vashishta, F. Shimojo, A. van Duin, W. A. Goddard III, R. Biswas and D. Srivastava, *Comput. Mater. Sci.* 38 (4), 642–652 (2007).

12. Recommended Practice API RP 571, *Damage Mechanisms Affecting Fixed Equipment in the Refining Industry*. (American Petroleum Institute, Washington, DC, 2003).

13. Z. Y. Chen, F. Cui and R. G. Kelly, *J. Electrochem. Soc.* 155, C360–C368 (2008).

14. Z. Y. Chen and R. G. Kelly, *J. Electrochem. Soc.* 157, C69–C78 (2010).

15. P. V. Strekalov, *Prot. Met.* 34 (6), 501–519 (1998).

16. A. Bouzoubaa, B. Diawara, V. Maurice, C. Minot and P. Marcus, *Corros. Sci.* 51, 2174–2182 (2009).

17. C. J. Boxley, J. J. Watkins and H. S. White, *Electrochem. Solid-State Lett.* 6 (10), B38–B41 (2003).

18. R. Cohen-Adad and J. W. Lorimer, *Solubility Data Series*. Edited by J. W. Lorimer (International Union of Pure and Applied Chemistry, Oxford, 1991), Vol. 47.

19. O. Forsen, J. Aromaa and M. Tavi, *Corros. Sci.* 35, 297–301 (1993).

20. P. M. Natishan, W. E. O'Grady, F. J. Martin, R. J. Rayne, H. Kahn and A. H. Heuer, *J. Electrochem. Soc.* 158 (2), C7–C10 (2010).

21. P. M. Natishan, W. E. O'Grady, E. McCafferty, D. E. Ramaker, K. Pandya and A. Russell, *J. Electrochem. Soc.* 146, 1737–1740 (1999).

22. M. A. Pletnev, S. G. Morozov and V. P. Alekseev, *Prot. Met.* 36, 202–208 (2000).

23. A. M. Sedenkov, *Prot. Met.* 22, 118–119 (1986).

24. V. A. Shimbarevich and K. L. Tseitlin, *Prot. Met.* 15, 455–457 (1979).

25. (a) V. A. Shimbarevich and K. L. Tseitlin, *Prot. Met.* 17, 144–148 (1981). (b) H.-H. Strehblow, *Werkst. Korros.* 35, 437–448 (1984). (c) H.-H. Strehblow and P. Marcus, *Mechanisms of Pitting Corrosion in Corrosion Mechanisms in Theory and Practice*. Edited by P. Marcus (CRC Press, 2012), pp. 349–417.

26. A. S. Couper and J. W. Gorman, *Corrosion/70*, 67 (1970).

27. A. L. Cummings, F. C. Veatch and A. E. Keller, *Mater. Perform.* 37, 42–48 (1998).

28. A. J. Devey, R. Grau-Crespo and N. H. de Leeuw, *J. Phys. Chem. C* 112, 10960–10967 (2008).

29. E. F. Ehmke, *Corrosion/75*, 6 (1975).

30. E. F. Ehmke, *Mater. Perform.* 14, 20–28, (1975).

31. E. F. Ehmke, *Corrosion/81*, 59 (1981).

32. J. Gutzeit, *Mater. Prot. Perform.* 7, 17–23 (1968).

33. C. Scherrer, M. Durrieu and G. Jarno, *Corrosion/79*, 27 (1979).

34. T. Skei, A. Wachter, W. A. Bonner and H. D. Burnham, *Corrosion* 9, 163–172 (1953).

35. E. Sosa, R. Cabrera-Sierra, M. T. Oropeza and I. González, *Corrosion* 58 (8), 659–669 (2002).

36. E. Sosa, R. Cabrera-Sierra, M. E. Rincón, M. T. Oropeza and I. González, *Electrochim. Acta* 47, 1197–1208 (2002).

37. G. M. Waid and R. T. Ault, *Corrosion/79*, 180 (1979).

38. S. M. Wilhelm and D. Abayarathna, *Corrosion* 50 (2), 152–159 (1994).

39. K. Yamakawa and R. Nishimura, *Corrosion* 55 (1), 24–30 (1999).

40. M. Zamanzadeh, R. N. Iyer and H. W. Pickering, *Corrosion/90*, 208 (1990).

41. C. S. Brossia and R. G. Kelly, *Corrosion* 54, 145–154 (1998).

42. J. S. Lee, M. L. Reed and R. G. Kelly, *J. Electrochem. Soc.* 151, B423–B433 (2004).

43. B. K. Nash and R. G. Kelly, *Corros. Sci.* 35, 817–825 (1993).

44. L. Stockert and H. Böhni, *Mater. Sci. Forum* 44–45, 313–328 (1989).

45. R. D. McCright, *Stress Corrosion Cracking and Hydrogen Embrittlement of Iron Base Alloys*. (NACE, Unieux–Firminy, 1977), pp. 306–325.

46. R. C. Newman, *Stem Corrosion cracking Mechanisms in Theory and Practice*. Edited by P. Marcus (CRC press, 2012), pp 499–544.

47. S. Persaud, A. Carcea and R. C. Newman, in *Corrosion 2013*. Edited by C. D. Taylor (NACE, Orlando, FL, 2013).

48. H. H. Pham and T. Cagin, *Acta Mater.* 58, 5142–5149 (2010).

49. P. Vashishta, R. K. Kalia, A. Nakano, E. Kaxiras, A. Grama, G. Lu, S. Eidenbenz, A. F. Voter, R. Q. Hood, J. A. Moriarty and L. H. Yang, *J. Phys. Conf. Ser.* 78, 12036–12042 (2007).

50. C. Y. Lee, J. A. McCammon and P. J. Rossky, *J. Chem. Phys.* 80, 4448–4455 (1984).

51. *Collins English Dictionary—Complete & Unabridged* (HarperCollins Publishers, Accessed Online, 2013).

52. H. V. Atkinson, *Oxid. Met.* 24, 177–197 (1985).

53. R. H. Doremus, *J. Appl. Phys.* 100, 101301 (2006).

54. I. Milas, B. Hinnemann and E. A. Carter, *J. Mater. Chem.* 21 (5), 1447 (2011).

55. T. Nakagawa, H. Nishimura, I. Sakaguchi, N. Shibata, K. Matsunaga, T. Yamamoto and Y. Ikuhara, *Scr. Mater.* 65, 544–547 (2011).

56. O. Runevall and N. Sandberg, *J. Phys. Condens. Matter* 23 (34), 345402 (2011).

57. C. A. C. Sequeira and D. M. F. Santos, *Nucl. Eng. Des.* 241, 4903–4908 (2011).

58. I. Schuster and C. Lemaignan, *J. Nucl. Mater.* 189, 157–166 (1992).

59. X. M. Bai, A. F. Voter, R. G. Hoagland, M. Nastasi and B. P. Uberuaga, *Science* 327 (5973), 1631–1634 (2010).

60. M. Christensen, T. M. Angeliu, J. D. Ballard, J. Vollmer, R. Najafabadi and E. Wimmer, *J. Nucl. Mater.* 404, 121–127 (2010).

61. C. D. Taylor, *Chem. Phys. Lett.* 469, 99–103 (2009).

62. C. D. Taylor, *Corrosion* 68, 591–599 (2012).

63. E. E. Ebenso, T. Arslan, F. Kandemirli, N. Caner and I. Love, *Int. J. Quantum Chem.* 110 (5), 1003–1018 (2009).

64. E. E. Ebenso, T. Arslan, F. Kandemirli, I. Love, C. Öğretır, M. Saracoğlu and S. A. Umoren, *Int. J. Quantum Chem.* 110 (14), 2614–2636 (2010).

65. K. Asakura, *Catal. Today* 157, 2–7 (2010).

66. J. Hafner, *Comput. Phys. Commun.* 177, 6–13 (2007).

67. H. J. Grabke, *ISIJ Int.* 36 (7), 777–786 (1996).

68. C. D. Taylor, *Int. J. Corrosion* 2012, 204640 (2012).

69. E. Deltombe, N. de Zoubov and M. Pourbaix, *Atlas of Electrochemical Equilibria in Aqueous Solution.* (NACE, Houston, TX, 1974).

70. H.-H. Strehblow and P. Marcus, Fundamentals of Corrosion, in *Corrosion Mechanisms in Theory and Practice*. 3rd edition, Edited by P. Marcus (CRC Press, 2012), pp. 1–104.

71. (a) J. O. M. Bockris, A. K. N. Reddy and M. Gamboa-Aldeco, 2A (2000).
    (b) Handbook of Chemistry and Physics. (CRC Press, Cleveland, OH, 1977–1978).

72. E. Protopopoff and P. Marcus, *J. Vac. Sci. Technol. A* 5, 944–947 (1987).

73. E. Protopopoff and P. Marcus, in *Corrosion: Fundamentals, Testing and Protection* (ASM International, 2003), Vol. 13A, pp. 17–30.

74. J. Rossmeisl, J. K. Nørskov, C. D. Taylor M. J. Janik and M. Neurock, *J. Phys. Chem. B* 110, 21833 (2006).

75. C. D. Taylor, R. G. Kelly and M. Neurock, *J. Electrochem. Soc.* 153, E207–E214 (2006).

76. C. D. Taylor, R. G. Kelly and M. Neurock, *J. Electrochem. Soc.* 154 (3), F55–F64 (2007).

77. H. J. T. Ellingham, *J. Soc. Chem. Ind. Lond.* 63, 125 (1944).

78. A. Anderko and P. J. Shuler, *Comput. Geosci.* 23, 647–658 (1997).

79. K. R. Cooper and R. G. Kelly, *Corros. Sci.* 49, 2636–2662 (2007).

80. N. J. Laycock, M. H. Moayed and R. C. Newman, *J. Electrochem. Soc.* 145, 2622–2628 (1998).

81. N. J. Laycock, S. P. White, J. S. Noh, P. T. Wilson and R. C. Newman, *J. Electrochem. Soc.* 145, 1101–1108 (1998).

82. N. J. Laycock, J. S. Noh, S. P. White and D. P. Krouse, *Corros. Sci.* 47, 3140 (2005).

83. W. Sun and S. Nesic, *Corrosion* 2007, 076551 (2007).

84. M. S. Venkatraman, I. S. Cole and B. Emmanuel, *Electrochim. Acta* 56, 8192–8203 (2011).

85. P. B. Balbuena, K. P. Johnston and P. J. Rossky, *J. Phys. Chem.* 100, 2706–2715 (1996).

86. G. Barabino, C. Gavotti and M. Marchesi, *Chem. Phys. Lett.* 104, 478–484 (1984).

87. P. S. Crozier, R. L. Rowley and D. Henderson, *J. Chem. Phys.* 114, 7513–7517 (2001).

88. D. Dominguez-Ariza, C. Hartnig, C. Sousa and F. Illas, *J. Chem. Phys.* 121 (2), 1066–1073 (2004).

89. D. L. Price and J. W. Halley, *J. Chem. Phys.* 102 (16), 6603–6612 (1995).

90. E. Spohr, *Solid State Ion.* 150, 1–12 (2002).

91. N. J. Laycock and R. C. Newman, *Corros. Sci.* 40, 887–902 (1998).

92. S. Nesic, *Chem. Eng. Sci.* 61, 4086–4097 (2006).

93. S. A. Wasileski, C. D. Taylor and M. Neurock, *Device and Materials Modeling in PEM Fuel Cells* (2009).

94. G. Jerkiewicz, J. J. Borodzinski, W. Chrzanowski and B. E. Conway, *J. Electrochem. Soc.* 142, 3755–3763 (1995).

95. C. D. Taylor, M. Neurock and J. R. Scully, *J. Electrochem. Soc.* 158 (3), F36–F44 (2011).

96. T. P. Radhakrishnan and L. L. Shrier, *Electrochim. Acta* 11, 1007–1021 (1966).

97. R. D. McCright, T. Zakroczymski and Z. Szklarska-Smialowska, *J. Electrochem. Soc.* 11, 2548–2552 (1973).

98. E. Protopopoff and P. Marcus, *Surface effects on Hydrogen entry into Metals, in Corrosion Mechanisms in Theory and Practice*, 3rd edition. Edited by P. Marcus (CRC Press, 2012), pp. 105–148.

99. G. A. Di Bari and J. V. Petrocelli, *J. Electrochem. Soc.* 112, 99–104 (1965).

100. U. Narkiewicz and W. Arabczyk, *Langmuir* 15, 5790–5794 (1999).

101. J. Oudar and P. Marcus, *Appl. Surf. Sci.* 5, 48 (1979).

102. P. Marcus, A. Teissier and J. Oudar, *Corros. Sci.* 24, 259 (1984).

103. A. B. Anderson and T. V. Albu, *J. Electrochem. Soc.* 147, 4229–4238 (2000).

104. T. Bligaard, J. K. Nørskov, S. Dahl, J. Matthiesen, C. H. Christensen and J. Sehested, *J. Catal.* 224 (1), 206–217 (2004).

105. F. Gao and D. W. Goodman, *Annu. Rev. Phys. Chem.* 63, 265–286 (2012).

106. J. Greeley, *ECS Trans.* 16 (2), 209–213 (2008).

107. J. K. Nørskov, F. Abild-Peterson, F. Studt and T. Bligaard, *Proc. Natl. Acad. Sci.* 108, 937–943 (2011).

108. R. A. van Santen and M. Neurock, *Molecular Heterogeneous Catalysis: A Conceptual and Computational Approach.* (Wiley VCH, Weinheim, 2006).

109. E. Gileadi, *Chem. Phys. Lett.* 393, 421–424 (2004).

110. E. Gileadi, *Isr. J. Chem.* 48, 121–131 (2008).

111. E. Gileadi, *J. Solid State Electrochem.* 15, 1359–1371 (2011).

112. D. D. Macdonald, *J. Electrochem. Soc.* 139 (12), 3434–3449 (1992).

113. (a) P. Marcus, *Advances in Localized Corrosion* (NACE, Houston TX, 1990), pp. 289. (b) K. J. Vetter and H.-H. Strehblow, *Ber. Bunsenges. Phys. Chem.* 74, 1024 (1970).

114. (a) R. S. Lillard, G. F. Wang and M. I. Baskes, *J. Electrochem. Soc.* 153, B358–B364 (2006). (b) K. E. Heusler and Z. Elektrochem. *Ber. Bunsenges. Phys. Chem.* 62, 582 (1958).

115. J. W. Halley, B. Johnson, D. Price and M. Schwalm, *Phys. Rev. B* 31 (12), 7695–7709 (1985).

116. J. W. Halley, A. Mazzolo, Y. Zhou and D. Price, *J. Electroanal. Chem.* 450, 273–280 (1998).

117. J. W. Halley and D. Price, *Phys. Rev. B* 35 (17), 9095–9102 (1987).

118. J. W. Halley, B. B. Smith, S. Walbran, L. A. Curtiss, R. O. Rigney, A. Sutjianto, N. C. Hung, R. M. Yonco and Z. Nagy, *J. Chem. Phys.* 110 (13), 6538–6552 (1999).

119. D. L. Price and J. W. Halley, *Phys. Rev. B* 38, 9357–9367 (1988).

120. B. B. Smith and J. W. Halley, *J. Chem. Phys.* 101, 10915–10924 (1994).

121. S. Walbran, A. Mazzolo, J. W. Halley and D. L. Price, *J. Chem. Phys.* 109, 8076 (1998).

122. A. B. Anderson and T. V. Albu, *J. Am. Chem. Soc.* 121, 11855–11863 (1999).

123. A. B. Anderson and T. V. Albu, *Electrochem. Commun.* 1, 203–206 (1999).

124. P. Biedermann, E. Torres and A. Blumenau, in *212th Meeting of the Electrochemical Society* (The Electrochemical Society, Washington D.C., 2007).

125. R. A. Sidik and A. B. Anderson, *J. Electroanal. Chem.* 528, 69–76 (2002).

126. F. U. Renner, A. Stierle, H. Dosch, D. M. Kolb, T.-L. Lee and J. Zegenhagen, *Nature* 439, 707–710 (2006).

127. D. M. Artymowicz, J. Erlebacher and R. C. Newman, *Philos. Mag.* 89, 1663–1693 (2009).

128. J. Morales, P. Esparza, S. Gonzalez, L. Vazquez, R. C. Salvarezza and A. J. Arvia, *Langmuir* 12, 500–507 (1996).

129. (a) F. Wiame, B. Salgin, J. Swiatowska-Mrowiecka, V. Maurice and P. Marcus, *J. Phys. Chem. C* 112, 7540–7543 (2008). (b) H.-H. Strehblow, *Passivity of Metals in Advances in Electrochemical Science and Engineering*. Edited by R. C. Alkire and D. M. Kolb (Wiley VCH, Weinheim, 2003), pp. 271–374. (c) H.-H. Strehblow, V. Maurice and P. Marcus, *Passivity of Metals in Corrosion Mechanisms in Theory and Practice*, 3rd edition, Edited by P. Marcus (CRC Press, 2012), pp. 235–325.

130. I. Diez-Perez, F. Sanz and P. Gorostiza, *Curr. Opin. Solid State Mater. Sci.* 10, 144–152 (2006).

131. U. R. Evans, *Metallic Corrosion, Passivation and Protection*. (Edward Arnold, London, England, 1937).

132. S. Hendy, B. Walker, N. Laycock and M. Ryan, *Phys. Rev. B* 67, 085407–085416 (2003).

133. B. MacDougall and M. J. Graham, *Growth and Stability of Passive Films, in Corrosion mechanisms in Theory and Practice*. Edited by P. Marcus (Marcel Dekker, Inc., New York, 2002), pp. 189–216.

134. P. M. Natishan, H. S. Isaacs, M. Janik-Czachor, V. A. Macagno, P. Marcus and M. Seo *Passivity and its Breakdown*. (The Electrochemical Society, Pennington, NJ, 1998).

135. S. Song, W. Song and Z. Fang, *Corros. Sci.* 31, 395–400 (1990).

136. J. W. Schultze and M. M. Lohrengel, *Electrochim. Acta* 45, 2499–2513 (2000).

137. P. Marcus, V. Maurice and H.-H. Strehblow, *Corros. Sci.* 50, 2698–2704 (2008).

138. R. Reigada, F. Sagues and J. M. Costa, *J. Chem. Phys.* 101, 2329–2337 (1994).

139. V. S. Sastri, *Corrosion Inhibitors: Principles and Applications*. (Wiley, New York, NY, 1998).

140. M. Kang, *J. Dispers. Sci. Technol.* 27, 587–597 (2006).

141. S. G. Zhang, W. Lei, M. Z. Xia and F. Y. Wang, *J. Mol. Struct. (THEOCHEM)* 732, 173–182 (2005).

142. J. M. Costa and J. M. Lluch, *Corros. Sci.* 24, 929–933 (1984).

143. M. M. Islam, B. Diawara, P. Marcus and D. Costa, *Catal. Today* 177, 39–49 (2011).

144. A. R. Bailey, *Met. Rev.* 6, 101–142 (1961).

145. T. Misawa, *Corros. Sci.* 18, 199–216 (1978).

146. T. Shahrabi, R. C. Newman and K. Sieradzki, *J. Electrochem. Soc.* 140, 348–352 (1993).

147. P. S. Sidky, *J. Nucl. Mater.* 256, 1–17 (1998).

148. D. Tromans, *Acta Metall. Mater.* 42 (6), 2043–2049 (1994).

149. G. S. Duffo and S. B. Farina, *Corros. Sci.* 47, 1459–1470 (2005).

150. S. B. Farina and G. S. Duffo, *Corros. Sci.* 46, 2255–2264 (2004).

151. F. Friedersdorf and K. Sieradzki, *Corrosion* 52, 331–336 (1996).

152. V. Y. Gertsmann and S. M. Bruemmer, *Acta Mater.* 49, 1589–1598 (2001).

153. S. Jain, N. D. Budiansky, J. L. Hudson and J. R. Scully, *Corros. Sci.* 52, 873–885 (2010).

154. A. King, G. Johnson, D. Engelberg, W. Ludwig and J. Marrow, *Science* 321, 382–385 (2008).

155. E. M. Lehockey, A. M. Brennenstuhl and I. Thompson, *Corros. Sci.* 46, 2383–2404 (2004).

156. X. Liu, G. S. Frankel, B. Zoofan and S. I. Rokhlin, *Corros. Sci.* 49, 139–148 (2007).

157. T. Kubo, Y. Wakashima, K. Amano and M. Nagai, *J. Nucl. Mater.* 132, 1–9 (1985).

158. R. Lillard, *Electrochem. Solid-State Lett.* 6, B29–B31 (2003).

159. R. Lillard, G. Wang and M. Baskes, *J. Electrochem. Soc.* 153, B358-B364 (2006).

160. B. Harmon, *J. Phys. Conf. Ser.* 16, 273–276 (2005).

161. J. P. Perdew and A. Ruzsinszky, *Int. J. Quantum Chem.* 110 (15), 2801–2807 (2010).

162. K. Burke, *J. Chem. Phys.* 136 (15), 150901 (2012).

163. L. Pauling and E. B. Wilson Jr., *Introduction to Quantum Mechanics with Applications to Chemistry.* (Dover Publications, Mineola, NY, 1985).

164. M. C. Payne, M. P. Teter, D. C. Allan, T. A. Arias and J. D. Joannopoulos, *Rev. Mod. Phys.* 64 (4), 1046–1097 (1992).

165. M. P. Allen and D. J. Tildesley, *Computer Simulation of Liquids.* (Oxford Science Publications, Oxford, UK, 1989).

166. A. Szabo and N. S. Ostlund, *Modern Quantum Chemistry: Introduction to Advanced Electronic Structure.* (Dover Publications, Mineola, 1996).

167. I. Prodan, G. Scuseria and R. Martin, *Phys. Rev. B* 76 (3) (2007).

168. R. C. Albers, N. E. Christensen and A. Svane, *J. Phys. Condens. Matter* 21, 343201 (2009).

169. P. Hohenberg and W. Kohn, *Phys. Rev.* 136, 864B (1964).

170. R. G. Parr and W. Yang, *Density–Functional Theory of Atoms and Molecules.* (Oxford University Press, Oxford, 1994).

171. A. E. Mattsson, R. Armiento, J. Paier, G. Kresse, J. M. Wills and T. R. Mattsson, *J. Chem. Phys.* 128 (8), 084714 (2008).

172. J. Perdew, A. Ruzsinszky, G. Csonka, O. Vydrov, G. Scuseria, L. Constantin, X. Zhou and K. Burke, *Phys. Rev. Lett.* 100 (13) (2008).

173. C. J. Cramer and D. G. Truhlar, *Phys. Chem. Chem. Phys.* 11 (46), 10757–10816 (2009).

174. V. R. Cooper, *Phys. Rev. B* 81 (16) (2010).

175. K. Lee, É. D. Murray, L. Kong, B. I. Lundqvist and D. C. Langreth, *Phys. Rev. B* 82 (8) (2010).

176. M. L. Rossi and C. D. Taylor, *J. Nucl. Mater.* 433, 30–36 (2013).

177. A. Krishtal, D. Geldof, K. Vanommeslaeghe, C. V. Alsenoy and P. Geerlings, *J. Chem. Theory Comput.* 8 (1), 125–134 (2012).

178. S. Grimme, *J. Comput. Chem.* 27 (15), 1787–1799 (2006).

179. J. Klimes, D. R. Bowler and A. Michaelides, *J. Phys. Condens. Matter* 22 (2), 022201 (2010).

180. A. E. Mattson, P. A. Schultz, M. P. Desjarlais, T. R. Mattson and K. Leung, *Model. Simul. Mater. Sci. Eng.* 13, R1–R31 (2005).

181. O. K. Anderson and O. Jepsen, *Phys. Rev. Lett.* 53, 2571–2574 (1984).

182. A. B. Anderson, *J. Chem. Phys.* 62, 1187–1188 (1975).

183. C. R. Herbers, C. Li and N. F. A. van der Vegt, *J. Comput. Chem.* 34, 1177–1188 (2013).

184. C. D. Taylor, *Chem. Phys. Lett.* 469, 99–103 (2009).

185. P. C. Gehlen, J. R. Beeler and R. J. Jaffee, (Plenum Press, New York, NY, 1972).

186. M. J. Stott and E. Zaremba, *Phys. Rev. B* 22, 1564 (1980).

187. C.D. Taylor, *Phys. Rev. B* 80 (2) (2009).

188. M. Baskes, *Phys. Rev. B* 46, 2727–2742 (1992).

189. M. I. Baskes, S. G. Srinivasan, S. M. Valone and R. G. Hoagland, *Phys. Rev. B* 75, 94113–94128 (2007).

190. M. I. Baskes and R. A. Johnson, *Model. Simul. Mater. Sci. Eng.* 2 (1), 147–163 (1994).

191. M. I. Baskes, S. G. Srinivasan, S. M. Valone and R. G. Hoagland, *Phys. Rev. B* 75, 94113 (2007).

192. M. S. Daw and M. I. Baskes, *Phys. Rev. B* 29 (12), 6443–6453 (1984).

193. S. M. Foiles, M. I. Baskes and M. S. Daw, *Phys. Rev. B* 33, 7983–7991 (1986).

194. B.-J. Lee and M. I. Baskes, *Phys. Rev. B* 62 (13), 8564–8567 (2000).

195. F. H. Stillinger and T. A. Weber, *Phys. Rev. B* 31 (8), 5262–5271 (1985).

196. J. C. Shelley, G. N. Patey, D. R. Berard and G. M. Torrie, *J. Chem. Phys.* 107, 2122–2141 (1997).

197. J. D. Gale, *J. Chem. Soc. Faraday Trans.* 93, 629–637 (1997).

198. F. Streitz and J. Mintmire, *Phys. Rev. B* 50 (16), 11996–12003 (1994).

199. J. Chen and T. J. Martínez, *Chem. Phys. Lett.* 438 (4–6), 315–320 (2007).

200. J. H. Rose, J. R. Smith, F. Guinea and J. Ferrante, *Phys. Rev. B* 29, 2963 (1984).

201. M. J. Buehler, A. C. T. van Duin and W. A. Goddard III, *Phys. Rev. Lett.* 96, 95505–95508 (2006).

202. A. C. T. van Duin, S. Dasgupta, F. Lorant and W. A. Goddard III, *J. Phys. Chem. A* 105, 9396–9409 (2001).

203. Q. Zhang, T. Cagin, A. van Duin, W. A. Goddard, Y. Qi and L. G. Hector, *Phys. Rev. B* 69, 045423 (2004).

204. H. Y. Wei, S. Z. Luo, G. P. Liu, Z. L. Xiong and H. T. Song, *Acta Phys.-Chim. Sin.* 24 (11), 1964–1968 (2008).

205. S. E. Wonchoba, W.-P. Hu and D. G. Truhlar, *Phys. Rev. B* 51, 9985–10002 (1995).

206. J. G. Yu, K. M. Rosso and S. M. Bruemmer, *J. Phys. Chem. C* 116 (2), 1948–1954 (2012).

207. A. Bouzoubaa, D. Costa, B. Diawara, N. Audiffren and P. Marcus, *Corros. Sci.* 52 (8), 2643–2652 (2010).

208. M. Legrand, B. Diawara, J.-J. Legendre and P. Marcus, *Corros. Sci.* 44, 773–790 (2002).

209. O. K. Rice, *Statistical Mechanics, Thermodynamics and Kinetics*. (Freeman, New York, NY, 1967).

210. R. V. Pappu, R. K. Hart and J. W. Ponder, *J. Phys. Chem. B* 102, 9725–9742 (1998).

211. F. Jensen, *Introduction to Computational Chemistry*. (Wiley, New York, NY, 1999).

212. C. D. Taylor, M. J. Janik, M. Neurock and R. G. Kelly, *Mol. Simul.* 33 (4–5), 429–436 (2007).

213. R. A. van Santen, M. Neurock and S. G. Shetty, *Chem. Rev.* 110 (4), 2005–2048 (2010).

214. S. Glasstone, K. J. Laidler and H. Eyring, *The Theory of Rate Processes*. (McGraw-Hill, New York, NY, 1942).

215. G. Henkelman, B. P. Uberuaga and H. Jonsson, *J. Chem. Phys.* 113, 9978 (2000).

216. G. Henkelman, B. P. Uberuaga and H. Jonsson, *J. Chem. Phys.* 113, 9901 (2000).

217. D. Sheppard and G. Henkelman, *J. Comput. Chem.* 32, 1769–1771 (2011).

218. D. Sheppard, R. Terrell and G. Henkelman, *J. Chem. Phys.* 128, 134106 (2008).

219. D. Sheppard, P. Xiao, W. Chemelewski, D. D. Johnson and G. Henkelman, *J. Chem. Phys.* 136, 074103 (2012).

220. S. Nose, *J. Chem. Phys.* 81, 511 (1984).

221. S. Nose, *Prog. Theor. Phys. Suppl.* 103, 1 (1991).

222. M. S. Daw and M. I. Baskes, *Phys. Rev. Lett.* 50, 1285–1288 (1983).

223. S. Serebrinsky, E. A. Carter and M. Ortiz, *J. Mech. Phys. Solids* 52, 2403–2430 (2004).

224. M. Wen, X.-J. Xu, Y. Omura, S. Fukuyama and K. Yokogawa, *Comput. Mater. Sci.* 30, 202–211 (2004).

225. J.-M. Delaye and D. Ghaleb, *Phys. Rev. B* 61, 14481–14494 (2000).

226. B. J. Teppen, K. Rasmussen, P. M. Bertsch, D. M. Miller and L. Shafer, *J. Phys. Chem. B* 101, 1579–1587 (1997).

227. E. Spohr, G. Toth and K. Heinzinger, *Electrochim. Acta* 41, 2131–2144 (1996).

228. A. F. Voter, F. Montalenti and T. C. Germann, *Annu. Rev. Mater. Res.* 32, 321–346 (2002).

229. D. Perez, B. P. Uberuaga, Y. Shim, J. G. Amar and A. F. Voter, in *Annual Reviews in Computational Chemistry*. Edited by R. A. Wheeler (Elsevier, The Netherlands, 2009), Vol. 5.

230. M. E. J. Newman and G. T. Barkema, *Monte Carlo Methods in Statistical Physics*. (Oxford University Press, Oxford, UK, 1999).

231. M. P. Allen and D. J. Tildesley, *Computer Simulation of Liquids*. (Oxford University Press, Oxford, UK, 1987).

232. A. F. Voter, *Radiation Effects in Solids*. Edited by K. E. Sickafus and E. A. Kotomin (Springer, NATO Publishing Unit, Dordrecht, The Netherlands, 2005).

233. J. R. Kitchin, K. Reuter and M. Scheffler, *Phys. Rev. B* 77, 075437 (2008).

234. A. V. Ruban and I. A. Abrikosov, *Rep. Prog. Phys.* 71, 046501 (2008).

235. S. Miertus, E. Scrocco and J. Tomasi, *Chem. Phys.* 55 (1), 117–129 (1981).

236. S. Desai and M. Neurock, *Phys. Rev. B* 68, 75420–75426 (2003).

237. C. D. Taylor, S. A. Wasileski, J. W. Fanjoy, J.-S. Filhol and M. Neurock, *Phys. Rev. B* 73, 165402–165417 (2006).

238. S. Y. Park, J. H. Kim, B. K. Choi and Y. H. Jeong, *Met. Mater. Int.* 13, 155–163 (2007).

2

MOLECULAR MODELING OF STRUCTURE AND REACTIVITY AT THE METAL/ENVIRONMENT INTERFACE

CHRISTOPHER D. TAYLOR

Strategic Research & Innovation, DNV GL, Dublin, OH, USA
Fontana Corrosion Center, Department of Materials Science and Engineering,
The Ohio State University, Columbus, OH, USA

2.1 INTRODUCTION

It is a common point of view that the chemical nature and structure of the surface of a metal, which is in contact with an electrolyte, are decisive in the kinetics of electrochemical reactions that proceed on this surface.

M. A. Pletnev, 2000

Corrosion is a strongly coupled process that consists of three elements: chemical reaction, electron transfer, and mass transport. The *common point of view*, alluded to in the quotation by Pletnev above, applies to the former two subprocesses [1]. Thus, in cases in which mass transport is not the limiting factor, the surface structure and reactivity become preeminent in corrosion control.

Corrosion in aqueous environments is an electrochemical process: hence, coupled anodic and cathodic reactions take place at unique sites distributed across the materials/environment interface [2, 3]. The reactions themselves involve transfer of electrons or ions—often both—across an electrochemical double layer [4]. For this reason, the mechanisms via which corrosion proceeds can be strongly influenced by perturbations in the surface and interfacial environment.

Molecular Modeling of Corrosion Processes: Scientific Development and Engineering Applications, First Edition.
Edited by Christopher D. Taylor and Philippe Marcus.
© 2015 John Wiley & Sons, Inc. Published 2015 by John Wiley & Sons, Inc.

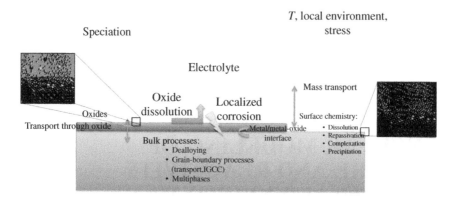

FIGURE 2.1 Depending on the material and the environment, a number of candidate processes may control the corrosion properties of a material. These include processes inside the bulk material, such as dealloying, grain-boundary segregation and transport, or the precipitation of secondary phases; processes at the metal/metal-oxide interface such as point defect creation and annihilation; processes internal to the oxide, such as point defect migration and short-circuit diffusion; reactivity at the oxide/environment interface, such as cation dissolution; localized corrosion that may involve alternative chemical processes, such as oxide dissolution; processes at the metal/solution interface, such as bare metal dissolution, repassivation, complexation, and precipitation; and processes in the environment itself that may play a role, such as speciation. Even this complex picture omits several important subprocesses, including phase transformations and spalling of corrosion product films. Reproduced with permission from Taylor [5].

Although it is common to see corrosion processes reduced to a candidate set of chemical reactions, in many cases, such sets should be considered as placeholders: representative entities that subsume a host of microkinetic processes such as mass transport, surface adsorption and desorption, and bond-making/bond-breaking chemical reactions. Pletnev's *chemical nature and structure* of the materials/environment interface are, in fact, rather complex quantities, as the schematic in Figure 2.1 indicates [6]. Taking a point of view that begins with the environment, the elements that will lead to corrosion are initiated by speciation of ions and neutral molecules in the electrolyte or vapor phase, processes that are moderated by factors such as temperature, partial pressures, and pH. Other externalities that control corrosion include local stresses, the geometry of macro- and microstructures, and fluid flow (i.e., mass transport). The first line of defense for most structural materials is the passive oxide film that provides kinetic control over general (i.e., uniform) corrosion. The degree of control provided is a function of the transport properties of point defects across the metal/metal-oxide and metal-oxide/environment interfaces, as well as through the film itself. Transport across the latter has been demonstrated to be a function of the propensity of *short-circuit* diffusion paths, such as grain boundaries. Localized corrosion may initiate at such weakest links in the oxide phase or at stress points that fracture the oxide film. Under these conditions, active dissolution may take place, in which case metal dissolution kinetics compete with repassivation, complexation, and precipitation. Neutral or ionic adsorption may also take place under these conditions, including the formation of self-assembled monolayers. The final set of aspects that can factor into a thorough analysis of the corrosion properties of a material include the bulk properties: ingrowth of species such as hydrogen or carbon via grain boundaries can lead to intergranular failure, for example, and hence, the microstructure and near-surface composition can play a decisive role.

Scientists working in the fields of surface science and solid-state physics have made it possible to directly simulate a variety of such interfacial reactions with atomistic resolution using electronic structure methods or interatomic potentials, finding particular applications in the design and modeling of heterogeneous catalysts [7–9], but also, more recently, in corrosion science also [5]. Whereas heterogeneous catalysis might be described as the deployment of a material to manipulate its environment, events in a corrosion scenario constitute an inverse process: modification and loss of integrity of a material due to its interaction with its environment. The deterioration of a solid-state material as a consequence of its interactions with a fluid phase, however, involves processes that are typically slower and involve higher-energy transition states than the catalysis of molecular transformations by a substrate [10]. This is, in general, a very good thing. However, the goal of corrosion science is to understand these slow processes and design for their mitigation by the implementation of environmental or materials control.

Although it is not, at the present time, possible to include all of the elements contained within Figure 2.1 within a comprehensive molecular dynamics simulation of a corrosion event (although for some particular mechanisms and materials, this has been attempted) [11, 12], it is possible to deconstruct certain elements deemed to be critical within a given application. For instance, in cases of localized corrosion or corrosion cracking, one may focus on the chemistry occurring at a bare metal/solution interface [13–15]. In cases where oxide growth is relevant, one can focus on ensembles that consist of defect states in the oxide material [16, 17]. It is also possible to look at dissolution processes from oxide surfaces or to interrogate questions regarding bulk or surface thermodynamic stability [5, 18, 19]. The field of computational materials science is currently growing such that, by coupling continuum kinetic or fracture mechanics models with atomistic scale codes or phase-field theory, it may soon be possible to realistically simulate increasingly comprehensive portions of Figure 2.1 [11, 12, 20].

Atomistic modeling techniques can, at present, be grouped into four main categories: molecular modeling, static band-structure calculations, molecular dynamics, and Monte Carlo methods. Monte Carlo methods are reviewed in more detail in separate chapters of this volume [21–23]. Molecular modeling has its origins in chemistry and molecular orbital theory [23]. Molecular orbital theory is used to solve the electronic structure problem that defines the bonding relations between the atoms, as well as the interactions between molecules that may be associated or reacting. The molecular wave function that is produced in these calculations in turn allows access to molecular properties, such as the highest occupied molecular orbital–lowest unoccupied molecular orbital (HOMO–LUMO, respectively) gap, bond lengths and bond angles, and energetics of reactant, intermediate, transition, and product states. Extensions of molecular orbital theory allow, for example, the modeling of cationic states of metals when dissolved in solution phase [24, 25], the study of metallic nanoparticles [26], investigations of bonding of water and oxygen to metal atoms and clusters [27, 28], and the evaluation of molecular properties of chemical inhibitors [29, 30]. Semiempirical and empirical versions of the theory also exist that can provide a more rapid screening of molecules for desirable physical or chemical properties.

Static band-structure calculations have recently flourished due to developments in the accuracy and computational efficiency of density functional theory calculations [31]. These calculations were originally developed to evaluate the properties of solids, in a similar way to how molecular orbital theory calculations are used to evaluate the properties of molecules. From a reaction point of view, the electronic and atomistic properties of reactant, product, intermediate, and transition states can also be evaluated using these methods, allowing insights to be gained regarding rate-limiting steps, for example, in mass-transport processes such as solid-state diffusion [32–34] or phase transformations [35]. These techniques can be extended to look at surfaces as well, as has been applied heavily in the fields of heterogeneous catalysis [7] and electrocatalysis [8] and, more recently, to problems in corrosion [18, 36–40]. This technique, as well as molecular modeling, is limited primarily by the computational cost of performing such calculations. Ensembles containing small numbers of atoms, either isolated as a cluster in space or in cells with periodic boundary conditions, must therefore be carefully constructed to imitate the system one is trying to model, and the appropriate caveats should be given regarding artifacts that may be induced by these idealized conditions. Furthermore, the inability to readily incorporate dynamic effects means that reactions that require significant thermal reorganization—such as the formation of solvation shells around ions during a dissolution event—may be difficult to capture reliably. Overall, however, these methods can be powerfully applied to interrogate detailed, atomistically resolved reaction mechanisms and mass-transport processes relevant to corrosion. Their dependence upon electronic structure evaluation makes them the most flexible and reliable, in terms of fundamental, deterministic physics, but at the same time, makes them exceedingly expensive in terms of computational cost and user time.

Molecular dynamics simulations can be used to simulate reaction events, particularly those involving high-energy or thermally stimulated processes. While simulations using electronic structure techniques to evaluate the forces on atoms and their thermal trajectories can be performed, the cost of such methods is usually far too high to enable useful timescales to be reached within the scope of the simulation. Ab initio simulations can presently be performed up to no more than 10–100 ps at most. Therefore, modelers employing molecular dynamics will typically apply interatomic potential techniques. These techniques employ algorithms for evaluating forces on atoms using only their relative positions in space without a self-consistent computation of the electron structure at each step, for example, the embedded atom method [41–43]. Based on this technique, one can simulate the evolution of corrosion morphologies; diffusion processes in materials; the behavior of defects in materials, such as grain boundaries; and the rates at which oxidation or other processes occur. Examples include the oxidation of aluminum and zirconium metals [11, 17], the structure of the metal–water interface [44–52], and the evolution of radiation damage effects in nanolayered materials [53].

In the following sections, a variety of applications of the above techniques to provide an improved understanding of the structure and reactivity of the materials/environment interface are reviewed in detail. The implications of the studies for our understanding of the relevance of surface chemical transformations to corrosion are also discussed. Finally, a summary is made of the types of materials/environment modeling that has been performed to date and what avenues may be explored in the near future.

2.2 STRUCTURE AND REACTIVITY OF WATER OVER METAL SURFACES

The smallest and simplest first-principles theoretical model for the metal/aqueous interface consists of a single metal atom in contact with a single molecule of water. However, the electronic structure of a single metal atom is rather distinct from that of a metal atom residing on the surface of a metal. On the other hand, the electronic character of a molecule of water, however, is not very different from that of water contained in a condensed phase. For this reason, much of our initial understanding of the adsorption and reactivity of water on a metal surface has been derived from first-principles calculations performed on single water molecule adsorption to metal clusters [54–56] or periodic metallic slabs [57–59]. Investigation of the molecular structure of water on metal surfaces provides a first approach to understanding more complex processes at the metal/environment interface, such as events taking place during active corrosion. By characterization of bond lengths, the electronic interactions between substrate and environment, and the strengths with which water adsorbs (i.e., the binding energy), one can develop an appreciation for how readily other species, chloride, for example, may attach to the metal surface.

Michaelides et al. performed a methodical study of water monomer adsorption on various close-packed (111) transition and noble metal surfaces [57]. It was found that water chemisorbs to metallic close-packed surfaces in a *horizontal orientation*, that is, the OH bonds are very slightly tilted out of the plane parallel to the metal surface. For incompletely filled d-shell transition metals, the empty d_{z^2} states on the metal surface can be populated by the $3a_1$ and $1b_1$ frontier molecular orbitals of a closed-shell water molecule, creating a relatively strong water chemisorption. When, on the other hand, the d_{z^2} state is fully occupied, as in the more noble late transition metals, a four-electron chemisorption bonding interaction is initiated. In this bonding state, those electrons that would otherwise occupy antibonding states (thus preventing bonding in a simple molecular/atomic scenario) are relieved by undergoing transfer into the Fermi level conduction band of the metal substrate. Michaelides' band-structure calculations determined the optimum metal–oxygen bond lengths, ranging from 2.25 Å for Cu to 3.02 Å for Au. The adsorption energies, ΔE_{ads}, were comparable to hydrogen bond strengths: 0.13 eV for Au and 0.42 eV for Rh. The relative binding energies between hydrogen bonding in water and water–metal chemisorption can be employed to assess whether or not water will be expected to form ordered structures when it encounters a metal surface. In particular, Meng et al. established a hydrophilicity (or wettability) parameter for this very purpose [60]:

$$w = \frac{E_{HB}}{\Delta E_{ads}} \qquad (2.1)$$

When $w \leq 1$, water–metal interactions are dominant, and water molecules will wet the metal surface. When $w \gg 1$, on the other hand, water–water interactions are more significant, resulting in cluster formation rather than surface wetting. For Au(111) w is close to 3, whereas on Pt(111) $w \sim 1$. In line with this theoretical prediction, formation of ordered structures on Pt(111) has been confirmed using low-energy electron diffraction [61, 62].

The question may be asked how is this behavior modified at the electrochemical interface. That is, water molecule chemisorption may be sensitive to the presence of local electric fields or variations in the applied electrochemical potential as a result of the molecular dipole. The response of water molecules to such a field has been simulated by Sanchez [63]. Variations in the local charge density on a cluster model for Ag(111) were used to investigate this phenomenon. Polarization of the surface to ±15 μC/cm² was observed to produce a *standing structure* in which water molecules are bound via the oxygen atom, with the hydrogen atoms directed away from the surface in the normal direction. Consequently, the phenomenon known as dielectric saturation ensues, in which there exists a reduced dielectric constant at the interface, and, hence, electrostatic interactions across the metal/aqueous interface become enhanced.

The next logical question concerns the cluster of water molecules—that is, do two or more water molecules on a metal surface become more or less than the sum of their parts?

(a)   (b)

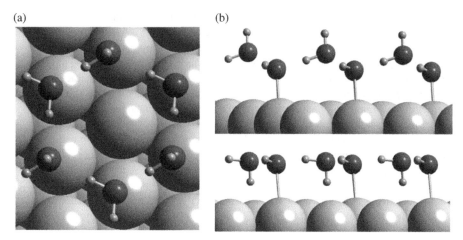

FIGURE 2.2   Several studies of the metal–water interface have revealed that water has a tendency to structure near the metal surface. A commonly observed motif is the bilayer structure, shown here both in top view (a) and in the side view (b) in two separate orientations. Reproduced with permission from Taylor and Neurock [65]. © Elsevier.

Due to the similarity in binding strengths between hydrogen bond and metal–water interactions, intricate structures are found to emerge. One common motif that has emerged from both computational and surface science studies of metal–water interfaces is the hexagonal bilayer arrangement. The classical example is the wetting layer formed by water as it adsorbs onto Ru(0001) by Doering and Madey [64]. Water orients in a hexagonal arrangement consisting of two water layers that are closely coordinated with the surface having only a small separation of about an ångstrom. The top view of the bilayer is shown in Figure 2.2a, and the side view in Figure 2.2b. A number of theoretical studies of bilayer structures on transition and noble metal surfaces were performed following this observation [66–71]. Formation of hydrogen-bonding networks can be advantageous for electrochemical events that occur at the interface, as proton transport becomes rather facile. The connectedness of the hydrogen-bonded network, on the other hand, suggests that there will be some barrier to the approach of ions and other reactive species to the surface, as they would have to interrupt the network leading to substantial molecular reorganization. Variations on the intact bilayer structure include the dissociated water bilayer in which hydrogen transfers to the surface, leaving an OH species in the bilayer region [72–75] as well as other distinct two-dimensional hexagonal patterns based on an ice-like configuration [76]. The bilayer configuration is a function of both thermodynamic tendency and the kinetics of formation. The wettability parameter provides one means for inferring the relative tendencies for water molecules to organize at the interface.

Fluctuations in the structure of these water bilayers and the formation of multibilayers were modeled using density functional theory calculations for a Cu(111) surface [77]. The activation barrier for inversion of a bilayer between the configurations shown in Figure 2.2b is 0.16 (forward) or 0.3 eV (reverse). The lower barrier corresponds to water directly coordinated to the metal substrate with no hydrogen bonds engaged to the surrounding water matrix (Fig. 2.2b, lower image), since the metal stabilizes the inversion transition state by engaging with both the oxygen and hydrogen atoms of the H_2O in the inversion, whereas the severance of hydrogen bonds to nearby water molecules tends to increase the energy of the transition state.

Such studies provide a detailed glimpse at the steps that control molecular transformation at interfaces, but over the time scale of a corrosion event, many thousands or even millions of such reorganizations must be encountered. Hence, dynamic effects needs be considered. For this purpose, ensemble sampling methods, such as Monte Carlo [78–85], molecular dynamics, and integral equation techniques [86–92], have been used to estimate the time-averaged structural and electrical properties of the metal/aqueous interface. The general findings from these studies support the formation of icelike networks near the interface, consistent with the studies reported earlier on molecular systems. These configurations however are dynamic, consistent with the kind of inversion barriers computed for the water bilayers in contact with a metal surface that are on the same order as a hydrogen bond [44, 46, 51, 74, 75, 93–101]. Such fluctuations lead to variant structures that may consist of four-, five-, and six-membered rings. Molecular dynamics simulations have indicated that such structures may persist at the interface for timescales of up to 12.5 ps [94, 96].

Molecular dynamics simulations and Monte Carlo simulations of the metal–water interface have been performed with varying degrees of complexity (from approximation of the metal as a nonpolarizable, rigid wall [83, 84, 98, 99, 101] through to quantum mechanical representations), yet all methods consistently point toward the existence of very particular water structures that coalesce at the interface. This result is indicative of the fact that hydrogen bonding alone provides a strong enough thermodynamic impetus to impose structural requirements on interfacial H_2O, resulting in an icelike boundary condition when the bulk symmetry is terminated [99, 100, 102–104]. In such simulations, the structure of interfacial water may be characterized by oscillating density profile, having two main peaks that represent an interfacial bilayer.

Ab initio molecular dynamics simulations of the metal/aqueous interface were pioneered by Halley and coworkers [44–47]. Their papers, which incorporate successively higher levels of theoretical complexity [105–107], couple the physics of the electronic interactions that control bonding and/or chemisorption between the electrode and the molecules in solution with the internal solvent dynamics. The oxygen density distributions, derived at potentials positive and negative of the potential of zero charge, compare favorably with Toney's X-ray scattering experiments [108]. At cathodic potentials (negative surface charge), the water molecules were shown to be repelled from the surface (Fig. 2.3), and at positive potentials (positive surface charge), the first distribution peak is closer to the electrode. These distinctions emerge due to electrostatic effects between the electric field at the interface and the dipole of interfacial water.

Fully ab initio molecular dynamics simulations by Desai and Neurock were applied to simulate the dynamics and chemistry of water over the Pt/Ru bimetallic surface, important for catalysis [110, 111]. This ab initio model was capable of simulating arbitrary bond-breaking and bond-formation events across the O–H bonds, according to the dynamic fluctuations emerging in the electronic structure. Microscopic events, therefore, could be simulated by this approach, including the adsorption and activation of water molecular preferentially at the Ru sites to form adsorbed hydroxyl submonolayers. The complementary proton produced from dissociation diffused away into the solution phase via proton transfer across local hydrogen-bonding networks. Hydroxide transport across the surface was expedited by fast proton transfer throughout the interfacial water bilayer (Fig. 2.4). The hydrogen ion that forms in the near-surface region upon dissociation was shown to exist primarily as a Zundel $H_5O_2^+$ state and also as $H_9O_4^+$. The metal-mediated dissociation of H_2O on ruthenium was stabilized in solution as the surrounding water molecules stabilized the partial charge on the transition state by about 15 kJ/mol.

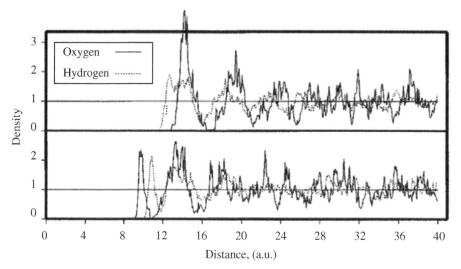

FIGURE 2.3 Hydrogen and oxygen correlation functions for (a) negative surface charge and (b) positive surface charge as a function of distance from the metallic slab. Reproduced with permission from Price and Halley [109]. © American Institute of Physics.

One may now extend such detailed study of the molecular/water interfaces to shed light on processes more directly relevant to corrosion. For instance, specific details regarding the reactivity of water molecules at the metal–water interface have been modeled with varying degrees of sophistication [27, 28, 112–116]. The dissociation of H_2O at the surface into products presents an important first probe reaction and has been related to the initiation of passivity of fresh metal surfaces when first exposed to aqueous solution:

$$H_2O_{(ads)} + e \rightarrow H_{(ads)} + OH^-_{(aq)} \qquad (2.2)$$

$$H_2O_{(ads)} \rightarrow H^+_{(aq)} + OH_{(ads)} + e \qquad (2.3)$$

$$OH_{(ads)} \rightarrow O_{(ads)} + H^+ + e \qquad (2.4)$$

Various cluster and periodic slab quantum mechanical calculations have been performed to determine the adsorption structure and energy for both water and its dissociation products over different metal surfaces. Anderson performed some of the first studies of this nature using the semiempirical ASED-MO method [112] to model the homolytic dissociation of H_2O on Pt(111) and Pt(100) at potentials of 0, 0.5, and 1.5 V relative to the potential of zero charge [113]. While the calculated energies were much higher than the more recent values provided by Michaelides et al. [57] (1.31 eV for H_2O), they show that, qualitatively, adsorption is stronger at more anodic potentials. In addition, a basis was provided for this shift in terms of the electronic structure: anodic potentials lower the d-band of Pt such that there is greater overlap between the d-orbitals at the surface and the lone-pair orbitals of H_2O, which strengthens the overall bonding interaction and reduces the occupation of antibonding states. This description coincides with Hoffman's four-electron attractive bond model recounted previously [117]. Of even more relevance to structural materials, Anderson also

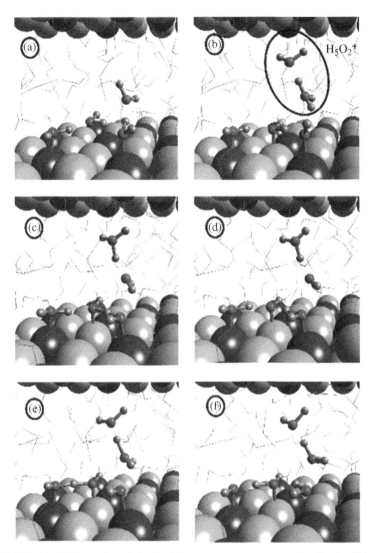

FIGURE 2.4 (a)–(b) Water dissociation initially occurs at the more reactive Ru site. This event is subsequently followed by proton transfer through the aqueous phase (c)–(f) and consequently the exchange of a proton between the surface H_2O species, such that the OH moiety migrates to a Pt site. Reproduced with permission from Desai and Neurock [111]. © American Physical Society.

modeled the behavior of iron clusters bonded to single molecules of water, hydronium, and hydroxyl species [27, 28, 114]. These theoretical calculations further illuminated the influence of the band structure on the stability of each of the adsorbates. The shifts in the d-band structure of surface Fe that result from the adsorption of OH destabilize the surface FeOH complex, weakening the Fe–Fe interactions, thus preparing the bonding environment for the dissolution of FeOH. Later extensions of this *local reaction center* model have been successful in estimating transfer coefficients and equilibrium potentials for water dissociation and oxygen reduction [63, 102–104, 118–120].

The ASED-MO model mimics the response of an interfacial system to electrochemical potentials by directly raising or lowering the energy levels of the atomic orbitals. Alternatively, one may model the structure and adsorption of water and its dissociated products upon clusters of metal atoms by varying the number of electrons available to the cluster (creating a charged surface) or by applying an electric field to the system Hamiltonian. Using the latter technique, Patrito et al. noted that elongation of Pt–OH adsorption bond lengths for the hydrated OH species, as well as a reorientation of the water dipoles, occurred in the presence of increasingly cathodic electric fields [115, 116]. Using charge accounting techniques, the adsorption of OH was also shown to involve a partial charge transfer from the OH species to the metal (0.3 electrons, as opposed to the conventional picture of one-electron transfer). Thus, the simplistic view of electron transfer exhibited in Equations 2.2 through 2.4 may in fact be more complicated and require the transfer of a partial number of electrons, while some electron density is retained by the local adsorption environment. At the same time, it should be noted that electron accounting from quantum mechanical calculations rarely conforms exactly to the notions of traditional chemical redox theory.

By directly applying an excess or deficit surface charge to a periodic slab model of the Pd(111)/H_2O interface, Neurock and Filhol directly modeled the electrochemical response of extended multilayers of water and its activation products to changes in the electrochemical potential [121]. This work marked the first direct implementation of an electrical double layer into a first-principles band-structure model for the simulation of chemistry at the electrochemical interface. By changing the number of electrons available to the metal surface, the attraction of water to the slab could be tuned, such that at negative surface charge densities, water was repelled and assumed the inward orientation (hydrogen atoms directed toward the metal) and at positive surface charge densities, water was attracted and assumed an outward orientation (hydrogen atoms directed toward the aqueous phase), as shown in Figure 2.5a. Further polarization of the interface in the anodic direction leads to the dissociation of water to form an adsorbed hydroxide, whereas polarization in the cathodic direction leads to the dissociation of water to form surface hydrogen. The energies of each of these phases can be determined following the decoupling scheme introduced earlier and described in the papers by Filhol and coworkers [121, 122]. The phase diagram derived by Neurock and Filhol for water and its dissociation products over Pd(111) is shown in Figure 2.5b. An outstanding challenge in this model is to couple the accurate evaluation of surface chemistry with dynamic restructuring of the aqueous phase. Only when a sufficiently thermodynamic ensemble is represented will calculations of this nature provide quantitative relations between electrochemical phase transformations and the potentials at which these transformations occur.

Calculations performed in a similar way for water and its reactive intermediates at the Ni(111)/water interface showed that the range of potentials for which water is inert over Pd(111) (see Fig. 2.5b) is nonexistent on Ni(111) [123]. The varying activity of metals for the dissociation of H_2O can, in fact, be inferred from the studies of bilayer wetting layers by Michaelides and coworkers [67], who calculated the energy to dissociate the hexagonal wetting layer of water into layers containing hydroxyl and atomic hydrogen on the surface.

This first-principles electrochemical model was directly applied to investigate the corrosion and destabilization of copper. A series of geometry optimizations was performed for a model Cu(111)–H_2O interface, with the potentials made successively more anodic at each step [77]. The model was, therefore, set up in an equivalent way to a voltammetry experiment, although kinetic processes were not explicitly taken into account. When the resulting geometries were plotted, it could be seen that, at various values for the surface charge density, phase

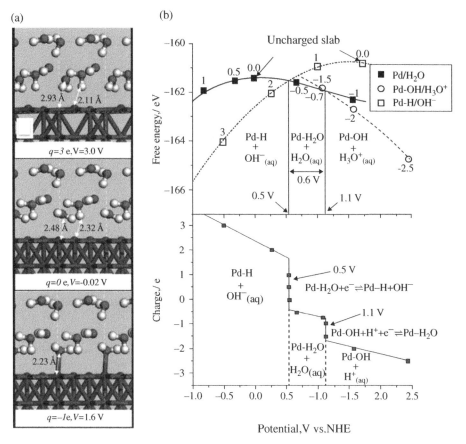

FIGURE 2.5 (a) Structural response of water to changes in the applied electrochemical potential. (b) *Upper*: energetic response of water to changes in the applied potential, with phase diagrams indicated for hydroxyl and hydrogen formation. *Lower*: the discontinuities in surface charge occurring upon water activation. Reproduced with permission from Filhol and Neurock [121]. © Wiley.

transitions on the surface were predicted. These began with the simple water dissociation reactions, described earlier, but ultimately led to destabilization of the metallic copper surface, culminating in place exchange between the electron-rich oxygen species and the electron-depleted copper atoms. This series of phase transformations is illustrated in Figure 2.6.

Taken altogether, the stepwise charging of the model system and subsequent relaxation of the atomic structure prompted the following structural changes (Fig. 2.6):

a. Chemisorption of H_2O to the atop site of a Cu atom:

$$H_2O_{(aq)} \rightarrow H_2O_{(atop)} \tag{2.5}$$

b. Migration of H_2O from the atop site to a bridge site:

$$H_2O_{(atop)} \rightarrow H_2O_{(bridge)} \tag{2.6}$$

(a)   (b)   (c)   (d)   (e)   (f)   (g)

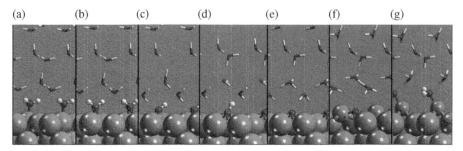

FIGURE 2.6 Charge-induced reaction sequence for the activation and migration of H_2O (a–c), OH (c–e), and O (e–g) on Cu(111), followed by oxidation of Cu (f–g) effected by adsorbed oxygen.

c. Deprotonation of the bridging H_2O to form $OH_{(ads)} + H^+_{(aq)} + e^-$:

$$H_2O_{(bridge)} \rightarrow HO_{(bridge)} + H^+_{(aq)} + e^- \tag{2.7}$$

d. Migration of the bridging OH to the threefold hollow site, consistent with the emergence of a second type of OH resonance in the surface-enhanced resonance spectrum recorded by Niaura [124]:

$$HO_{(bridge)} \rightarrow HO_{(3\,fold)} \tag{2.8}$$

e. Deprotonation of $OH_{(ads)}$ to form $O_{(ads)} + H^+_{(aq)} + e^-$:

$$HO_{(3\,fold)} \rightarrow O_{(3\,fold)} + H^+_{(aq)} + e^- \tag{2.9}$$

f. Surface reconstruction by eruption of a Cu adatom from the surface layer. Adsorbed oxygen acts as a tether:

$$Cu_{(metal)} + O_{(3\,fold)} \rightarrow Cu_{(bridge)} - O_{(bridge)} \tag{2.10}$$

g. Complete displacement of the copper adatom by oxygen, resulting in a Cu_2O-like adlayer, in which O is adsorbed at the vacancy site left by the Cu atom, and this partially adsorbed Cu atom is now coordinated to H_2O, according to the reaction:

$$H_2O_{(l)} + Cu_{(bridge)} - O_{(bridge)} \rightarrow H_2O - Cu_{(ads)} + O_{(ads)} + vacancy \tag{2.11}$$

This work was then generalized to a larger range of transition metals using a simplified version of the technique that does not require polarization of the interfacial cells, but simply shifts the free energy of electrochemical reactions by an energy term $nF\Delta U$ [15]. This method works most simply when the transfer of an electron and proton occurs simultaneously, as in the reaction schemes presented earlier, as these energies can be related to the energy of H_2 for a reference potential of the standard hydrogen electrode (SHE). Over the years, various methods have been employed to incorporate this effect, as compared by Rossmeisl et al. [125]

To summarize, the activation of water over metal surfaces requires that one follow the potential-dependent behavior of adsorbed water as well as its activation products, which include adsorbed hydrogen, hydroxyl, and oxygen. For some metals such as nickel, water is reactive over all electrochemical potentials. The surface of these metals contains either adsorbed hydroxyl, hydrogen species, or, in some cases, both. This can impact other potential surface reactions due to changes in the surface coverage as well as the formation of local acidic or basic intermediates at the metal/aqueous interface. Finally, anodic potentials directly deplete the surface charge density, weakening the metal–metal bonding, making these atoms ultimately more vulnerable toward water dissociation and the incorporation of water to form the first monolayers of the oxide phase.

2.3 MOLECULAR MODELING OF CHEMISORBED PHASES UNDER COMPETING ADSORPTION CONDITIONS

Specific adsorption of ions and neutral species other than water may also exercise control over the corrosive chemistry occurring at a bare metal surface exposed to solution. Sulfates, chlorides, nitrates, nitrites, neutral hydrocarbons, and borates, for example, can inhibit or accelerate corrosion processes or form self-assembled monolayers that modify the surface properties. A particular example in which multiple species compete for the dominant surface chemistry is the problem of sour-water corrosion. *Sour water*, as experienced in the oil and gas industry, may contain various mixtures of ammonia, amines, organic acids, chlorides, sulfides, hydrogen sulfides, carbonates, and cyanides. Failure mechanisms resulting from this environment include ammonium chloride corrosion, ammonium bisulfide corrosion, high-temperature sulfide corrosion, cyanide-accelerated corrosion, or stress corrosion cracking that may be brought on by hydrogen, chloride, sulfide, or carbonate [57]. For any given set of thermodynamic and flow conditions, corrosion occurs via surface reactions operative at the protective or semiprotective scale or the exposed or partially exposed surface of the metal itself [1, 126, 127].

First-principles thermodynamic calculations can be useful in this context by comparing the relative strength of chemisorption for different ions at the metal/environment interface and delineating the relevant thermodynamic conditions. First-principles thermodynamics involves the extrapolation of internal energies determined at 0 K via electron structure calculation to finite temperature free energies through the incorporation of vibrational, rotational, and translational enthalpic and entropic contributions as well as configurational effects.

Such a study was performed for the different ionic and neutral species involved in ammonium chloride corrosion: ammonium chloride corrosion occurs when ammonia and hydrogen chloride crystallize *in situ* as hygroscopic NH_4Cl particulates, interacting with condensed water vapor to form highly saturated chloride solutions attended by aggressive corrosion behavior [57, 128, 129]. In such cases, the oxide scale is compromised, and repassivation potentially hindered by competing adsorption processes on the metal surface or an accelerated rate of metal dissolution. Investigations of mild steel passivity in NH_4Cl media by Shimbarevich and Tseitlin, for example, revealed that increasing NH_4Cl content for a given NH_3 concentration can cause enhanced corrosion, pointing to the relevance of chloride adsorption in the acceleration of anodic dissolution [130]. Likewise, increasing NH_3 content (and thus the basicity) potentially elevated the importance of OH adsorption, diminishing chloride presence and so hastening the onset of passivity [131–133].

To illumine these hypotheses, adsorption geometries and energies of environmental species on this surface were computed by performing first-principles slab calculations. In particular, adsorption energies were calculated for NH_x, OH_x, H, and Cl species over a Fe(110) representative of the kinds of mild steel surfaces that may be exposed during active dissolution. The first-principles thermodynamic corrections were included, as obtained from statistical mechanics and speciation considerations. To correct for the effect of a saturated solution of ammonium chloride, the solubility product was taken from the literature [134], as was the equilibrium constant for the pKa of NH_4 [135]. An appropriate set of thermodynamic cycles was also employed to correct for the effects of ion/molecular solvation as well as the half-cell potentials for adsorptions that proceed via electron transfer (see Fig. 2.7). The first-principles thermodynamics model was constructed with the help of standard tables obtained from the literature: enthalpic and entropic contributions of the gas and solution-phase preadsorption molecules of the adsorbed species, solvation energies, and electrochemical potentials for the $Cl_2/2Cl^-$ and $H_2/2H^+$ half-cell reactions [137–141]. The latter quantities at room temperature were taken from the table of standard reduction potentials [142]. Corrections to the Gibbs free energies of a species X were then made via the term

$$\Delta G^{conc}(X) = RT \ln X \tag{2.12}$$

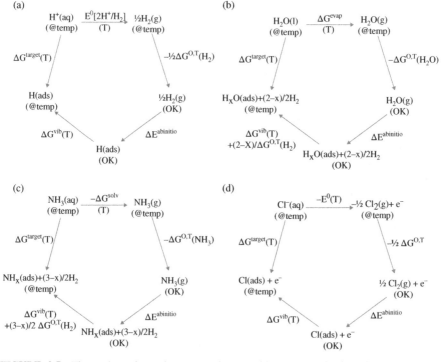

FIGURE 2.7 Thermodynamic cycles are used to combine computed adsorption energies for molecular/ionic species from first-principles calculations, with thermodynamic data available in the literature, to determine the overall free energy of adsorption of solution-phase species to the mild steel surface. Cycles are shown for Equations (a) 2.13, (b) 2.16, (c) 2.15, and (d) 2.14. Reproduced with permission from Taylor [136]. © National Association of Corrosion Engineers.

Similar assumptions have been made previously in the construction of Pourbaix diagrams for iron [143].

Once thermodynamic corrections were made based upon the standard reduction potentials, solvation data, etc., the free energy may be plotted as a function of the electrochemical potential by taking into account the number of electrons being consumed or released as a result of the particular adsorption process being considered [125]. These reactions include

$$H^+_{(aq)} + e^- \rightarrow H_{(ads)} \tag{2.13}$$

$$Cl^-_{(aq)} \rightarrow Cl_{(ads)} + e^- \tag{2.14}$$

$$NH_{3(aq)} \rightarrow NH_{x(ads)} + (3-x)H^+_{(aq)} + (3-x)e^- \tag{2.15}$$

$$H_2O_{(l)} \rightarrow OH_{x(ads)} + (2-x)H^+_{(aq)} + (2-x)e^- \tag{2.16}$$

The adsorption energies for the high symmetry adsorption states of the various species that compete for surface activity on Fe(110) are reproduced in Table 2.1. In Figure 2.8, the preferred surface adsorption geometries for selected species are shown. Note from Table 2.1 that adsorption strength can vary widely depending on the exact position assumed by the adsorbate on the metal surface. Furthermore, whereas O binding is stronger than both OH and H_2O, the lowest energy binding structure for the ammonia-based species is NH. This variation underscores the need for a first-principles analysis to provide both a quantitative and qualitative evaluation of the surface processes that may proceed during corrosion electrochemistry on a metal surface.

Application of the thermodynamic cycles described earlier allows the Gibbs free energies for the various adsorptions to be compared to one another at a given temperature, chosen here to be 25°C. The relative adsorption of the ammonia-, water-, hydrogen-, and chloride-derived species was compared at the equilibrium pH associated with saturated NH_4Cl (pH 4.45). Plotting the Gibbs free energy as a function of electrochemical potential reveals the shifting nature of the surface as electrochemical conditions are varied (Fig. 2.9).

TABLE 2.1   First-principles adsorption energies for H, NH_x, OH_x, and Cl species on Fe(110) prior to zero-point energy correction, given in eV

Species	ATOP	BRIDGE1	BRIDGE2	HOLLOW
H	0.037	−0.51	0.677	**−0.685**
Cl	−2.316	−2.618	−0.089	**−2.811**
NH_3	**−0.733**	−0.508	−0.347	2.012
NH_2	−0.081	−0.658	**−0.686**	1.659
NH	0.905	−0.38	**−1.005**	−0.945
N		0.802	**−0.487**	−0.32
H_2O	**−0.383**	−0.245	−0.078	0.916
OH	−0.266	0.889	**−1.004**	−0.426
O	0.564	−0.541	−1.042	**−1.058**

Table reproduced with permission from Taylor [136]. © National Association of Corrosion Engineers. Reference states are the bare Fe(110) surface and (a) H_2 for H, (b) Cl_2 for Cl, (c) NH_3 for NH_x with H_2 energies used for hydrogen balance, and (d) H_2O for OH_x with H_2 energies used for hydrogen balance. Preferred (lowest energy) sites are highlighted in bold face.

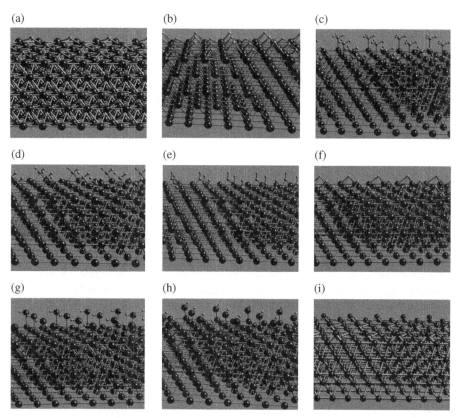

FIGURE 2.8 Adsorption sites for (a) hydrogen, (b) chloride, (c) ammonia, (d) NH_2, (e) NH, (f) N, (g) water, (h) hydroxyl, and (i) oxygen on Fe(110) surface as optimized using first-principles calculations.

The proper interpretation of Figure 2.9 is as follows. The derivative of the free energy with respect to electrochemical potential (dG/dV) is the amount of charge transferred, q, from the adsorbate to the surface. The free energy is given with respect to a reference state, which was taken in this study to be the environment of stable species: chloride, water, ammonia, and hydronium. Correspondingly, oxygen adsorption is a two-electron process, hydroxyl adsorption is a one-electron process, and chloride electron donates one electron. Furthermore, adsorption reactions that do not exchange electrons with the surface appear as horizontal lines, whereas adsorption reactions that do exchange electrons with the surface have a nonzero slope, equal to the number of electrons exchanged. Reduction reactions have a positive slope; oxidation reactions possess a negative slope. The intersection of the lines shows where two states are in equilibrium. The lowest energy state at each given potential corresponds to the state that dominates the exposed mild steel surface.

At 298 K, there are four surface phases predominant on the close-packed iron surface: H, Cl, O, and N (Fig. 2.9). For clarity, not all possible adsorption phases are shown, just the lowest energy phases. In pH-moderated *sour-water* conditions, where the pH is controlled to 9.0, the hydrogenated surface is stable at potentials below −710 mV SHE, and the hydrogen evolution reaction is likely to proceed. Between −710 and −170 mV SHE, the surface is covered by a prepassive/passive oxygen layer. Focusing only on one-electron

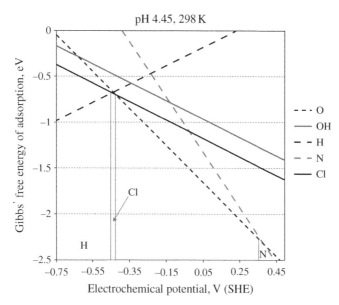

FIGURE 2.9 Free energy diagrams as a function of the electrochemical potential under equilibrium NH$_4$Cl(satd.) pH 4.45 at 25 °C (298 K). Reproduced with permission from Taylor [136]. © National Association of Corrosion Engineers.

processes, chloride adsorption is stronger than hydroxyl adsorption, with the implication that chloride may block oxide nucleation as it proceeds first via formation of iron hydroxyls [112]. The open-circuit potential for this system has been measured at −0.65 V, and the passivation potential at −0.3 V [144]. These values line up with the oxygen-covered region of the potential space (−710 mV through to −170 mV), with passivation occurring toward the more anodic end of this region. Oxygen adsorption is critical to a heterogeneous growth model for the inner oxide film, consistent with other modeling works [13, 15], as well as high-resolution observation of anodic oxidation processes on other metals [124, 145, 146].

Beyond −170 mV SHE, the oxygen layer is displaced by nitrogen layer, which seems also likely to have a protective effects [147–149]. The prediction of a nitrogen-passivated surface sheds some light on the role of nitrogen in inhibiting both pitting and general corrosion [147–151]. Iron and iron-based alloys treated with nitrogen in various ways are enriched with nitrogen at the metal/metal-oxide interface. This enrichment leads to faster passivation, thinner and more corrosion-resistant oxide films, and pitting resistance, which has been linked to enhanced repassivation rates [150]. The theoretical phase diagrams shown in Figure 2.9 indicate that nitrogen can adsorb onto iron surfaces at more strongly anodic potentials, which suggests that if the oxide film is compromised during pitting, nitrogen can then adsorb on the surface, potentially forming nitrides, that may then be covered by outer-layer oxides.

If pH-moderating species are not present, the NH$_4$Cl equilibrium produces a pH of 4.45. This shift in pH raises the free energy of the O and OH phases, opening a window between −450 and −420 mV SHE in which chloride becomes the dominant surface phase on mild steel [1]. Chloride access to the surface provides an opportunity for accelerated dissolution via surface complexation or interruption of the passivation process on Fe. This picture is consistent with the corrosion measurements made in ammonium chloride media made by Pletnev et al., Shimbarevich and Tseitlin, and Alvarez et al. [1, 130, 131, 152] and validates a number of the hypotheses regarding the ammonium chloride corrosion mechanism. Pletnev et al. found that chloride accelerated the corrosion of mild steel in the active-dissolution region and that the effect of chloride reached a plateau at concentrations greater than 1 M [1]. Pletnev suggested that this was consistent with a surface binding mechanism, whereby FeClOH could be produced, and that this effect would plateau once the surface concentration of Cl was saturated at one monolayer [1]. The theoretical results confirm that during active dissolution of iron and at acidic pH, chloride adsorbs onto iron, excluding OH, O, and NH_x species. Likewise, the specific adsorption of chloride has been speculated to hinder pit repassivation and, at the same time, increase the dissolution rate [152].

This example shows that more complex systems can be examined using first-principles thermodynamics beyond the sorption of water on metallic surfaces. Future extensions of this type of approach should push to describe the competitive adsorption of environmental species on oxide surfaces, as well as the adsorption of ions and neutral species at surface defects, which often catalyze the corrosion process.

2.4 COADSORPTION OF IONS AT THE INTERFACE AND PROMOTION OF HYDROGEN UPTAKE

Displacement is only one of the mechanisms via which species in the environment can compete for surface activity. An alternative can be found in the phenomenon of coadsorption. Coadsorption can open pathways for unique chemical reactivity (i.e., catalysis) but can also modify the kinetics and thermodynamics of stand-alone reactions, through site blocking or through surface electronic effects. An example of this effect relevant to corrosion is the promotion of hydrogen uptake by coadsorbing anions [153]. Sulfur is an example of a coadsorbed species that produces this effect. During hydrogen evolution, adsorbed hydrogen can follow two possible reaction pathways: recombination to form molecular H_2 and evolve hydrogen gas and absorption into the metal surface. The rate of absorption of H atoms on the surface expressed as a flux is often far less than 10% of the hydrogen deposition rate on most metals such as Fe and Ni [154–156]. This flux is significantly elevated when certain other elements are present, such as sulfur, arsenic, and cyanide, for instance.

A through-surface model for this promotion has been provided by Zolfaghari and coworkers [157] and elaborated by Protopopoff and Marcus [158]. According to this model, the chemical potential of hydrogen in the vicinity of a promoter is increased, effectively lowering the activation energy for hydrogen uptake. Such a model can be readily investigated using a first-principles analysis. Taylor, Scully and Neurock performed such an analysis and found that, indeed, adsorbed atoms and molecules are surrounded by a *ring of influence*, in which the chemical potential of hydrogen is locally elevated [159]. At the same time, they discovered two additional *zones* of influence (Fig. 2.10). These three zones are:

1. A free zone, in which the promoter has no effect on the adsorption and uptake rate of hydrogen atoms
2. A promoter zone, in which the adsorbed promoter increases the chemical potential of hydrogen sufficiently to significantly impact the reaction kinetics
3. An exclusion zone, for those sites closest to the promoter species where the adsorption of H may not occur due to strong repulsion

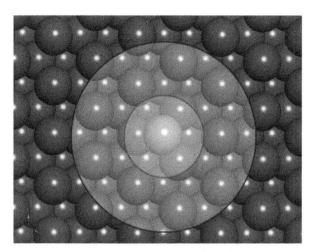

FIGURE 2.10   A *promoter* atom, as shown here in gray adsorbed at the threefold site on a fcc (111) surface, may have three zones of influence on neighboring adsorbates: an inner *exclusion* zone, consisting of sites that have a sufficiently raised chemical potential so as to exclude adsorption (inner circle); a *promoter* zone, consisting of sites with slightly elevated chemical potential that leads to enhanced reactivity (outer circle); and a *promoter-free* zone, at which sites the chemical potential is unaffected by the *promoter* (unshaded sites, outside of the outer circle). Reproduced with permission from Taylor et al. [159]. © Electrochemical Society.

Following the deconstruction of the coadsorption into zone, one can then write an adsorption isotherm governing hydrogen uptake. Coupling this with the Volmer–Heyrovsky model for hydrogen uptake versus hydrogen evolution, an expression for the relative rate of hydrogen uptake can be derived. The relative uptake rate, for a particular coverage of the promoter species, θ_P, is given by the equation

$$r' = 1 - \sigma_p \theta_P - \sigma_x \theta_P + \left(\frac{\sigma_p \theta_P}{\alpha}\right)\left(\frac{\theta_H^P}{\theta_H^f}\right) \tag{2.17}$$

where $\alpha = e^{-g p/RT}$ is the promotion factor, determined by the interaction energy g^P that corresponds to the local chemical potential elevation induced by the promoter; θ_H^f is the actual coverage of hydrogen on the surface; θ_H^p is the coverage of hydrogen in the promotion zone; σ_p is the number of sites promoted per promotion atom adsorbed; σ_x is the number of sites excluded by the adsorption; and θ_P is the coverage of the promoter species. Obviously, appropriate limits on θ_P must be obeyed such that r' remains greater than 0. Following this expression, there exists a peak in the permeation rate of hydrogen in steels at an intermediate concentration of promoters (because at some point the exclusion zone dominates over the promotion effect—the negative and positive terms in Eq. (2.17), respectively). This same peak was shown for a variety of promoters in the review by Radhakrishnan and Shrier [160].

In addition to illuminating the qualitative concept of blocking and promotion zones, the first-principles calculations also provided quantitative metrics for the size of the promoter zone, θ_P, and the interaction energy, g^P. To do this, S–H interactions were studied in a number of different configurations (Fig. 2.11). The coadsorption energies are given in Table 2.2.

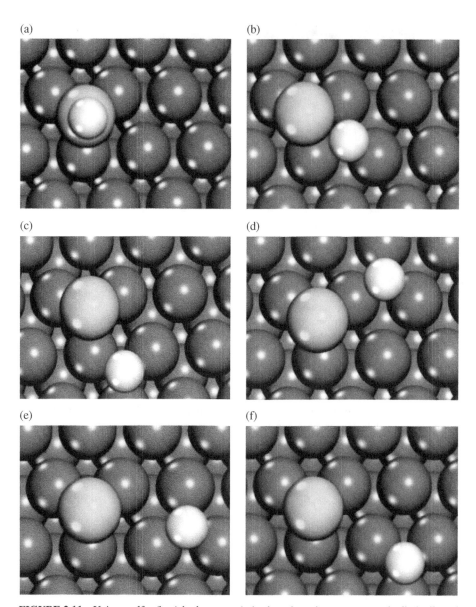

FIGURE 2.11 Unique sulfur (hcp)-hydrogen optimized coadsorption states up to the limit allowed by the confines of the $2\sqrt{3} \times 2\sqrt{3}$ periodic simulation cell: (a) SH ads, (b) nearest neighbor, (c) next-nearest neighbor, (d) next^2-nearest neighbor, (e) next^3-nearest neighbor, and (f) next^4-nearest neighbor. Reproduced with permission from Taylor et al. [159]. © Electrochemical Society.

Sulfur binds to the nickel surface with a strong binding energy of −5.23 eV. The binding energy for subsequent adsorption of hydrogen can vary between +0.15 and −0.52 eV depending on the relative position of the adsorption site. Hydrogen adjacent to S has an elevated potential of +0.15 eV; thus, S provides a strong chemical potential gradient that opposes the occupation of hydrogen. As hydrogen is moved to sites that share only a single

TABLE 2.2 Binding energies, ΔE, with reference to H₂(g) and interatomic distances (d(Ni–H), d(Ni–S), and d(S–H)) for adsorption states of hydrogen and sulfur, embedded sulfur, and coadsorption states of hydrogen, hydrogen–sulfur, and hydrogen-embedded sulfur on Ni(111), respectively

Adsorption state	ΔE/eV	d(Ni–H)/Å	d(S–H)/Å
H_{hcp}	−0.53	1.72	
H_{fcc}	−0.53	1.72	
S_{hcp}–H			
Exclusion	+0.15	1.80	2.20
Promotion	−0.40	1.73	2.87
Free	−0.48	1.71	3.14
Free	−0.52	1.70	3.89
Free	−0.49	1.72	4.34
HS	+0.69	2.17 (Ni–S)	1.37

Reproduced with permission from Taylor et al. [159]. © Electrochemical Society.

Ni atom with the coadsorbed sulfur, the chemical potential is only mildly elevated relative to the free zone. HS may also form; however, the nickel surface provides a strong driving force for dissociation (>1 eV).

Given these inputs, Taylor, Scully, and Neurock plotted r' as a function of θ_P for selected values of the hydrogen overpotential η (Fig. 2.12a). One observes a linear increase in the promotion rate as the promoter coverage is increased, moving to a peak at $\theta_P = 0.1$. This threshold value corresponds to the area at which the promoter region is maximized. The promotion rate then linearly drops to zero as θ_P continues to increase. The surface promotion terminates here at $\theta_P = 0.25$ as a result of the site blocking. The general shape of the curve is also consistent with experimental investigations of the hydrogen permeation rate [160, 161].

The peak values for r' are plotted with respect to θ_P as a function of η in Figure 2.12b. This maximum value can be found by the expression

$$r'_{max} = \frac{1}{\alpha}\left(\frac{\sigma_p}{\sigma_p + \sigma_x}\right) \tag{2.18}$$

At high $\eta > 300$ mV, there is no rate enhancement, as the three-state model leads to a negative slope regardless of coverage. However, for overpotentials, $\eta < 300$ mV, a significant rate enhancement can be observed for values of θ_P close to 0.1 ML. Although previous studies did not report surface coverage, a qualitatively similar optimal concentration was seen for promotion. Assuming that coverage scales with bulk concentration by a MacLean-type isotherm, the connection is clear. Moreover, the promotion effect increased with hydrogen overpotential at optimal promoter concentrations in experimental permeation studies.

2.5 DISSOLUTION OF METAL ATOMS

Thus far, we have charted in this review a progression that began with the structure and reactivity of water at a metallic surface and moved on to the adsorption of other ions and neutral species and next to the response of a system to coadsorption and surface factors that control hydrogen embrittlement. As a final instance of how first-principles modeling can illuminate

(a)

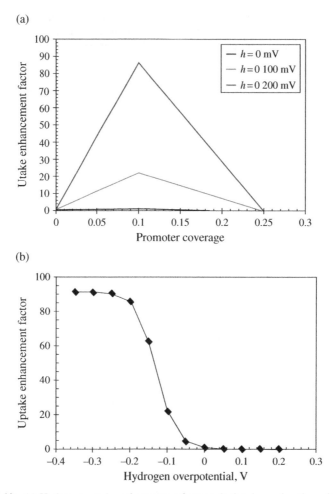

(b)

FIGURE 2.12   (a) Hydrogen uptake enhancement factor calculated as a function of the promoter coverage (in ML), for three different values of the hydrogen adsorption overpotential, η used in text. (b) Maximum rate enhancement at the critical value of θ_P as a function of the hydrogen overpotential. Reproduced with permission from Taylor et al. [159]. © Electrochemical Society.

the surface chemistry of corrosion, we consider the dissolution reaction. Loss of metal from the surface during a corrosion event is typically expressed by a chemical equation of the type

$$M_{(s)} \rightarrow M^{z+}_{(aq)} + Ze^{-} \tag{2.19}$$

The above chemical equation highlights a tension between the forces of metal–metal bonding and the formation of bonds with species present in the aqueous solution. An analysis of the complexity contained within this tension was recently made through a series of papers published by Gileadi [10, 162, 163]. To understand how theoretically challenging this concept is, consider that molecular water is bound to transition and noble metal close-packed surfaces by energies no more than a few tenths of an eV [64], whereas the surface cohesive energies are an order of magnitude greater [164]. The reason that such dissolution events do occur, on a regular and (from a corrosion mitigation perspective) inconveniently frequent basis, is that electric fields at the interface (of order 0.1 V/Å) provide a gradient sufficiently strong enough to stabilize the formation and extraction of that metal atom in its ionic form [162].

One way to interrogate this process via atomistic simulation is to begin by simply considering the role of metal–metal bonding in controlling corrosion morphologies. This issue has been addressed previously to some extent via kinetic Monte Carlo simulations of the dealloying of binary alloy systems and of crystallographic pitting [165–167]. Motivated by the development of robust metal alloy waste forms a study of iron–technetium binary systems revealed that local coordination and composition both play a role in control of the cohesive energies of surface metal atoms [6]. A study of copper nanoparticles compared metal atoms on curved surface to those on close-packed and stepped surfaces, revealing that, although surface cohesive energies correlate approximately with the number of nearest neighbors, significant scatter exists in the correlation due to the fact that individual topological features also should be taken into account [164].

Because it is fundamentally challenging to directly model the interplay between metal–metal cohesion and metal–solvent attraction using typical materials dynamics simulation techniques (such as embedded atom potentials) [168–171], we decided to build a model system that could explore aspects of the dissolution reaction using highly controlled, density functional theory calculations [172]. The model we built to study this system consisted of a Cu nanoparticle, with a single Cu adatom adsorbed on the surface in contact with a hemispherical *nanodroplet* of close to a dozen water molecules (see Fig. 2.13a). The metal particle was designed to be large enough to provide a fair representation of metal–metal bonding and electronic band structure, and the solution model is sufficient to model one to two layers of an emerging ion's solvation sheath. The configuration of the solvent molecules was obtained by performing a quantum mechanical molecular dynamics simulation at 300 K and then cooling the system to an annealed configuration, which was governed by the hydrogen bonding between water molecules and their interactions with the metal particle. We then performed a series of geometry optimizations, similar to what had been performed for water molecules on Cu(111) surfaces in the study mentioned earlier, but rather than varying the system charge state, we varied the height of the metal atom above the particle surface stepwise.

To understand some of the mechanistic details of the study, we performed the simulation under three separate sets of conditions (no water, water, and water plus an applied electric field, thus simulating a double layer). We plotted the charge states of the adatom as a function of the height above the surface (or, equivalently, as a function of the dissolution process) and the associated energies (Fig. 2.13b and c, respectively).

In the absence of solvent, the energy of the adatom appears to rise in a quasiharmonic way, which will at some point vary from harmonicity and tend asymptotically to the value for the surface binding energy of the adatom. There is some weak charging of the surface adatom. When the solvent molecules are present, the charge on the adsorbed copper atom is significantly more cationic. The Bader method yields a charge of +0.36 for the copper adatom in the solvent phase, even before dissolution occurs. Thus, water molecules appear to *precharge* adsorbed metal atoms on the surface, as they form a partial solvation shell around the adatom. As the bond is broken (due to raising of the surface metal atom), the energy increases, as does the charge. Furthermore, the copper atom begins to form more and more bonds with the surrounding water molecules, eventually obtaining a rudimentary solvation shell.

(a)   (b)

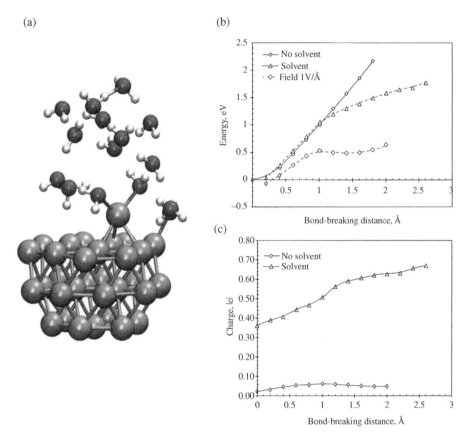

(c)

FIGURE 2.13 (a) Initial solvent–metal particle configuration, (b) electronic structure energies for dissolution/bond breaking of a copper atom in proximity to a (111) copper surface, and (c) charges determined by the method of Bader for the copper adatom being *dragged* away from the (111) copper surface. Reproduced with permission from Taylor [172]. © Elsevier.

The early part of the energy–distance curve, for distances less than 1 Å, is not so different from the case of the vacuum, or *no solvent*. It appears that *the energy in the early stages of dissolution is mostly concerned with the strong effect of breaking bonds with the metal surface*. In the latter part of the curve, for distances greater than 1 Å, the formation of bonds with the solvent environment alleviates the high energies incurred when the metal/surface bond is broken, in agreement with the *make-before-break* argument as posed in the manuscript by Gileadi [162]. The development of what appears to be a +1 charge on the Cu ion may be an indicator that the +1 state develops first, and then transition to +2 takes place as an outer-layer, electron-transfer event. The application of the electric field further lowers the energy trajectory of the dissolving ion.

2.6 SUMMARY AND PERSPECTIVES

The metal/environment interface poses a fascinatingly complex theoretical problem, which, as illustrated in the aforementioned case studies, has only just begun to be interrogated using advanced atomistic modeling and simulation techniques. The context of corrosion provides a particularly interesting backdrop, as the role of the metal itself becomes more than a simple catalyst or substrate, but its identity and potential for transformation is integral to the phenomena under investigation. In addition to templating the water environment, through encouragement of the formation of water bilayers, controlling water dissociation chemistry, or providing an electrode surface for the electrochemical potential, the metal surface is also a surface to which both neutral and ionic species in the electrolyte may stick and participate in chemical reactions of their own. Furthermore, interactions between these adsorbates may modify their individual chemistries, as seen in the promotion of hydrogen uptake by sulfur in the theoretical work described earlier. Finally, surface interactions can promote disruption of the metal lattice itself, as shown in the final case study, in which adsorption and electric fields may both play a role in modifying the potential energy surface for metal adatom dissolution. The aforementioned case studies show that atomistic modeling powerfully provides insight as to the thermodynamics and kinetics of each of these processes, as well as provide a means to observe how nature optimizes the geometries of the atoms and molecules as they organize at the electrochemical interface. The previous examples are just a subset of the kinds of simulations currently being performed for corrosion-related problems in the materials science literature. We anticipate many new applications developing that go beyond the specifications outlined here, as more groups adopt atomistic modeling within their research toolkit and new algorithms and synergies with continuum or mesoscale techniques are developed.

Despite these tremendous capabilities, which are only just being now applied to many of the outstanding problems in corrosion science, there remain several key challenges: there is still some way to go for the computational science community to be able to represent the kinds of length and timescales associated with macro- and mesoscopic corrosion processes. Some of these challenges include:

1. Capturing adequate dynamic timescales for the rearrangement of solvent particles around atoms undergoing dissolution
2. Capturing not only details of the metal/environment interface but also the metal/oxide and oxide/environment interfaces
3. Incorporating the important effects of alloying components in both the metallic and oxide states, in both solid solution and as particle inclusions
4. Including the role of defect states, such as step edges, terraces, roughened surfaces, grain boundaries, and dislocations, as potentially the most active sites at which corrosion events are going to occur

The development of such advanced toolkits, using either state-of-the-art potentials such as the reactive force field [173] or multiscale modeling projects [11, 12, 20], is an active focus in our current portfolio of research projects.

A final point should be made regarding experimental validation. First-principles modeling techniques are subject to variation depending upon the granularity and accuracy of the physics approximations used in their construction. For this reason, it is not uncommon to find variations, for example, in binding energies of molecules and materials calculated using different methods of 30–50 kJ/mol. Similarly, structural features, such as bond lengths and lattice parameters, can also be subject to variation depending upon the physics model selected. For these reasons, it is imperative that an equal, if not greater, effort continue to be applied in the characterization of interfacial phenomena using high-resolution, surface-specific probes (e.g., scanning tunneling microscopy and surface-enhanced Raman spectroscopy). Improving resolution in both timescales and length scales is critical to providing the guidance needed for the improvement of the physics models that underlie the kinds of simulations described in this review. As is the case in most research areas, advances in the understanding of corrosion processes can occur when a synergy is applied between the theoretical and experimental sciences.

REFERENCES

1. M. A. Pletnev, S. G. Morozov and V. P. Alekseev, *Prot. Met.* 36, 202–208 (2000).

2. U. R. Evans, *Metallic Corrosion, Passivation and Protection*. (Edward Arnold, London, England, 1937).

3. C. Wagner and W. Traud, *Z. Elektrochem.* 44, 52 (1938).

4. J. O. M. Bockris, A. K. N. Reddy and M. Gamboa-Aldeco, *Modern Electrochemistry*. (Kluwer Academic/Plenum Publishers, New York, 2000).

5. C. D. Taylor, *Int. J. Corrosion* 2012, 204640 (2012).

6. C. D. Taylor, *J. Metall* 2011, 954170 (2011).

7. J. K. Nørskov, F. Abild-Peterson, F. Studt and T. Bligaard, *Proc. Natl. Acad. Sci.* 108, 937–943 (2011).

8. S. A. Wasileski, C. D. Taylor and M. Neurock, *Device and Materials Modeling in PEM Fuel Cells*. Edited by S. J. Paddison (Springer-Verlag, Berlin, 2009).

9. R. A. van Santen and M. Neurock, *Molecular Heterogeneous Catalysis: A Conceptual and Computational Approach*. (Wiley-VCH, Weinheim, 2006).

10. E. Gileadi, *J. Solid State Electrochem.* 15, 1359–1371 (2011).

11. T. J. Campbell, G. Aral, S. Ogata, R. K. Kalia, A. Nakano and P. Vashishta, *Phys. Rev. B* 71, 205413–205426) (2005).

12. S. Serebrinsky, E. A. Carter and M. Ortiz, *J. Mech. Phys. Solids* 52, 2403–2430 (2004).

13. C. D. Taylor, R. G. Kelly and M. Neurock, *J. Electrochem. Soc.* 153, E207 (2006).

14. C. D. Taylor, R. G. Kelly and M. Neurock, *J. Electrochem. Soc.* 154 (3), F55–F64 (2007).

15. C. D. Taylor, R. G. Kelly and M. Neurock, *J. Electrochem. Soc.* 154 (12), F217–F221 (2007).

16. G. Jomard, T. Petit, A. Pasturel, L. Magaud, G. Kresse and J. Hafner, *Phys. Rev. B* 59 (6), 4044–4052 (1999).

17. S. K. R. S. Sankaranarayanan and S. Ramanathan, *J. Phys. Chem. C* 112, 17877–17882 (2008).

18. J. Greeley, *ECS Trans.* 16 (2), 209–213 (2008).

19. E. Kim, P. F. Weck, C. D. Taylor, O. Olatunji-Ojo, X.-Y. Liu, E. Mausolf, G. D. Jarvinen, K. R. Czerwinski, *J. Electrochem. Soc.* 161, C83–C88 (2014).

20. R. Spatschek, E. Brener and A. Karma, *Philos. Mag.* 91 (1), 75–95 (2011).

21. R. M. Martin, *Electronic Structure: Basic Theory and Practical Methods*. (Cambridge University Press, Cambridge, UK, 2004).

22. E. Kaxiras, *Atomic and Electronic Structure of Solids*. (Cambridge University Press, Cambridge, UK, 2003).

23. I. N. Levine, *Quantum Chemistry*, 5th edn. (Prentice Hall, Upper Saddle River, NY, 2000).

24. P. B. Balbuena, K. P. Johnston and P. J. Rossky, *J. Phys. Chem.* 100, 2706–2715 (1996).

25. D. Dominguez-Ariza, C. Hartnig, C. Sousa and F. Illas, *J. Chem. Phys.* 121 (2), 1066–1073 (2004).

26. P. F. Weck, E. Kim, F. Poineau and K. R. Czerwinski, *Phys. Chem. Chem. Phys.* 11 (43), 10003–10008 (2009).

27. A. B. Anderson and N. C. Debnath, *J. Am. Chem. Soc.* 105, 18–22 (1983).

28. A. B. Anderson and N. K. Ray, *J. Phys. Chem.* 86, 488–494 (1982).

29. E. E. Ebenso, T. Arslan, F. Kandemirli, N. Caner and I. Love, *Int. J. Quantum Chem.* 110 (5), 1003–1018 (2009).

30. E. E. Ebenso, T. Arslan, F. Kandemirli, I. Love, C. Öğretır, M. Saracoğlu and S. A. Umoren, *Int. J. Quantum Chem.* 110 (14), 2614–2636 (2010).

31. M. C. Payne, M. P. Teter, D. C. Allan, T. A. Arias and J. D. Joannopoulos, *Rev. Mod. Phys.* 64 (4), 1046–1097 (1992).

32. I. Milas, B. Hinnemann and E. A. Carter, *J. Mater. Chem.* 21 (5), 1447 (2011).

33. O. Runevall and N. Sandberg, *J. Phys. Condens. Matter* 23 (34), 345402 (2011).

34. D. A. Andersson, B. P. Uberuaga, P. V. Nerikar, C. Unal and C. R. Stanek, *Phys. Rev. B* 84 (5) (2011).

35. C. D. Taylor, T. Lookman and R. S. Lillard, *Acta Mater.* 58, 1045–1055 (2010).

36. G. Lu and E. Kaxiras, *Phys. Rev. Lett.* 94, 155501–155504 (2005).

37. G. Lu, D. Orlikowski, I. Park, O. Politano and E. Kaxiras, *Phys. Rev. B* 65, 64102–64109 (2002).

38. G. Lu, Q. Zhang, N. Kioussis and E. Kaxiras, *Phys. Rev. Lett.* 87, 95501–95504 (2001).

39. J. Greeley and J. K. Nørskov, *Electrochim. Acta* 52, 5829–5836 (2007).

40. A. Bouzoubaa, D. Costa, B. Diawara and P. Marcus, *Corros. Sci.* 52, 2643 (2010).

41. M. I. Baskes and R. A. Johnson, *Model. Simul. Mater. Sci. Eng.* 2 (1), 147–163 (1994).

42. S. M. Foiles, M. I. Baskes and M. S. Daw, *Phys. Rev. B* 33, 7983–7991 (1986).

43. B.-J. Lee and M. I. Baskes, *Phys. Rev. B* 62 (13), 8564–8567 (2000).

44. J. W. Halley, A. Mazzolo, Y. Zhou and D. Price, *J. Electroanal. Chem.* 450, 273–280 (1998).

45. J. W. Halley, B. B. Smith, S. Walbran, L. A. Curtiss, R. O. Rigney, A. Sutjianto, N. C. Hung, R. M. Yonco and Z. Nagy, *J. Chem. Phys.* 110 (13), 6538–6552 (1999).

46. D. L. Price and J. W. Halley, *J. Chem. Phys.* 102 (16), 6603–6612 (1995).

47. B. B. Smith and J. W. Halley, *J. Chem. Phys.* 101, 10915–10924 (1994).

48. E. Spohr, *Chem. Phys.* 141, 87–94 (1990).

49. E. Spohr, *Solid State Ion.* 150, 1–12 (2002).

50. E. Spohr, *Electrochim. Acta* 49, 23–27 (2003).

51. E. Spohr and K. Heinzinger, *Chem. Phys. Lett.* 123, 218–221 (1986).

52. E. Spohr, G. Toth and K. Heinzinger, *Electrochim. Acta* 41, 2131–2144 (1996).

53. X. M. Bai, A. F. Voter, R. G. Hoagland, M. Nastasi and B. P. Uberuaga, *Science* 327 (5973), 1631–1634 (2010).

54. H. Ruuska and T. A. Pakkanen, *J. Phys. Chem. B* 108, 2614–2619 (2004).

55. M. W. Ribarsky, W. D. Luedtke and U. Landman, *Phys. Rev. B* 32 (2), 1430–1433 (1985).

56. A. Ignaczak and J. A. N. F. Gomes, *J. Electroanal. Chem.* 420, 209–218 (1997).

57. A. Michaelides, V. A. Ranea, P. L. de Andres and D. A. King, *Phys. Rev. Lett.* 90, 216102–216106 (2003).

58. S. Wang, Y. Cao and P. A. Rikvold, *Phys. Rev. B* 70, 205410–205414 (2004).

59. D. Sebastiani and L. D. Site, *J. Chem. Theory Comput.* 1, 78–82 (2005).

60. S. Meng, E. G. Wang and S. Gao, *J. Chem. Phys.* 119, 7617–7620 (2003).

61. S. Haq, J. Harnett and A. Hodgson, *Surf. Sci.* 505, 171–182 (2002).

62. G. Zimbitas, S. Haq and A. Hodgson, *J. Chem. Phys.* 123, 174701–174709 (2005).

63. A. B. Anderson, N. M. Neshev, R. A. Sidik and P. Shiller, *Electrochim. Acta* 47, 2999–3008 (2002).

64. D. L. Doering and T. E. Madey, *Surf. Sci.* 123, 305–337 (1982).

65. C. D. Taylor, M. Neurock, *Curr. Opin. Solid State Mater. Sci.* 9(2), 49 (2005).

66. H. Ogasawara, B. Brena, D. Nordlund, M. Nyberg, A. Pelmenschikov, L. G. M. Pettersson and A. Nilsson, *Phys. Rev. Lett.* 89, 276102–276105 (2002).

67. A. Michaelides, A. Alavi and D. A. King, *Phys. Rev. B* 69, 113404–113407 (2004).

68. A. Michaelides, A. Alavi and D. A. King, *J. Am. Chem. Soc.* 125, 2746–2755 (2003).

69. S. Meng, L. F. Xu, E. G. Wang and S. Gao, *Phys. Rev. Lett.* 89 (17), 176104–176107 (2002).

70. S. Meng, E. G. Wang and S. Gao, *Phys. Rev. B* 69, 195404–195416 (2004).

71. S. Meng, E. G. Wang, C. Frischkorn, M. Wolf and S. Gao, *Chem. Phys. Lett.* 402, 384–388 (2005).

72. J. Weissenrieder, A. Mikkelsen, J. N. Andersen, P. J. Feibelman and G. Held, *Phys. Rev. Lett.* 93, 196102–196105 (2004).

73. P. J. Feibelman, *Science* 295, 99–102 (2002).

74. P. S. Crozier, R. L. Rowley and D. Henderson, *J. Chem. Phys.* 114, 7513–7517 (2001).

75. I.-C. Yeh and M. L. Berkowitz, *Chem. Phys. Lett.* 301, 81–86 (1999).

76. J. Cerda, A. Michaelides, M.-L. Bocquet, P. J. Feibelman, T. Mitsui, M. Rose, E. Fomin and M. Salmeron, *Phys. Rev. Lett.* 93, 116101–116104 (2004).

77. C. D. Taylor, Ph.D. Dissertation, University of Virginia, 2006.

78. D. Boda, K.-Y. Chan and D. Henderson, *J. Chem. Phys.* 109, 7362–7371 (1998).

79. J. C. Shelley, G. N. Patey, D. R. Berard and G. M. Torrie, *J. Chem. Phys.* 2122–2141 (1997).

80. L. Zhang, H. T. Davis and H. S. White, *J. Chem. Phys.* 98, 5793–5799 (1992).

81. J. P. Valleau and A. A. Gardner, *J. Chem. Phys.* 86, 4162–4170 (1987).

82. A. A. Gardner and J. P. Valleau, *J. Chem. Phys.* 86, 4171–4176 (1987).

83. B. Jonsson, *Chem. Phys. Lett.* 82, 520–525 (1981).

84. N. I. Christou, J. S. Whitehouse, D. Nicholson and N. G. Parsonage, *J. Chem. Soc. Faraday Symp.* 16, 139–149 (1981).

85. G. M. Torrie and J. P. Valleau, *Chem. Phys. Lett.* 65, 343–346 (1979).

86. A. Kramer, M. Vossen and F. Forstmann, *J. Chem. Phys.* 106 (7), 2792–2800 (1997).

87. M. Vossen and F. Forstmann, *J. Chem. Phys.* 101, 2379–2389 (1994).

88. C. W. Outhwaite and M. Molero, *Electrochim. Acta* 36, 1685–1687 (1991).

89. G. M. Torrie, P. G. Kusalik and G. N. Patey, *J. Chem. Phys.* 88, 7826–7840 (1988).

90. C. W. Outhwaite and L. B. Bhuiyan, *J. Chem. Soc. Faraday Trans.* 79, 707–718 (1983).

91. F. Vericat, L. Blum and D. Henderson, *J. Chem. Phys.* 77, 5808–5815 (1982).

92. A. A. Kornyshev, W. Schmickler and M. A. Vorotyntsev, *Phys. Rev. B* 25, 5244–5256 (1982).

93. P. Vassilev, R. A. van Santen and M. T. M. Koper, *J. Chem. Phys.* 122, 54701–54712 (2005).

94. S. Izvekov, A. Mazzolo, K. VanOpdorp and G. A. Voth, *J. Chem. Phys.* 114 (7), 3248–3257 (2001).

95. A. Klesing, D. Labrenz and R. A. van Santen, *J. Chem. Soc. Faraday Trans.* 94, 3229–3235 (1998).

96. K. Raghavan, K. Foster, K. Motakabbir and M. L. Berkowitz, *J. Chem. Phys.* 94, 2110–2117 (1991).

97. A. Wallqvist, *Chem. Phys. Lett.* 165, 437–442 (1990).

98. C. Y. Lee, J. A. McCammon and P. J. Rossky, *J. Chem. Phys.* 80, 4448–4455 (1984).

99. G. Barabino, C. Gavotti and M. Marchesi, *Chem. Phys. Lett.* 104, 478–484 (1984).

100. A. Sonnenschein and K. Heinzinger, *Chem. Phys. Lett.* 102, 550–554 (1983).

101. M. Marchesi, *Chem. Phys. Lett.* 97, 224–230 (1983).

102. A. B. Anderson and T. V. Albu, *J. Am. Chem. Soc.* 121, 11855–11863 (1999).

103. A. B. Anderson and T. V. Albu, *Electrochem. Commun.* 1, 203–206 (1999).

104. Y. Cai and A. B. Anderson, *J. Phys. Chem. B* 108, 9829–9833 (2004).

105. J. W. Halley, B. Johnson, D. Price and M. Schwalm, *Phys. Rev. B* 31 (12), 7695–7709 (1985).

106. J. W. Halley and D. Price, *Phys. Rev. B* 35 (17), 9095–9102 (1987).

107. D. L. Price and J. W. Halley, *Phys. Rev. B* 38, 9357–9367 (1988).

108. M. F. Toney, J. N. Howard, J. Richer, G. L. Borges, J. G. Gordon, O. R. Melroy, D. G. Wiesler, D. Yee and L. B. Sorensen, *Nature* 368, 444–446 (1994).

109. D. L. Price, J. W. Halley, *J. Chem. Phys.* 102, 6603 (1995).

110. S. Desai and M. Neurock, *Electrochim. Acta* 48, 3759–3773 (2003).

111. S. Desai and M. Neurock, *Phys. Rev. B* 68, 75420–75426 (2003).

112. A. B. Anderson, *Int. J. Quantum Chem.* 49, 581–589 (1993).

113. S. Seong and A. B. Anderson, *J. Phys. Chem.* 100, 11744–11747 (1996).

114. A. B. Anderson, *Surf. Sci.* 105, 159–176 (1981).

115. E. M. Patrito and P. Paredes-Olivera, *Surf. Sci.* 527, 149–162 (2003).

116. P. Paredes Olivera, A. Ferral and E. M. Patrito, *J. Phys. Chem. B* 105, 7227–7238 (2001).

117. R. Hoffmann, *Rev. Mod. Phys.* 60 (3), 601–628 (1988).

118. A. B. Anderson and T. V. Albu, *J. Electrochem. Soc.* 147, 4229–4238 (2000).

119. A. B. Anderson, *Electrochim. Acta* 48, 3743–3749 (2003).

120. R. A. Sidik and A. B. Anderson, *J. Electroanal. Chem.* 528, 69–76 (2002).

121. J.-S. Filhol and M. Neurock, *Angew. Chem. Int. Ed. Engl.* 45, 402–406 (2006).

122. C. D. Taylor, S. A. Wasileski, J. W. Fanjoy, J.-S. Filhol and M. Neurock, *Phys. Rev. B* 73, 165402 (2006).

123. C. D. Taylor, S. A. Wasileksi, J. W. Fanjoy, J.-S. Filhol and M. Neurock, *Phys. Rev. B.* 73, 165402 (2006).

124. G. Niaura, *Electrochim. Acta* 45, 3507–3519 (2000).

125. J. Rossmeisl, J. K. Nørskov, C. D. Taylor and M. Neurock, *J. Phys. Chem. B* 110, 21833 (2006).

126. Pletnev et al. write "now it is a common point of view that the chemical nature and structure of the surface of a metal, which is in contact with an electrolyte, are decisive in the kinetics of electrochemical reactions that proceed on this surface."

127. P. Marcus, *Electrochim. Acta* 43, 109–118 (1998).

128. R. A. White, *Materials Selection for Petroleum Refineries and Gathering Facilities.* (NACE International, Houston, TX, 1998).

129. O. Forsen, J. Aromaa and M. Tavi, *Corros. Sci.* 35, 297–301 (1993).

130. V. A. Shimbarevich and K. L. Tseitlin, *Prot. Met.* 15, 455–457 (1979).

131. V. A. Shimbarevich and K. L. Tseitlin, *Prot. Met.* 17, 144–148 (1981).

132. B. Ingham, S. C. Hendy, N. J. Laycock and M. P. Ryan, *Electrochem. Solid-State Lett.* 10, C57–C59 (2007).

133. A. K. N. Reddy, M. G. B. Rao and J. O. M. Bockris, *J. Chem. Phys.* 42, 2246–2248 (1965).

134. R. Cohen-Adad and J. W. Lorimer, *Solubility Data Series.* Edited by J. W. Lorimer (IUPAC, Oxford, 1991), Vol. 47.

135. R. E. Mesmer, W. L. Marshall, D. A. Palmer, J. M. Simonson and H. F. Holmes, *J. Solut. Chem.* 17 (8), 699–718 (1988).

136. C. D. Taylor, *Corrosion (NACE)* 68, 591 (2012).

137. A. Ben-Naim and Y. Marcus, *J. Chem. Phys.* 81 (4), 2016–2027 (1984).

138. Y. Marcus, *J. Chem. Soc. Faraday Trans.* 82, 233–242 (1986).

139. Y. Marcus, *J. Chem. Soc. Faraday Trans.* 87, 2995–2999 (1987).

140. Y. Marcus, *J. Chem. Soc. Faraday Trans.* 83, 339–349 (1987).

141. P. J. Linstrom and W. G. Mallard, NIST Chemistry WebBook, NIST Standard Reference Database Number 69, http://webbook.nist.gov/chemistry/ Ed: P. J. Linstrom and W. G. Mallard (retrieved September 21, 2009).

142. A. J. Bard, R. Parsons and J. Jordan, *Standard Potentials in Aqueous Solutions.* (IUPAC, Marcel Dekker, New York, 1985).

143. B. Beverskog and I. Puigdomenech, *Corros. Sci.* 38 (12), 2121–2135 (1996).

144. A. M. Sedenkov, *Prot. Met.* 22, 118–119 (1986).

145. V. Maurice, H.-H. Strehblow and P. Marcus, *Surf. Sci.* 458, 185–194 (2000).

146. B. MacDougall and M. J. Graham, *Corrosion Mechanisms in: Theory and Practice.* Edited by P. Marcus (Marcel Dekker, Inc., New York, 2002), pp. 189–216.

147. C. R. Clayton, G. P. Halada and J. R. Kearns, *Mater. Sci. Eng. A* 198, 135–144 (1995).

148. U. Kamachi Mudali, B. Reynders and M. Stratmann, *Mater. Sci. Forum* 185–188, 723–730 (1995).

149. S. Song, W. Song and Z. Fang, *Corros. Sci.* 31, 395–400 (1990).

150. H. J. Grabke, *ISIJ Int.* 36 (7), 777–786 (1996).

151. M. K. Lei and X. M. Zhu, *J. Electrochem. Soc.* 152, B291–B295 (2005).

152. M. G. Alvarez and J. R. Galvele, *Corros. Sci.* 24, 27–48 (1984).

153. J. O. M. Bockris, J. McBreen and L. Nanis, *J. Electrochem. Soc.* 112 (10), 1025–1031 (1965).

154. J. R. Scully and P. J. Moran, *J. Electrochem. Soc.* 135 (6), 1337–1348 (1988).

155. D. G. Enos, A. J. J. Williams, G. G. Clemena and J. R. Scully, *Corrosion* 54, 389–402 (1998).

156. R. S. Lillard, D. G. Enos and J. R. Scully, *Corrosion* 56, 1119–1132 (2000).

157. A. Zolfaghari, G. Jerkiewicz, W. Chrzanowski and A. Wieckwoeski, *J. Electrochem. Soc.* 146, 4158-4165 (1999).

158. E. Protopopoff and P. Marcus, *J. Vac. Sci. Technol. A.* 5, 944 (1987).

159. C. D. Taylor, M. Neurock and J. R. Scully, *J. Electrochem. Soc.* 158, F36-F44 (2011).

160. T. P. Radhakrishnan and L. L. Shrier, *Electrochim. Acta* 11, 1007–1021 (1966).

161. R. D. McCright, *Stress Corrosion Cracking and Hydrogen Embrittlement of Iron Base Alloys* (Unieux-Firminy, France, 1977), pp. 306–325.

162. E. Gileadi, *Chem. Phys. Lett.* 393, 421–424 (2004).

163. E. Gileadi, *Israel J. Chem.* 48, 121–131 (2008).

164. C. Taylor, M. Neurock and J. Scully, *J. Electrochem. Soc.* 155, C407–C414 (2008).

165. R. S. Lillard, G. F. Wang and M. I. Baskes, *J. Electrochem. Soc.* 153, B358 (2006).

166. J. Erlebacher, *J. Electrochem. Soc.* 151, C614–C626 (2004).

167. D. M. Artymowicz, J. Erlebacher and R. C. Newman, *Philos. Mag.* 89 (21), 1663–1693 (2009).

168. W. Schmickler, K. Pötting and M. Mariscal, *Chem. Phys.* 320, 149–154 (2006).

169. M. Baskes, S. Srinivasan, S. Valone and R. Hoagland, *Phys. Rev. B* 75, 94113–94128 (2007).

170. P. Biedermann, E. Torres and A. Blumenau, *212th Meeting of the Electrochemical Society* (The Electrochemical Society, Washington, D.C., 2007).

171. S. Valone and S. Atlas, *Philos. Mag.* 86, 2683–2711 (2006).

172. C. Taylor, *Chem. Phys. Lett.* 469, 99 (2009).

173. A. C. T. van Duin, S. Dasgupta, F. Lorant and W. A. Goddard III, *J. Phys. Chem. A* 105, 9396–9409 (2001).

3

PROCESSES AT METAL–SOLUTION INTERFACES: MODELING AND SIMULATION

NOELIA B. LUQUE[1], WOLFGANG SCHMICKLER[1], ELIZABETH SANTOS[1,2] AND PAOLA QUAINO[3]

[1]*Institute of Theoretical Chemistry, Ulm University, Ulm, Germany*
[2]*Faculdad de Matemática, Astronomia y Física, IFEG-CONICET, Universidad Nacional de Córdoba, Córdoba, Argentina*
[3]*PRELINE, Universidad Nacional del Litoral, Santa Fé, Argentina*

3.1 INTRODUCTION

Metal–solution interfaces are of obvious importance to corrosion, but they are particularly difficult to model. By definition, the interface comprises that part of the system in which the intensive variables of the two adjoining phases differ from their respective bulk values, and even in concentrated solutions this implies a thickness of the order of 15–20 Å. This is too large to be modeled solely by density functional theory (DFT), which surface scientists often use as a panacea for the metal–gas interface. In addition, the two adjoining phases are of very different nature: metals are usually solid at ambient temperatures, and their properties do not differ too much from those at 0 K, so that DFT, or semiempirical force fields like the embedded atom method, are good methods for their investigation. By contrast, the molecules in solutions are highly mobile, and thermal averaging is indispensable. Therefore, the two parts of the interface usually require different models, and an important part of the art consists in their combination.

Despite these inherent difficulties, much progress has been achieved during the past decade, not least due to the ever-increasing powers of computers. In this chapter, we focus on two processes of relevance to corrosion, on which our own group has worked: surface mobility

Molecular Modeling of Corrosion Processes: Scientific Development and Engineering Applications, First Edition.
Edited by Christopher D. Taylor and Philippe Marcus.
© 2015 John Wiley & Sons, Inc. Published 2015 by John Wiley & Sons, Inc.

enhanced by an electric field or by co-adsorbed chlorine atoms, and hydrogen evolution on pure metals and nanostructures. Although the methods by which we treat these processes are rather different, they have in common that they consist of a combination of several techniques: DFT, force fields, and kinetic Monte Carlo (KMC) for surface mobility and quantum statistics and DFT for electron transfer and electrocatalysis. We believe that these methods can also be applied to other, as yet little explored areas of corrosion research.

The rest of this chapter is organized as follows. We start with the mobility of metal surfaces, which we investigated by the KMC technique. This requires the rates of all possible processes as input, which we obtain by a combination of DFT and a semiempirical potential. The first application is to a Ag(100) electrode, which is quite mobile even at room temperature. Interestingly, the mobility increases when the electrode potential is raised, which we explain by field–dipole interactions. Explicitly, we consider island shapes and dynamics, step fluctuations, and Ostwald ripening for this surface. In contrast to silver, a clean Au(100) surface is not mobile at ambient temperatures; however, the adsorption of chloride ions enhances the mobility. We explain the underlying mechanism and present results for Ostwald ripening.

Hydrogen evolution is our second major topic. We start by presenting the model Hamiltonian, on which our work is based, and explain how this can be combined with DFT calculations. Explicit results are presented for a series of single-crystal and several nanostructured electrodes; they compare well with experimental data.

3.2 SURFACE MOBILITY

The mechanism by which metal atoms migrate on a metal surface is of fundamental importance in processes such as corrosion, crystal growth, epitaxy, catalyst degradation, faceting, sintering, annealing, evaporation, condensation, and chemical surface reactions. Migration of individual atoms on solid surfaces is among the most fundamental processes in surface science. It takes place by a series of displacements that cross the minimum in the potential barrier between adjacent binding sites.

A detailed understanding of these basic processes is important in corrosion prevention or in the development of new materials by controlling their growth at the atomic level. Theoretical calculations ranging from first-principles approaches, to effective-medium and embedded-atom methods are able to provide insight concerning the underlying physics involved in atom motion and cluster nucleation on metal surfaces. In this part of the chapter, we shall review some of the latest computational results on KMC obtained in our own group. There is a large amount of papers where the MC method has been used to study catalytic reactions; those are not covered in the chapter. Here, we focus on the mobility on the surface.

The basic concepts of KMC are well documented in the literature [1–7]. There are various of KMC algorithms; we have used the most popular one which is called the "*n*-fold way," which is also known as the residence–time algorithm or the Bortz–Kalos–Liebowitz (BKL) [8] algorithm, or just the KMC algorithm.

In principle, KMC can give a faithful representation of the dynamics of the system, and this is a fascinating advantage. However, in order to have a good accuracy in the dynamical development, the *rates of all the possible events* must to be known in advance, which is one of the major disadvantages of this method.

KMC has been used in several areas of physics, not only in surface diffusion, but also in molecular beam epitaxy growth, chemical vapor deposition growth, polymers, electrochemical nucleation and growth, and so on. There is an enormous amount of articles and reviews already in each area of physics where KMC was used. Several applications to surface growth can be found in a review from 2005 [9 and reference therein].

In the following, we will present results on the surface self-mobility of metals obtained in our own group in cooperation with our partners from Ulm and Jlich [10, 11]. We have organized this part of the chapter as follows: In Section 3.3, we briefly discuss the principles that underlie the KMC method, and the way the inputs for the simulation were obtained. In the next section we discuss the effect of an electric field on electrodes of silver (Section 3.4); this part contains two subsections on step fluctuations and the analysis of island shapes. In Section 3.5, we deal with Ostwald ripening in two systems: the first is Ag in the presence of electric field and the second is Au in the presence of chloride atoms.

3.3 KMC: DETAILS IN THE MODEL AND SIMULATION TECHNIQUE

3.3.1 The Model

Like all computer simulations, KMC must be based on a consistent microscopic model of the processes involved. In addition, the transitions must be Poisson processes, and the different processes have to be independent. As pointed out earlier, the implementation of KMC requires the rates of all relevant processes.

In the simulations addressed later, three processes were studied: (i) step fluctuations, (ii) island shape fluctuations, and (iii) Ostwald ripening of an island inside a vacancy island. The corresponding configurations are shown in Figure 3.1. In each case, the electrode surface, Ag(100) or Au(100), was represented by an arrangement of $N_x \times N_y$ sites, using periodic boundary conditions in both directions. Each lattice point corresponds to a site for a silver (or gold) atom. For the sizes of the simulation cells, we refer to the original publications [10, 11].

Within the lattice, interactions with the nearest neighbors have been considered—with the exception of the exchange mechanism, which will be discussed later. In an elementary step, an atom can move to a nearest neighbor site through hopping over a bridge site. Every lattice site can be occupied or unoccupied, therefore $2^{10} = 1024$ diffusion processes have to be treated. Because of the symmetry of the fcc(100) surface, this number reduces to 544.

We have labeled all possible processes by a binary notation that is explained in Figure 3.2. According to Müller and Ibach [12], the exchange mechanism, in which an atom leaves the terrace and is simultaneously replaced by an adatom, is unfavorable on Ag(100) and was therefore not considered for silver, but this process was implemented later for the self-diffusion of Au on Au(100); for more details, see Section 3.6.

3.3.2 Energy Calculations for Silver

Since DFT calculations for such a large number of processes are prohibitive, the potential energy curves along the diffusion pathways were calculated with the embedded atom method (EAM) [13]. This method takes into account many-body effects, so the description of the metallic bonding is much better than with simple pair potentials. Within this method, the total energy E_{tot} of a system is the sum of the energies of the individual atoms E_i.

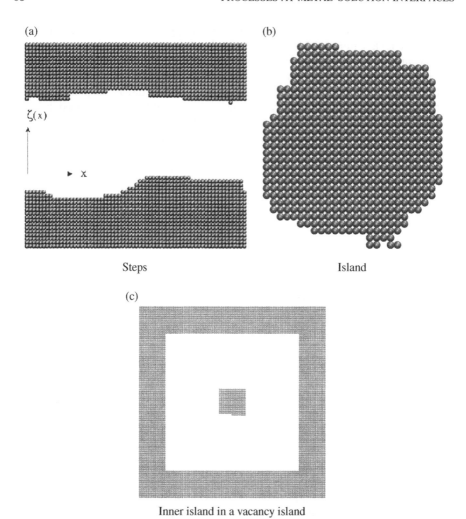

(a)

$\zeta(x)$

x

Steps

(b)

Island

(c)

Inner island in a vacancy island

FIGURE 3.1 Different initial configurations for processes on an fcc(100) surface. (a) A typical stepped surface used in the simulations with $N_x = 64$ and $N_y = 57$. (b) A typical island used in the simulations with $N_x = N_y = 105$. (c) A typical inner island in a vacancy island on a fcc(100) surface used in the simulations with $N_x = N_y = 105$. Reproduced with permission from Pötting et al. [10] and Luque et al. [14]. © 2009, 2010, Elsevier.

$$E_{\text{tot}} = \sum_i E_i \quad \text{with} \quad E_i = \frac{1}{2}\sum_{i,j}\varphi_{ij}(r_{ij}) + \sum_i F_i(\rho_i) \qquad (3.1)$$

The attractive part of the potential is given by the embedding function F_i, which depends on the local electronic density ρ_i and represents the energy to embed an atom i in the electronic density at the site where the atom i is located. $\rho_i(r_{ij})$ is calculated as the superposition of the individual electronic densities. The repulsion between the ion cores is represented by a pair potential φ_{ij}, which depends only on the distance between the cores. The EAM contains

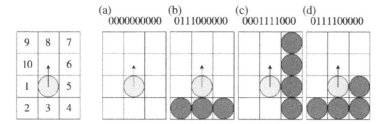

FIGURE 3.2 Nomenclature for surface diffusion. Occupied sites are labeled "1", and unoccupied "0" in the order that is shown in the picture on the left. The main part shows four possible processes and their *configuration number*: (a) free diffusion, (b) detachment from a step, (c) diffusion along a step, and (d) detachment from a kink. Reproduced with permission from Pötting et al. [10], © 2009, Elsevier.

parameters that have been fitted to experimental data like elastic constants, lattice constants, sublimation heats, and so on.

The potential curves for the various processes were obtained with a five layers slab in a 20×20 unit cell. The migrating adatom was relaxed perpendicular to the surface at each calculated point along the diffusion path, while all other atoms were fixed at the experimental nearest neighbor distance of 2.89 Å. The activation energy E_{act} was evaluated as the difference between the saddle point and the initial minimum in the energy curve along the reaction coordinate. The vibrational frequencies v_0 in the initial states were estimated by a harmonic approximation around the local minima [10]. They do not vary much between the processes, and they are all in the classical range $hv_0 \ll kT$. They can therefore be identified with the pre-factor in the diffusion rate. A more exact statistical treatment tends to give values that are somewhat higher [15], but since this would apply equally to all processes, this should not affect our results.

3.3.3 Dipole Moments

To study the influence of an electric field on the diffusion rates, the local dipole moments associated with all important processes must be known. These cannot be obtained from the EAM, so we have calculated them from DFT. Obviously, it is not possible to calculate the changes in the dipole moments for all 544 processes, so we have limited our calculations to the most important processes. KMC offers a convenient way to determine which processes contribute most. In this way, we made a screening of all the processes and selected the basic ones that determine the rates and the fluctuations. We have identified 16 diffusion events that strongly contribute to the fluctuations. All of them were treated with DFT to obtain the local dipole moments at the initial and transition states. A list of the processes considered and the corresponding changes in the dipole moments is given in Table 3.1.

Within DFT, dipole moments of adsorbates are best obtained from the change $\Delta\Phi$ in the work function caused by the adsorbate:

$$\Delta\Phi = -e_0\mu n_s/\varepsilon_0 \tag{3.2}$$

where μ is the dipole moment perpendicular to the surface, and n_s is the number of adsorbates per unit area. The technical details of the calculations are given in Ref. [10].

PROCESSES AT METAL–SOLUTION INTERFACES

TABLE 3.1 Dipole moment differences $\Delta\mu$ (in E_0 Å) between transition and initial states for the considered diffusion pathways

Configuration	Pathway	$\Delta\mu$
0000000000	Adatom → adatom	0.0043
1100000011	Step adatom → step adatom	0.0051
0111000000	Step adatom → adatom	0.0000
0000001110	Adatom → step adatom	0.0051
1110000011	Kink → step-adatom	0.0140
0001111100	Step adatom → kink	0.0048
0111100000	Kink → adatom	0.0096
0000011110	Adatom → kink	0.0048
1111100000	Stepatom → adatom	0.0100
0000011111	Adatom → stepatom	0.0000
0011000000	Corner → adatom	0.0220
0000001100	Adatom → corner	0.0073
1100000000	Adatom → corner	0.003
0000000011	Corner → adatom	0.007
0000111000	Corner → step adatom	0.0150
0001110000	Step adatom → corner	0.0074

3.3.4 Effect of the Electric Field on the Diffusion Rates

An electric dipole $\vec{\mu}$ in an external electric field $\vec{E}$ acquires an energy $-\vec{\mu} \cdot \vec{E}$. We take the z-axis as perpendicular to the electrode surface and assume that E_z is the only nonvanishing component in the electric double layer; this implies that the external field is not greatly disturbed by the presence of monoatomic steps and migrating atoms. We therefore need to consider only the z components of the dipole moment. If for a given process the dipole moment $\mu^\dagger$ in the activated state differs from the moment μ_0 in the initial state, the energy of activation is modified by the presence of the field, and the rate can be written as follows:

$$\nu = \nu_0 exp - \frac{E_{act}^0 - \Delta\mu E_z}{k_B T}, \quad \Delta\mu = \mu^\dagger - \mu_0, \quad (3.3)$$

where ν_0 is the frequency factor, E_{act}^0 the activation energy in the absence of the field, T is the temperature, and k_B Boltzmann's constant.

On metal electrodes, in accordance with the Smoluchowski effect, defects like steps, adatoms, and vacancies generally have a positive dipole moment [12], which is larger in the activated than in the initial state of a process. The rate of such processes is enhanced by a positive field and therefore rises with increasing electrode potential. The field experienced by the defects [12, 16] is practically the unscreened field cix $E_z = \sigma/\varepsilon_0$, where σ is the surface charge density. In order to cover a range of field strengths compatible with experiments, simulations in the range $0 \leq E_z \leq 2 \times 10^{10}$ Vm^{-1} have been performed. Negative fields were not explored since they decrease the rates; and for vanishing field and $T = 300$ K, we are already at the upper range of the time that can be covered in the simulations.

3.3.5 Energy Calculations for Gold

The DFT results for the previous addressed 16 selected processes were used to train a reactive force field (RFF) [17], from which the rates of all 544 relevant processes were calculated for the self-diffusion of gold on bare Au(100) surface. In order to investigate the effect of chlorine on Ostwald ripening, the explicit influence of chlorine atom on top of a gold adatom was tackled by means of DFT results for the same 16 selected processes as before. The results will be presented in Section 3.6.

DFT calculations were performed with SeqQuest [18], employing localized basis sets represented by a linear combination of optimized double-ζ plus polarization-contracted Gaussian functions. Effects of Au core electrons were treated by norm-conserving pseudopotentials, including nonlinear core corrections.

The PBE flavor [19] of the generalized gradient approximation (GGA) was used as exchange correlation potential. Convergence in binding energies was obtained with a (5×5) unit cell, a Monkhorst–Pack k-point mesh of 2×2 and six-layer slabs. Geometry optimizations up to 0.01 eVÅ^{-1} have been performed for all atoms except the two bottom layers. In agreement with previous DFT calculations [20, 21], we find that adatom diffusion on Au(100) via the exchange mechanism possesses a lower activation energy than hopping processes [11].

3.4 ISLAND DYNAMICS ON CHARGED SILVER ELECTRODES

The dependence of the surface stability and the dynamics of surface structures on the potential or, equivalently, the electric field, are interesting topics in their own right. Indeed, early studies of the mobility of electrode surface structures showed a strong, typically exponential dependence on the electrode potential (see Refs. [22, 23]). At a first glance, this is surprising, since the metal surface is an equipotential, so that a change in the electrode potential does not affect the free energy of processes that occur on the surface only. However, as pointed out earlier, the electrode potential generates an electric field in the double layer, which interacts with local surface dipole moments. In accord with the Smoluchowski effect [24], steps and point defects on metal surfaces generally have an associated dipole moment that points away from the surface. If the dipole moment in the activated state of a process differs from that in the initial state, the field–dipole interaction enters exponentially into the rate [16]. Since the electric field is roughly proportional to the potential, this mechanism explains the experimental findings. The subject has been reviewed by Giesen [25] and by Ibach [26].

3.4.1 Mesoscopic Theory of Step Fluctuations

At the mesoscopic level, at scales substantially larger than the diameter of an atom, step fluctuations can be described by a theory equivalent to capillary wave theory—for more details, see Refs. [26–28]. It is valid for fluctuations that are sufficiently small that the boundary of the step can be described by a univalent function $\zeta(x)$ of the one-dimensional position vector x (see Fig. 3.1a). We resolve the step position into its Fourier components:

$$\zeta(x) = \frac{1}{N}\sum_q A(q)e^{-ix}, \tag{3.4}$$

where $A(q)$ is the amplitude and q the wave number of the fluctuations. N is the number of sites of the straight, unperturbed boundary.

The fluctuations perform work against the step line tension γ, short-range fluctuations corresponding to higher wave numbers q have lower amplitudes, since for a given amplitude they create more boundary than long-range fluctuations. Statistical mechanics (see Eq. 4.92 of Ref. [26]) gives the relation:

$$ln\langle A(q)A^*(q)\rangle = ln\frac{k_BT}{L_x\widetilde{\gamma}} - 2ln\,q \tag{3.5}$$

where L_x is the step length and $\widetilde{\gamma}$ the stiffness coefficient or step stiffness. The latter is related to the step line tension γ through

$$\widetilde{\gamma} = \gamma + \frac{\partial^2\gamma}{\partial\theta^2}. \tag{3.6}$$

θ is the angle that the tangent of a step makes with the tangent of a step oriented along the (110) direction (kink-free step). Equation (3.5) offers a convenient way to determine the step stiffness from the thermal average of fluctuations.

Based on the stiffness values $\widetilde{\gamma}$, the kink energy ε_K for adatoms positioned at a kink site on the surface can also be estimated. The stiffness $\widetilde{\gamma}$ is related to the so-called diffusivity b^2 of a step by (see Eq. 4.95 of Ref. [26]):

$$\widetilde{\gamma} = \frac{k_B T}{b^2} \tag{3.7}$$

Assuming that the energy of a kink is proportional to its length, one may relate the diffusivity b^2 to the kink energy ε_K by (see Eq. 5.27 of Ref. [2]):

$$b^2 = \frac{2}{\left(e^{\varepsilon_K/k_B T} + e^{-\varepsilon_K/k_B T} - 2\right)} a_{nn} \tag{3.8}$$

where a_{nn} is the nearest neighbor distance. Since Equation (3.8) is exact in the range $exp\left(-\varepsilon_K/k_B T\right) \ll 1$, it is also a reasonable approximation up to $k_B T = \varepsilon_K/2$ even when the aforementioned assumption is not fulfilled. ε_K was obtained by fitting the results for the step stiffness to Equation (3.7).

3.4.2 Step Fluctuations

3.4.2.1 Stiffness Coefficient The temperature and the electric field are the main external system parameters that determine the step fluctuations. The range over which the fluctuations can be described in the framework of the capillary wave theory is limited by the fact that a lattice model is used. Depending on the metal, the temperature range and the applied electric field, the fluctuations can be too weak or too intense. If they are too weak, no reasonable statistics can be obtained; if they are too strong, the fluctuations are too large to be described by capillary wave theory.

Figure 3.3a shows the power spectrum for various temperatures at vanishing field. At the lowest temperature investigated, 350 K, the fluctuations were too rare to provide good statistics. Therefore, the fluctuations were not evaluated in this case. For higher temperatures, the data follow the mesoscopic law of Equation (3.5) for wave vectors in the range $lnq \leq -1$. A logarithmic plot of the power spectrum versus lnq gives a slope of -2, and the line stiffness can be obtained from the intercept. For higher values of q, the corresponding wavelengths are of the same order of magnitude as the lattice constant, and capillary wave theory no longer holds. The same effect has been observed in simulations for liquid–liquid interfaces [29]. The values of the line stiffness obtained are given in Table 3.2. They decrease with increasing temperature, so that the fluctuations become larger.

The effect of a positive electric field can be seen in Figure 3.3b; with increasing field, the fluctuations become stronger and the line stiffness weaker. This effect is particularly large at lower temperatures. Indeed, at 300 K, the fluctuations are significant only in the presence of an appreciable positive electric field.

Simulations for a field of 5×10^{10} Vm^{-1} were also performed, but the fluctuations were too large to be described by capillary wave theory. These findings are in line with the observations that two-dimensional metal structures on metal surfaces become more mobile with increasing electrode potential.

3.4.2.2 Kink Energy Figure 3.4a shows the stiffness coefficients (points) and the fitted curves (solid lines) for different electric fields. The kink energies obtained from the fitting

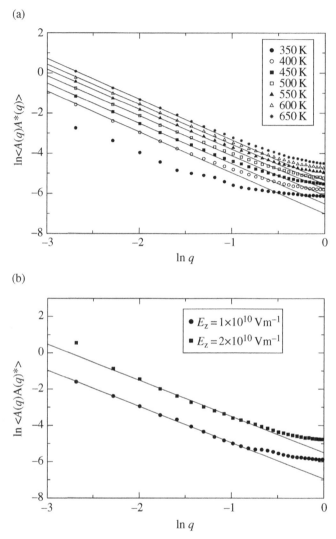

FIGURE 3.3 Power spectrum for Ag/Ag(100) (a) in the absence of an electric field and (b) in the presence of an electric field at 300 K. Reproduced with permission from Pötting et al. [10], © 2009, Elsevier.

are shown in Table 3.3. For higher electric fields, the fluctuations are too large, and the description of step fluctuations via capillary waves no longer applies.

3.4.3 Analysis of the Minimum Curvature of Island Shapes

Depending on the surface structure, the temperature, and the applied field, various island shapes can be observed. On fcc(100) and (111) surfaces, the equilibrium shape of islands show a quadratic or hexagonal profile for low temperatures and/or low fields, respectively.

TABLE 3.2 Stiffness coefficients $\tilde{\gamma}$ in eVÅ^{-1} as a function of temperature T (K) at different electric fields E_z (Vm^{-1}) for Ag(100)

			E_z		
T	0	1×10^9	5×10^9	1×10^{10}	2×10^{10}
300	0.473[a]	0.473[a]	0.295[a]	0.146	0.034
250	0.280[a]	0.287	0.162	0.096	–
400	0.206	0.200	0.149	0.089	–
450	0.142	0.134	0.115	0.071	–
500	0.105	0.105	0.091	0.063	–
550	0.084	0.082	0.076	0.051	–
600	0.072	0.071	0.064	0.042	–
650	0.065	0.058	0.049	0.040	–

Reproduced with permission from Pötting et al. [10], © 2009, Elsevier.
[a]Extrapolated values.

In the case of high temperatures and/or fields, the island contour either becomes circular or elliptical [25].

Following the analysis proposed by Giesen et al. [30], at any point on its perimeter an island in equilibrium has a constant chemical potential μ, which is related to the stiffness coefficient $\tilde{\gamma}$ and the local curvature. In addition, the stiffness coefficient depends on the step orientation θ (see Fig. 3.5) and the line tension γ. The authors derive a simple relation between the shape coordinates of islands and the energy parameters. Using the coordinate system and the island orientation shown in the figure, a point of minimum curvature exists at y, $x = 0$, and for reasons of symmetry, also at $-y$, $x = 0$. Line tension and step stiffness are then related by

$$\gamma = \tilde{\gamma} y y'' \tag{3.9}$$

where y'' denotes the second derivative.

3.4.4 Simulations of Islands

KMC simulations were performed over a temperature range of 300 to 650 K and electric fields in the range ($0 \leq E_z \leq 2 \times 10^{10}$ Vm^{-1}). The dynamics was followed over about 1×10^9 KMC time steps, and the shapes were sampled every 5×10^3 steps. As discussed earlier, the temperature and adatom diffusion are two of the most important parameters that influence the profile of an island. Figure 3.5a shows the island shape for 300 and 600 K at vanishing field. For 300 K, the island shape is almost square and thus close to the low temperature limit, while for higher temperatures the island shape approaches a circular shape. Similarly, the island becomes more circular and the line tension smaller with increasing field strength, that is, the shape fluctuations become stronger with increasing electrode potential (see Fig. 3.5b).

For fields lower than 0.5×10^9 Vm^{-1}, the dynamics of the step is essentially the same as for the uncharged surface. The embedded atom method itself gives a value for the kink energy at an uncharged surface identical to that obtained from the fluctuations, indicating that our treatment is consistent. This value is also close to the value of $\varepsilon_k = 0.082$ eV obtained by Stoltze [31], and $\varepsilon_k = 0.102$ eV by Nelson and Khare [32, 33] using effective medium theory.

(a)

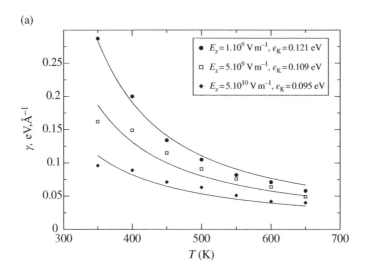

(b)

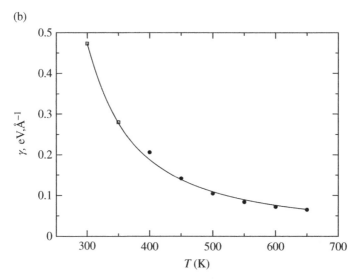

FIGURE 3.4 (a) $\tilde{\gamma}$ values obtained from the simulations at different fields, and the solid curves are obtained by fitting to Equation 1.7. (b) Extrapolation of the line stiffness at vanishing field to lower temperatures. Reproduced with permission from Pötting et al. [10], © 2009, Elsevier.

TABLE 3.3   Kink energies ε_k in eV as a function of the electric field E_z (Vm^{-1}) for Ag(100)

E_z	ε_k
1×10^9	0.121
5×10^9	0.109
1×10^{10}	0.095

Reproduced with permission from Pötting et al. [10], © 2009, Elsevier.

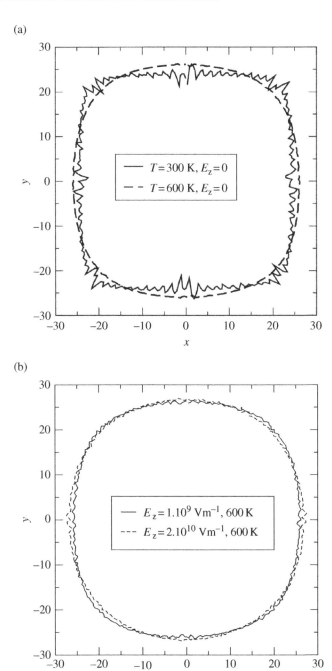

FIGURE 3.5 Ag island shapes on Ag(100), (a) at different temperatures in the absence of a field and (b) at different electric fields at 600 K. Reproduced with permission from Pötting et al. [10], © 2009, Elsevier.

TABLE 3.4   Line tension γ in eVÅ^{-1} as a function of temperature T (K) at different electric fields E_z (Vm^{-1}) for Ag(100)

			E_z		
T	0	1×10^9	5×10^9	1×10^{10}	2×10^{10}
300	0.109	0.109	0.082	0.046	0.014
500	0.061	0.105	0.081	0.047	–
600	0.052	0.071	0.050	0.036	–

Reproduced with permission from Pötting et al. [10], © 2009, Elsevier.

Equation (3.7) allows the extrapolation of the line stiffness from higher temperatures to lower ones, where the fluctuations are too rare to evaluate $\widetilde{\gamma}$ directly from the simulations (Table 3.2). As can be seen from Figure 3.4, the curves rise rapidly toward lower temperatures making the extrapolation somewhat uncertain. This explains why in Table 3.2 at 300 K the extrapolated value for $\widetilde{\gamma}$ is slightly larger for a low electric field than without a field.

Equation (3.9) offers the possibility to determine the line tension γ through the analysis of the minimum curvature of the island perimeter. The second derivative γ'' of the equilibrium shape at the point of minimum curvature has been evaluated through quadratic fitting to our simulation data. The line tension was then obtained from the previously determined line stiffness and the minimum curvature. The corresponding values of γ are listed in Table 3.4 for various temperatures and electric fields. For 300 K and small fields, the shape of the island is so close to a square that it is not possible to evaluate the line tension. Both the line tension and the line stiffness decrease with increasing temperature and/or electric fields. For higher temperatures and/or applied fields the line tension and the stiffness coefficients are similar, while they differ in the low-temperature regime.

3.5   OSTWALD RIPENING

Various mechanisms of island coarsening have been studied in ultra-high vacuum (UHV); they are well reviewed in the literature [25, 26, 34, 35]. For electrochemical systems, only few studies of island decay and Ostwald ripening have been reported [16, 23, 36]; thus far, they have been limited to gold islands on gold surfaces. These studies showed a strong, exponential increase of the surface mobility with increasing electrode potential, which can again be explained by the interaction of the double layer field with local dipole moments [16].

When two islands of different sizes are placed on a lattice, the ripening process depends critically on the positions of the islands, and there is no simple law for the time development. A better approach is to choose a well-defined geometry in which a single island sits in the center of a large vacancy island (see Fig. 3.1c); the symmetry of the system simplifies greatly the analysis. [25, 26]. In the course of time, the island at the center disappears, and the exterior island grows accordingly. For this geometry, approximate expressions for the rate of ripening have been derived in the literature, and will be discussed in the following text.

3.5.1   Ag/Ag(100): Field and Temperature Effect

Since most experimental data have been obtained for islands on Au(100), first simulations for this metal were performed. However, at temperatures below 500 K, the surface was practically immobile. Indeed, experimentally one also finds that in pure perchloric acids, ripening processes take place with an immeasurably slow rate (M. Giesen, private communication). Only in the presence of specifically adsorbed anions like sulfate or chloride is Au(100) sufficiently mobile to exhibit Ostwald ripening [16, 25, 37]. Therefore, for this surface, the presence of an anion should be taken into account; this is discussed in Section 3.6. In the rest of this section, Ag(100) is considered, which is much more mobile than Au(100).

Even on silver, the rates at ambient temperatures are low, and good statistics is difficult to obtain. Therefore, results for somewhat elevated temperatures are presented. Figure 3.6 shows a few typical transients for the size of the inner island at 400 K and various fields. As expected, the decay rate shows a significant increase with the applied field. Toward the very end, when the island is very small, fluctuations are particularly large.

Although such single runs give a good visual impression of the process, ensemble averages are more informative. For each reported field and temperature, we performed at

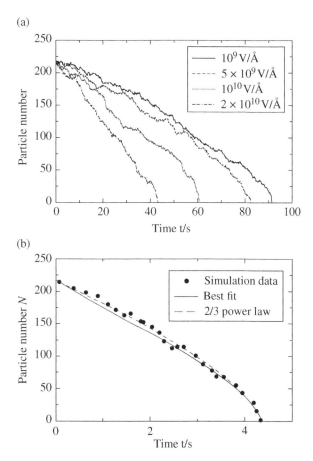

FIGURE 3.6 (a) Particle number of the inner island as a function of time at 400 K and various field strengths. (b) Average particle number of the inner island at 450 K and 0 field. The long dashes represent the 2/3 power law of Equation (3.11), and the full line was obtained by fitting the line tension to the simulation data. Reproduced with permission from Luque et al. [14]. © 2010, Elsevier.

TABLE 3.5 Lifetime t_f (s) of the Ag(100) inner island at different temperatures T and different electric fields E

2*T (K)	E_z (Vm^{-1})			
	0	5×10^9	1×10^{10}	2×10^{10}
400	88 ± 9	76 ± 7	65 ± 9	44 ± 3
450	4.4 ± 0.6	3.7 ± 0.5	3.2 ± 0.5	2.3 ± 0.3
500	0.39 ± 0.06	0.34 ± 0.04	0.29 ± 0.05	0.22 ± 0.03
550	0.055 ± 0.008	0.047 ± 0.006	0.041 ± 0.007	0.030 ± 0.004
600	0.010 ± 0.002	0.010 ± 0.002	0.008 ± 0.002	0.007 ± 0.001

Reproduced with permission from Luque et al. [14]. © 2010, Elsevier.

least $N = 21$ runs—sometimes more, if the fluctuations were too large. Table 3.5 shows the average lifetime of the inner island for various temperatures and fields. It decreases both with the field, as already noted earlier, and also strongly with temperature. This is another example of the similarity between a rise in temperature and in potential, which has been noted before in electrochemistry. Figure 3.6 shows a typical average decay curve. The slight curvature of the transient seems to indicate that the decay is limited by diffusion [26]. However, since the curvature is not a very clear criterion, we verified this directly by arbitrarily changing the rate constant for particle diffusion on the terrace only, and observed a faster decay rate with higher diffusion rate. The possibility to change rate constants of a particular process at will to investigate its importance in the overall process is a great advantage in simulations.

For diffusion limited decay and this symmetry, theory [25, 26] predicts the following rate law:

$$\frac{dN}{dt} = -2\pi D e^{-w_0/k_B T} \frac{e^{\zeta \Omega_s \gamma / r_i k_B T} - e^{-\zeta \Omega_s \gamma / r_0 k_B T}}{\ln(r_0/r_i)} \tag{3.10}$$

where D is the coefficient for diffusion on the terrace, w_0 is the work that is required to bring an atom from a kink site to a terrace site, Ω_s is the atomic area, γ is the line tension, and r_0 and r_i are the average radii of the outer and inner islands, respectively. Since the islands are not spherical but square, the average radii have to be multiplied by a scaling factor ζ, which for square geometry is $\zeta = 1.128$ [25, 26]. If the radii are sufficiently large, the two exponents containing the line tension γ can be expanded to first order. If, in addition, the time dependence of the logarithmic term in the denominator is neglected, a particularly simple decay law results for the inner island:

$$N(t) = \frac{N_0}{t_f^{2/3}} (t_f - t)^{2/3} \tag{3.11}$$

where N_0 is the initial number of particles, and t_f the time when the island disappears.

We have fitted the experimental decay curves to Equation (3.10) using the line tension γ as parameter. The fitted line represents the simulated data very well, but the line tensions obtained in this way (see Table 3.6) are larger by a factor of 2–3 than those obtained from island shapes [10]. Although the values obtained by fitting have a sizeable uncertainty, the discrepancy is outside the error bar. While this inconsistency is disturbing, it agrees with experimental findings: Values for the step line tension obtained from the decay of copper islands on Cu(111) [25, 26, 38] are also substantially higher than those obtained from island shapes [39]. The reason is difficult to discern, but it could be due to the fact that the adatom

TABLE 3.6   Step line tensions $\gamma \times 10^2$ (eVÅ^{-1}) obtained by fitting island decay curves at various temperatures T and electric fields E

T (K)	E_z (Vm^{-1})			
	0	5×10^9	1×10^{10}	2×10^{10}
400	2.4 ± 0.5	1.9 ± 0.5	1.9 ± 0.5	1.6 ± 0.5
450	2.1 ± 0.7	2.5 ± 0.7	2.2 ± 0.7	1.9 ± 0.6
500	1.9 ± 0.7	2.2 ± 0.7	1.3 ± 0.7	2.1 ± 0.7
550	1.6 ± 0.8	1.6 ± 0.7	1.6 ± 0.6	2.0 ± 0.5
600	1.6 ± 0.7	1.2 ± 0.7	2.2 ± 0.7	1.8 ± 0.7

Reproduced with permission from Luque et al. [14]. © 2010, Elsevier.

concentrations near the steps are not in equilibrium with the steps, as is assumed in the derivation of Equation (3.10). Other questionable assumptions are the simplified, circular island shape and a stationary mass center of the island.

3.6   THE EFFECT OF ADSORBED CL ATOMS ON THE MOBILITY OF ADATOMS ON AU(100)

Experimentally, Au(100) surfaces in pure HClO$_4$ are stable and show no adatom migration. However, even small amounts of adsorbed Cl atoms greatly enhance the surface mobility [36, 40]. A similar facilitation of diffusion on surfaces in vacuum by adsorbates has been studied experimentally and theoretically [41–45]. In order to understand this effect, we first investigated the effect of adsorbed Cl on the adatom migration barriers by DFT, and then used these results to simulate Ostwald ripening.

Since there is no evidence for *absorption* of Cl atoms into the Au surface [46, 47], only adsorption on the surface is considered. As the diffusing species in the case of chlorine adsorption a Au adatom with a chlorine atom on top was proposed to be the diffusing species in the case of chlorine adsorption [11]. Table 3.7 compares the energies of activation for the most important processes in the presence and absence of an adsorbed Cl atom. With few exceptions, the Cl atom significantly reduces the barrier.

As pointed in Section 3.3.1, the well-defined geometry of an island, which initially contained 218 atoms, situated in the center of a vacancy island of 75×75 atoms (Fig. 3.1c) was the initial configuration for the study of Ostwald ripening. In the absence of adsorbed Cl, the island was practically stable over the time of 1000 s (solid line, in Fig. 3.7). In the presence of chlorine, the island decayed completely with a decay rate of about 0.25 atoms s^{-1} (dashed line in Fig. 3.7).

The result of KMC simulations agrees well with experimental observations of island decay in 100 mM HClO$_4$ electrolyte with and without 1 mM HCl added. Without HCl added, no decay is observed (solid circles in Fig. 3.7). With 1 mM HCl, a considerable decay is observed. Figure 3.7 shows the experimental decay curve normalized to the size at the beginning of the observations as gray squares. The initial decay rate is about -1 atom s^{-1}.

Even with the fair agreement in the general trend, there is characteristic difference between simulations and experiment: the simulated island decays roughly linearly in time, whereas the experimentally observed decay rate increases as the island becomes smaller.

The thermodynamic theory of Ostwald ripening [26] distinguishes between two cases: the *detachment-limited* decay, which is linear in time, and the *diffusion-limited* decay, in

TABLE 3.7   Comparison of the activation energy of the most important process in Ostwald ripening obtained in the presence and absence of adsorbed Cl

		Activation energy (eV)	
Configuration	Pathway	With Cl	Without Cl
0000000000	Adatom → adatom	0.56	0.70
0000000000	Adatom →^{exch} adatom	0.34	0.55
1100000011	Step adatom → step adatom	0.43	0.42
0111000000	Step adatom → adatom	0.83	0.95
0000001110	Adatom → step adatom	0.58	0.69
1110000011	Kink → step-adatom	0.73	0.64
0001111100	Step adatom → kink	0.45	0.42
0111100000	Kink → adatom	0.56	0.91
0000011110	Adatom → kink	0.38	0.44
0011000000	Corner → adatom	0.91	0.94
0000001100	Adatom → corner	0.54	0.69
1100000000	Adatom → corner	0.41	0.44
0000000011	Corner → adatom	0.46	0.67
0000111000	Corner → step adatom	0.52	0.41
0001110000	Step adatom → corner	0.37	0.42

which the decay rate increases as the island becomes smaller. Diffusion-limited decay occurs when the barrier for the detachment of an atom to a kink site is equal or smaller than the barrier for diffusion.

According to Table 3.7, in the present system we are in the presence of the diffusion limited case. The experimentally observed curved shape of the decay curve is therefore expected. The simulation, on the contrary, appears to realize the detachment-limited case. The reason for the discrepancy is presumably that the continuum theory of Ostwald ripening

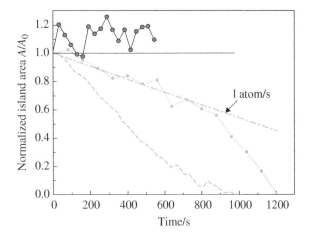

FIGURE 3.7 Comparison of the kinetic Monte Carlo simulation and experiments on the decay of islands on Au(100) surfaces. Solid and dashed lines are simulations without and with chlorine, respectively. Solid circles and squares are the experimental results without and with chlorine added to the electrolyte. Island areas are normalized to their initial size. The dash-dotted line shows approximately the initial rate of decay for the experimental island. Reproduced with permission from Mesgar et al. [48]. © 2013, Wiley.

assumes quasiequilibrium between the chemical potential of the islands and the local concentrations of adatoms on the terrace. However, equilibrium may not be reached in the KMC simulations of our small system. A possible reason for the minor difference between the continuum model and the simulation is that the system used in the simulation is too small for the continuum model to apply. Furthermore, the starting configuration in the simulation does not involve island equilibrium shapes but rather a single kink in the center island (Fig. 3.1c). As the equilibrium shape at 300 K would involve more kink sites its use as the starting configuration would presumably lead to a larger decay rate.

3.7 SOME CONCLUSIONS ON SURFACE MOBILITY

In all cases where the step fluctuations could be evaluated, the power spectrum followed the behavior predicted by the mesoscopic theory—the deviations observed for large wave vectors do not contradict the theory, since they correspond to atomic and not to mesoscopic dimensions. This shows that a consistent set of transition rates, both in the absence and presence of an electric field, was used. In particular, this implies that we have identified the most important processes that govern the fluctuations, and that the set of dipole moments that we calculated are consistent.

All in all, the simulations reviewed provide a consistent picture of the surface dynamics. With increasing field, the surface becomes more mobile, which entails larger step fluctuations and a decrease of the step stiffness. At the same time, the island shapes become more rounded and the coarsening faster. The same effects occur with increasing temperature. It has often been observed that in certain electrochemical experiments, the potential plays a similar role to that of the temperature in UHV. Thus, electrochemical desorption spectra obtained by a potential sweep bear a certain similarity to thermal desorption spectra in UHV.

It is therefore pleasing to observe that the electric field or, equivalently, the potential, has a similar effect on the surface mobility as does the temperature. The term *electrochemical annealing*, which is often used to denote the coarsening of electrode surfaces that occurs at high potentials [16], alludes to this fact.

Finally, we note that an enhanced mobility can also be caused by the coadsorption of anions. In the absence of specific adsorption, the islands on Au(100) are practically stable at ambient temperatures; this is in line with experimental observations that indicate that Ostwald ripening on this surface is induced by specific adsorption of anions. The surface of Ag(100) is much more mobile, and ripening occurs readily at uncharged and at positively charged surfaces.

In summary, the combination of large-scale KMC simulations with DFT calculations and analytically solvable thermodynamic models yield a good agreement with experiments on two-dimensional Ostwald ripening on Ag and Au metals in a bare fcc(100) surface or in chlorine containing electrolytes.

3.8 THEORY OF ELECTROCHEMICAL CHARGE TRANSFER REACTION

The processes considered so far were limited to the electrode surface; therefore, the electrode potential had only an indirect effect on the reaction rates, either through the double-layer field and its interaction with local dipole moments, or through the adsorption of anions, chloride being the prime example. We now turn to a theory of electrochemical reactions, in which charge is transferred through the interface, and the electrode potential affects directly the free energy of the reaction. Conventionally, one distinguishes between electron and ion transfer, but this seems a little artificial, since an ion, while it is being transferred through the interface, is discharged by electron transfer. In our view, a more important distinction is that between outer and inner sphere electron transfer. In the former process, both the reactant and the product are not adsorbed, but remain at a distance of several Ångstroms from the electrode surface, with which they exchange electrons. In inner sphere reactions, the reactant or the product, are adsorbed on—or even incorporated into—the electrode surface. Not only metal deposition and dissolution, but also electrocatalytic reactions like the Volmer reaction (hydrogen adsorption) or oxygen reduction belong to the latter class of reactions.

In our group, we have developed a theoretical framework that can be applied to both kinds of reactions. It is based on a model Hamiltonian incorporating concepts from theories of outer sphere electron transfer [49–51], Anderson–Newns theory [52, 53], and our own ideas. The model as was developed in the 1990s [54, 55]; and at that time, it was applied to various processes such as metal deposition/dissolution, anion adsorption, and outer-sphere electron transfer. However, this was at a time when DFT was not widely available, and several important system parameters had to be estimated, so that the applications had a qualitative character; nevertheless, they provided a basic understanding of these processes at the molecular level.

During the past few years, we have developed a method to combine our theory with DFT calculations [56, 57]. Basically, DFT provides several important parameters of our model, so that it has become possible to calculate free energy surfaces for the reactions and determine energies of activation. So far, our work has mainly been applied to electrocatalytic reactions, where we have focused on the effect of the electronic properties of the electrode on the reaction rate. However, the same procedure can be applied to metal deposition and dissolution, and we shall briefly return to this point at the end of this chapter. Here, we shall first present the theory on which our method is based, and then report on some of its applications to the hydrogen reaction.

3.8.1 A Model Hamiltonian for Electron and Ion Transfer Reactions at Metal Electrodes

We consider a reactant, labeled a, interacting with a metal surface. The model Hamiltonian consists of terms for the relevant electronic part of the system—the valence orbital on the reactant and the electronic states on the metal electrode—and terms for the solvent. We denote with the index a the reactant's orbital, by ε_a its energy, and by n_a the corresponding number operator; the metal states are labeled k, and thus have energies ε_k and number operators n_k. The valence orbital can take up two electrons; so in general, we have to consider two spin states, and the Coulomb repulsion between two electrons on the same orbital. This general case has been presented in Ref. [55], and is especially important when two electrons are transferred. In the one—electron processes that we consider here, both spin states are equally occupied in the region where electron transfer occurs. Therefore, we may limit ourselves to the spinless version of the Hamiltonian. For the general case, we refer to the cited literature.

The Hamiltonian can be written as the sum of several terms; in the spinless version, the electronic part is

$$H_{el} = \varepsilon_a n_a + \sum_k \varepsilon_k c_k + \sum_k \left[V_k c_k^+ c_a + V_k^* c_a^+ c_k \right] \tag{3.12}$$

The first two terms denote the reactant and the metal, the last term affects electron exchange between the metal and the reactant; c^+ denotes a creation and c an annihilation operator. Just like in Marcus (and polaron) theory, the solvent modes are divided into a fast part, which is supposed to follow the electron transfer instantly, and a slow part. The latter is modeled as a phonon bath; after transformation to a single, normalized reaction coordinate q, with corresponding momentum p, the corresponding part of the Hamiltonian is

$$H_{sol} = \lambda \left(q^2 + p^2 \right) + (z - n_a) 2\lambda q \tag{3.13}$$

where λ is the energy of reorganization and z is the charge number of the reactant when the state a is empty. If the terms for the interaction between reactant and electrode in Equation (3.12) are small and treated by first order perturbation theory, the model is equivalent to the Levich and Dogonadze theory [51]. In order to go beyond perturbation theory, we calculate the density of states of the hydrogen atom, treating the slow solvent modes as classical, from Green's function techniques [58]:

$$\rho_a(\varepsilon) = \frac{1}{\pi} \frac{\Delta(\varepsilon)}{\left[\varepsilon - (\varepsilon_a + \Lambda(\varepsilon) - 2\lambda q) \right]^2 + \Delta(\varepsilon)^2} \tag{3.14}$$

Here, $\Delta(\varepsilon)$ and $\Lambda(\varepsilon)$ are the chemisorption functions:

$$\Delta(\varepsilon) = \sum_k |V_k|^2 \pi \delta(\varepsilon - \varepsilon_k) \quad \Lambda(\varepsilon) = \frac{1}{\pi} P \int \frac{\Delta(\varepsilon')}{\varepsilon - \varepsilon'} d\varepsilon' \tag{3.15}$$

where P denotes the principal part. The total energy of the hydrogen atom and the slow part of the solvent modes, as a function of the solvent coordinate q, is then:

$$E(q) = \int_{-\infty}^{0} \rho_a(\varepsilon)\ \varepsilon\ d\varepsilon + \lambda q^2 + 2\lambda q \qquad (3.16)$$

and the occupancy of the orbital is given by

$$\langle n_a \rangle = \int_{-\infty}^{0} \rho_a(\varepsilon)\ d\varepsilon \qquad (3.17)$$

We divide the electronic states of the metal into a wide sp band and the d band. The former is represented by a wide, semielliptical band; usually, we have chosen its limit between -10 and 5 eV with respect to the Fermi level. This band is supposed to be the same for all metals, and to interact with the same strength with the hydrogen $1s$ orbital [59]. The coupling constant with the d band is assumed to be independent of energy; thus the k-dependent coupling constants V_k are replaced by a single constant V coupling the reactant to the d band.

To apply our formalism to hydrogen adsorption on a particular metal, we first calculate its d band by DFT. We then place a hydrogen atom at a given distance d from the electrode, allowing it to move freely in the direction parallel to the surface, and obtain both its energy $E_H(d)$ and the DOS of its $1s$ orbital. From the latter, we obtain the coupling constant V to the d band and the orbital energy ε_a as a function of the distance, by fitting the observed hydrogen DOS to Equation (3.14).

We use our model, as defined by the Hamiltonian, to interpolate between the electronic energies of the hydrogen atom E_H and the proton as a function of the solvent coordinate q. The electronic energy of the proton is 0, and it corresponds to an equilibrium value of $q = -1$ in our normalization of q. For a given distance, we calculate the electronic energy $E(q = 0)$ of the hydrogen atom from our model for $q = 0$ from the first part of Equation (3.16). The difference $\Delta E(q = 0)$ contains the multibody effects not contained in our model. We assume that this error is proportional to the occupancy:

$$\Delta E(q) = \Delta E(q = 0) \times \langle n_a \rangle \qquad (3.18)$$

This equation is obviously correct in the limits when $\langle n_a \rangle = 1$ and $\langle n_a \rangle = 0$. The linear interpolation is natural, and DFT itself is based on interpolations of this kind.

This defines the electronic energy as a function of the solvent coordinate q and the distance from the surface. Thus, the electronic part of the model is well defined in terms of parameters obtained from DFT. The electrode potential also enters into the electronic part: it shifts the energy ε_a with respect to the Fermi level of the metal.

The part that remains is the interaction with the solvent. Up to now, we have treated this in a comparatively simple manner, assuming that the energy of reorganization drops by a factor of 2 as the reactant approaches the surfaces [56]. This approximation is sufficient, as long as we focus on the catalytic properties of the electrode, comparing the electronic properties of various materials, as we do here. However, if we want to compare different processes, we need a better approximation. We shall return to this point toward the end of this chapter. In the next sections, we focus on hydrogen evolution/oxidation.

3.8.2 Principles of Electrocatalysis

In catalytic reactions, the electrode material plays a crucial role; thus, the rates of hydrogen evolution and oxidation vary by more than six orders of magnitude on the various metals. Put into everyday terms, on sp metals like mercury and cadmium, this reaction proceeds with the speed of an ant crawling on a leaf; on some transition metals, like platinum, it proceeds with the speed of a supersonic jet. Unfortunately, the prices of the metals follow their catalytic effectivity; and if mercury can be bought for the price of a scooter, platinum costs as much as a Ferrari race car.

Efforts to understand the catalytic activities of metals have a long history. Much of the early work was phenomenological, and correlations were established between the reaction rate and various properties such as work functions, hydrogen adsorption energies, and unfilled d orbitals [60–63]. All these attempts had limited success and contributed little to our understanding of electrocatalysis.

Before presenting our results on hydrogen evolution, we shall explain how a good catalyst enhances the reaction rate according to our model. All metals that are used as electrode material in aqueous solutions have an sp and a d band. The sp bands are broad, and behave very similar in all metals. Therefore, catalysis is effected by the d bands. To see how this works, we consider the inverse Volmer reaction: $H_{ad} \rightarrow H^+ + e^-$ as an example.

When the $1s$ orbital of hydrogen interacts with a metal surface, it is broadened and acquires a certain width and a corresponding density of states (DOS). This is the same effect as the lifetime broadening familiar from spectroscopy: An electron on the valence orbital can pass to the metal, and hence has only a finite lifetime in this state, which results in a broadened energy level. The width of the broadened orbital depends on the strength of the interaction with the electrode. The stronger the interaction, the shorter the lifetime and the broader the density of states. On all metals, the hydrogen orbital acquires a certain width because of its interaction with the sp band. However, the interaction with the d band can differ widely.

Figure 3.8 compares the situation on two different metal surfaces: Ag(111), which is a mediocre catalyst, and Pt(111), which is excellent. The initial state in the Volmer reaction is the adsorbed hydrogen atom. On both metals, the hydrogen $1s$ orbital is broadened by the interaction with the sp band, and lies below the metal d band, with which it interacts weakly because of the energy difference. The adsorbed hydrogen atom carries no charge, and practically does not interact with the solvent. For the desorption to occur, a thermal fluctuation must lift the valence orbital to the Fermi level, where it can transfer an electron to the metal. The resulting ion is solvated, and the valence level lifted to still higher energies. The energy of activation required for this process is largely determined by the difference in electronic energy between the initial state and the activated state, where the electron transfer occurs. On silver, the interaction of hydrogen with the d band is weak, because the d band lies well below the Fermi level, and the hydrogen DOS, at the activated state, is only little affected. By contrast, on platinum the d band spans the Fermi level, it interacts strongly with hydrogen, and at the activated state the DOS acquires a long tail extending well below the Fermi level, which significantly lowers the energy of activation. We may conclude that a good catalyst must have a d band that spans the Fermi level and interacts strongly with the reactant.

3.8.3 Hydrogen Electrocatalysis

On the basis of the principles presented earlier, we shall review a few representative examples of our work on the hydrogen evolution reaction. This is a reaction with two steps, and to

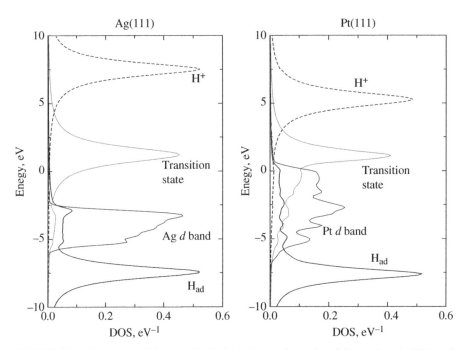

FIGURE 3.8 Mechanism of electrocatalysis for hydrogen desorption; left panel: at Ag(111) and right panel: at Pt(111). The Fermi level has been taken as the energy zero. Reproduced with permission from Santos et al. [64], figure 2. © 2012, Royal Society of Chemistry.

the rules for a good catalyst stated earlier, we may add the familiar principle of Sabatier: At equilibrium, the free energy for each step should be close to zero. The interplay between these three rules determines the reaction rate.

In particular, we shall present results for the two electrochemical steps of the hydrogen reaction mechanism (Volmer and Heyrovsky) on plain and nanostructured electrodes, and illuminate the role of the reaction intermediate. We also discuss in detail the influence of the position of the metal d band and its interaction with the $1s$ hydrogen orbital on the catalytic activity.

The first step in the hydrogen evolution reaction is always the adsorption of a hydrogen atom, which in acid solution occurs via:

$$H^+ + e^- \rightarrow H_{ad} \quad \text{Volmer reaction} \tag{3.19}$$

For the second step, there are two alternative pathways, chemical recombination:

$$H_{ad} + H_{ad} \rightarrow H_2 \quad \text{Tafel reaction} \tag{3.20}$$

or electrochemical desorption:

$$H^+ + H + e^- \rightarrow H_2 \quad \text{Heyrovsky reaction} \tag{3.21}$$

Therefore, the reaction evolves through an adsorbed intermediate (H_{ad}) on an active site on the metal electrode. This is one of the most fundamental reactions in electrochemistry; and during the past decades, much effort has been spent to clarify the mechanism of electrocatalysis, the behavior of the adsorbed intermediate, the effect of the electrode potential on the adsorbate, the role of the electrolyte, and so on. But still there is a lack of knowledge in understanding the nature of the hydrogen electrocatalysis.

We first applied our theory to the electrochemical adsorption of a proton from the solution onto the surface—the Volmer reaction of Equation (3.19)—on the densest surface planes of a good (Pt), a few mediocre (Au, Ag, and Cu) and a bad catalysts (Cd) [56, 65, 66] to see if we could reproduce the trend. The free energy surfaces for the Volmer reaction for four of these metals are shown in Figure 3.9, and the energies of activation are given in Table 3.8. To facilitate the interpretation, the densities of states of the d bands are shown

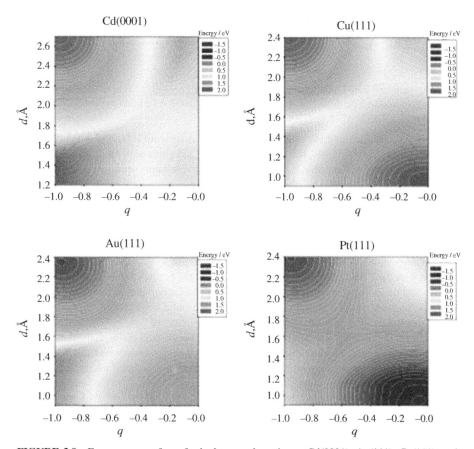

FIGURE 3.9   Free energy surfaces for hydrogen adsorption at Cd(0001), Au(111), Cu(111), and Pt(111). Reproduced with permission from Santos et al. [56]. © 2009, American Physical Society.

TABLE 3.8   Calculated energies of activation for hydrogen adsorption (Volmer step) on various metals at SHE

Metal	Cd	Cu	Ag	Au	Pt
ΔG_{act} (eV)	0.93	0.71	0.71	0.70	0.30

Reproduced with permission from Santos et al. [64], table 2. © 2012, Royal Society of Chemistry.

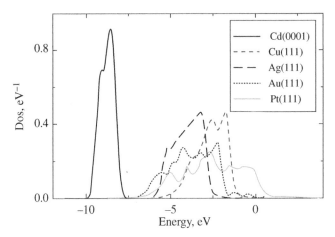

FIGURE 3.10   Surface d band DOS of the metals considered. The Fermi level has been taken as the energy zero. Reproduced with permission from Santos et al. [56]. © 2009, American Physical Society.

in Figure 3.10. All calculations have been performed for the standard hydrogen potential SHE. In all the free energy surfaces, the valley near $q = -1$ and situated at large distances corresponds to a solvated proton near the metal surface, the minimum near $q = 0$ right on the surface represents the adsorbed hydrogen atom. The two minima are separated by a barrier with a saddle point, whose height determines the activation energy ΔG_{act}.

In accord with experimental findings, the activation barrier is highest for Cd, because its d band lies too low the Fermi level to affect the activation energy; therefore it has no catalytic effect, and the reaction is mostly dominated by the sp band. On the coin metals Cu, Ag, and Au, the reaction has about the same activation energies. This is due to a compensation effect: the interaction with the d band increases down the column of the periodic table, because the atomic radius and hence the overlap increases; this lowers the energy of activation. On the other hand, the position of the d band becomes lower in the same order; according to our previous considerations, this raises the activation energy.

On all the three coin metals, the d band lies well below the Fermi level and both the bonding and the antibonding part of the hydrogen DOS caused by the interaction with the d band are filled, so that the d band does not contribute much to the adsorption bond that is dominated by the interaction with the sp band. Of all the metals considered, Pt is the only one whose d band extends over the Fermi level. Its interaction with H is strong, and therefore it has by far the lowest activation energy. At the equilibrium electrode potential, the energy of the adsorbed hydrogen is lower than that of the solvated proton. Therefore, adsorption sets in at potentials above the hydrogen evolution, so that one speaks of strongly adsorbed hydrogen (H_s). However, there is convincing experimental evidence that this is not the species that takes part in hydrogen evolution [67], but the intermediate is a more weakly adsorbed species H_w. So our calculations correspond to the deposition of strongly adsorbed hydrogen, which in the electrochemical literature is also denoted as hydrogen deposited at underpotential (see below). Experimentally, this reaction is so fast that it has not been possible to measure its rate. This is in line with the very low energy of activation that we reported.

Just like on platinum, there is sometimes more than one species of adsorbed hydrogen. This is generally the case for pure or nanostructured transition metals, which have a d band that straddles the Fermi level, and contributes to the bonding of hydrogen; hence, the

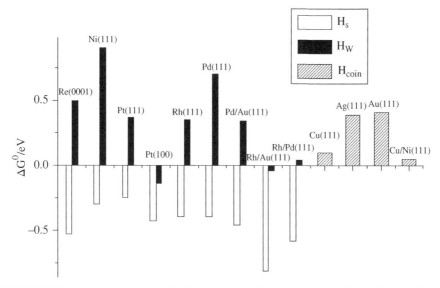

FIGURE 3.11 Adsorption free energies for both adsorbed hydrogen species. The values for H_s are for a quarter of a monolayer, and those for H_w are for a coverage of 1/9 in the presence of a monolayer of H_s. All values refer to the standard hydrogen potential (SHE). Reproduced with permission from Santos et al. [68], figure 1. © 2011, Wiley.

adsorption energies are much lower than on the coinage metals, where the d band lies well below the Fermi level. Often, the most strongly adsorbed species lies so low in energy that it is unfavorable for the second step; therefore it is important to consider at least two different adsorbed species: a strongly adsorbed species H_s, and a more weakly adsorbed species H_w.[1] Figure 3.11 shows the energies of both species for a fair number of systems; for comparison, we have also indicated the adsorption free energies for the coinage metals, where the lowest adsorption state has a positive adsorption free energy at SHE. Typically, the strongly adsorbed species H_s is adsorbed in the fcc hollow sites, the other on top, but there are exceptions. Also, the repulsion between the adsorbed hydrogen atoms typically induces a small shift in the position of H_s in the presence of H_w.

The values shown are for the equilibrium potential SHE; a first glance shows that according to Sabatier's principle, for some metals at that potential the strongly adsorbed species is more favorable from an energetic viewpoint, for others it is the weakly adsorbed one. Obviously, a conventional volcano plot in which the reaction rate is plotted versus the adsorption free energy of H_s only, as presented by Nørskov et al. [69], is misleading, since this state is often only a spectator. The adsorption energy shifts with the overpotential η; counting a cathodic overpotential as positive, so that the current increases with η, the shift is simply $-e_0\eta$, and thus favors the weakly adsorbed species, while the other is pushed even further below. In contrast, an anodic overpotential favors the strongly adsorbed state. Further details are given in [68].

---

[1] In the electrochemical literature, they are also referred to as H^{upd} and H^{opd}, respectively; for an explanation of this strange terminology, see Ref. [67].

3.8.4 Heyrovsky Reaction

As an example of a more complicated reaction, we consider the electrochemical recombination of hydrogen ($H^+ + e^- + H \rightarrow H_2$, Eq. (3.21)), which is also known as the Heyrovsky
reaction. In this case, we require DFT calculations for various positions of the approaching
and of the adsorbed hydrogen. Since only the approaching hydrogen changes its charge, we
have to interpolate the electronic energy for this atom only. So far, we have performed such
calculations for Ag(111), where the second step in hydrogen evolution is known to be the
electrochemical recombination. The free energy surface for this reaction really depends on
the positions of both hydrogen atoms and the solvent coordinates, but such a plot is difficult
to show. The saddle point for the reaction occurs when the adsorbed hydrogen has moved
from its equilibrium position, which is at a distance of 0.9 Å in the fcc hollow site, to a
position 1.4 Å above the surface. Figure 3.12 shows the free energy surface for the reaction
as a function of the position of the approaching hydrogen and the solvent coordinate. The
calculations have again been performed for the equilibrium potential of the hydrogen
evolution reaction.

The first step in hydrogen evolution is hydrogen adsorption, which is endergonic by
0.4 eV at equilibrium. Therefore the second step, electrochemical recombination,
is exergonic by this amount. At the saddle point, with the adsorbed fixed atom at 1.4 Å,
the reaction is slightly uphill. At the final state the approaching atom is at a distance of about
0.8 Å from the other one, which corresponds to the binding distance of the free molecule. The

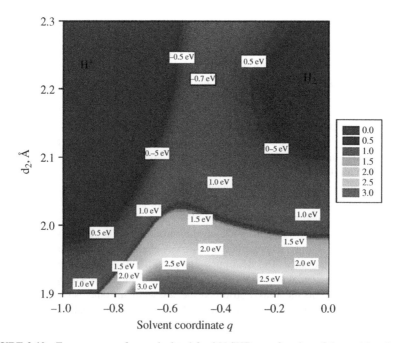

FIGURE 3.12 Free energy surface, calculated for 0 V SHE, as a function of the position d_2 of the
approaching hydrogen and of the solvent coordinate for the case where the adsorbed hydrogen atom is
at a distance of 1.4 Å. Reproduced with permission from Santos et al. [70], figure 8. © 2011, Royal
Society of Chemistry.

saddle point for the reaction is at about the same distance, indicating that the proton first approaches the hydrogen atom to the binding distance before electron transfer takes place. The newly formed hydrogen molecule gains energy as it leaves the surface.

3.8.5 Hydrogen at Nanostructured Electrodes

During the past few years, attention has been turned toward nanostructured surfaces, not only because some of them promise to be efficient catalysts, but also because they constitute an excellent field to explore the relation between structure and reactivity. A large variety of structures has been investigated experimentally and/or theoretically, ranging from metals covered by a monolayer to small clusters or even nanowires.

Here, we consider a particularly interesting case: a monolayer of palladium on Au(111), which is a significantly better catalyst than pure palladium [71, 72]. Pd on Au(111) forms a commensurate layer, which is expanded because the lattice constant of Au is larger than that of Pd by about 5%. This entails a shift of the d band center toward the Fermi level, which induces stronger adsorption bonds [73, 74]; in addition, there is a strong chemical effect of the gold substrate on the Pd overlayer [75]. Consequently, the energy of adsorption of the strongly adsorbed species is lower on Pd/Au(111) than on Pd(111)—see Figure 3.11. Other authors have taken this as the cause for the better catalytic activity toward hydrogen evolution [73]. However, an adsorbate with an energy of $\Delta G_{ad}^0 = -0.46\,eV$ lies too low to be a suitable intermediate, so from this argument alone Pd/Au(111) should be a *worse* catalyst than Pd(111). Our present calculations show clearly, that the weakly adsorbed species on Pd/Au(111) has a much lower energy than on Pd(111), and is therefore a favorable intermediate. From our theory we have calculated the free energy surface for hydrogen adsorption on the weak-adsorption site of Pd/Au(111) and obtained an energy of activation of about 0.57 eV, which is significantly lower than our value of 0.70 eV for Au(111), but not spectacularly low. This agrees with experiments, which show that a complete monolayer of Pd on Au(111) is a good, but not an excellent catalysts [71, 72]. For the dissociation of hydrogen over a Pd/Au(111) surface covered with a monolayer of H_s we obtained a value of about 0.8 eV, suggesting that chemical recombination is not favorable. Indeed, experimental data [76] indicate that either hydrogen adsorption or the electrochemical desorption according to Equation (3.21) determines the overall rate.

Islands of palladium on Au(111) are much better catalysts than a complete monolayer. Pandelov and Stimming [71] have systematically investigated hydrogen evolution for islands of various sizes. In general, they found that smaller islands are much more active. However, the observed current was neither proportional to the surface area, nor to the perimeter of the islands. To explore the matter further, we have calculated hydrogen adsorption on small palladium clusters of different shapes deposited on Au(111). Such clusters offer various adsorption sites, some of them with energies in the range of ±0.2 eV, which are more favorable than the strongly adsorbed species on the monolayer. The activation energies are only of the order of 0.2 eV higher than the adsorption energies, so that these sites are very active. This suggests the following interpretation: The rim of the islands contains very active sites, but the surface is quite active as well. Therefore, the measured current comes both from the surface and from the rim, and is therefore not proportional to the extension of either.

Another interesting system is a monolayer of copper on Ni(111) [77]. Both metals are comparatively cheap, and neither of them is a good catalyst by itself. The lattice constant of nickel is only about 2.5% shorter than that of copper, so there is no major stress, and

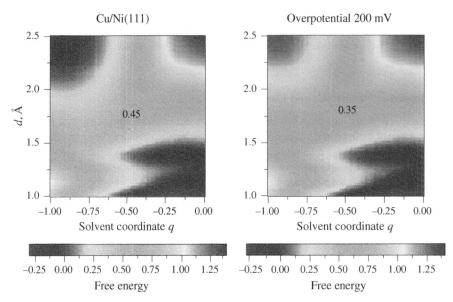

FIGURE 3.13   Free energy surfaces for the Volmer reaction at Cu/Ni(111) at equilibrium (left) and at a cathodic overpotential of 200 mV (right). Reproduced with permission from Gileadi [78], 699, 149, figure 4. Elsevier.

the d band of the adsorbed copper layer is not so much different from that of bulk copper. However, there is a strong chemical effect of nickel on the copper layer, and the interaction of copper with hydrogen is noticeably enhanced. As a consequence, the energy of activation for the Volmer reaction is about 0.45 eV at SHE, which makes this system a good catalysts—not as good as platinum, but much better than nickel or copper. Unfortunately, so far there are no experimental data for this system, so this remains a prediction.

As we mentioned earlier, in our model the electrode potential modifies the electronic energy ε_a of the reactant. This makes it possible to calculate free energy surface for various overpotentials. Figure 3.13b shows the surface for a cathodic overpotential of 200 mV. The activation energy is reduced by about 0.1 eV, indicating an transfer coefficient of $\alpha = 0.5$. For other systems, we typically found transfer coefficients of the order of 0.4–0.6 for the Volmer reaction, which is in line with typical experimental values.

3.8.6   Comparison With Experimental Data

We have performed calculations for a fair amount of systems, but the representative examples given earlier should suffice. Besides the Volmer and the Heyrovsky reaction discussed earlier, there are also a few systems, for which the chemical recombination (Tafel reaction) determines the rate. Since this does not involve electron transfer, it can be calculated by standard DFT; therefore, we have not discussed it in any detail.

Figure 3.14 summarizes those of our results that can be compared with experimental data. Considering the fact that our calculations contain no adjustable parameters, the agreement between our theory and experiment is very satisfactory.

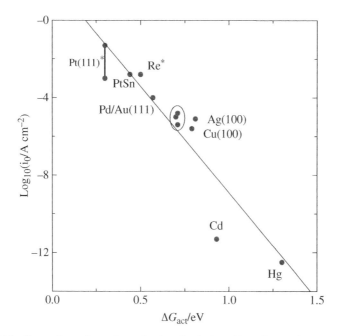

FIGURE 3.14 Comparison of experimental values for the logarithm of the exchange current density versus the free energy of activation calculated from our theory. For platinum, the lower value is from Markovic et al. [79], the higher value is from Chen and Kucernak [80] for polycrystalline Pt. The ellipse contains the (111) surfaces of the three coin metals in the order: Ag < Au < Cu. The asterisk denotes metals on which the chemical recombination determines the rate; for the other metals we have assumed hydrogen adsorption that determines the rate, but on Pd/Au(111) it may also be the electrochemical recombination [76]. The line is a least square fit. Reproduced with permission from Mesgar et al. [48]. © 2013, Wiley.

3.9 CONCLUSIONS AND OUTLOOK

In this overview, we considered two types of processes. The first type takes place only on the surface of the metal and does not involve charge transfer through the interface, the second type are proper electrochemical reactions and involve electron transfer. In the former, the electrodes potential has no direct effect, since the metal surface is an equipotential. We identified two indirect mechanisms: an effect of the double-layer field, which couples to local dipole moments, and the adsorption of ions, which depends on the potential. In particular, we showed that adsorbed chloride enhances the surface mobility of gold. The modeling of these processes was a straightforward combination of DFT with KMC. Perhaps, somewhat surprisingly, water played no role in these simulations. The interaction of water with silver and gold is weak, much weaker than the interaction of gold and silver atoms with their corresponding surfaces, and could therefore be neglected. On metals that interact more strongly with water, such as platinum, the situation is more complicated.

In charge transfer reactions, both the electrode potential and the solvent play important direct roles. This makes it impossible to model these reactions only with DFT, since the ensemble would have to contain the whole interface, which is at least 15 Å thick. By contrast, our approach contains all the important parts of the system from the start, and the parameters are obtained from DFT as far as possible. So far, the treatment of the solvent and its reorganization is comparatively simple, but is justified because we focus on the catalytic effect of the metal.

The same approach, using the simple solvation model, could be used to compare the deposition of one specific metal ion on various metal substrates. However, a more important problem is the realistic description of the whole process of metal deposition, including the desolvation of the metal ion as it approaches the surface. In principle, the latter aspect can be treated by molecular dynamics, and first results have already been obtained a few years ago [81]. What is missing is the incorporation of such simulations into a framework that contains all of the electronic interactions. In this way, we should be able to understand what has been termed the *enigma of metal deposition* [78]: Why is the deposition of certain metal ions so fast? The deposition of silver, for example, is one of the fastest electrochemical reactions known, even though the ion looses about 6 eV of solvation energy during the process. So we close this chapter on an optimistic note: we believe we now have the tools at hand to answer such fundamental questions.

ACKNOWLEDGMENTS

Financial supports by the Deutsche Forschungsgemeinschaft (Schm 344/34-1,2, SCHM 344/42-1,2, Sa 1770/1-1,2, and FOR 1376), and by the European Commission (through FP7 Initial Training Network under "ELCAT," Grant Agreement No. 214936-2) are gratefully acknowledged, and by an exchange agreement between the BMBF. CONICET PIP 112-201001-00411 and PICT-2008-0737 (ANPCyT) are gratefully acknowledged. We thank CONICET, Argentina, for continued support. A generous grant of computing time from the Baden–Wrttemberg grid is gratefully acknowledged.

REFERENCES

1. K.A. Fichthorn and W.H. Weinberg, *J. Chem. Phys.*, **95**, 1090 (1991).

2. K. Nordlund, http://beam.acclab.helsinki.fi/~knordlun/mc/mc8nc.pdf (2006).

3. U. Burghaus, J. Stephan, L. Vattuone and J.M. Rogowska, *A Practical Guide to Kinetic Monte Carlo Simulations and Classical Molecular Dynamics Simulations: An Example Book, Nova Science Publishers Inc., Hauppauge, NY, 2006.*

4. A.F. Voter, "Introduction to the Kinetic Monte Carlo Method," in *Radiation Effects in Solids, ed. by K.E. Sickafus, E.A. Kotomin and B.P. Uberuaga, Springer, Berlin, 2007.*

5. P. Kratzer, "Monte Carlo and kinetic Monte Carlo methods—a tutorial," in *Multiscale Simulation Methods in Molecular Sciences—Lecture Notes, NIC Series, ed. by J. Grotendorst, N. Attig, S. Blügel and D. Marx, Forschungszentrum Jülich, Jülich, 2009.*

6. K. Reuter, "First-principles kinetic Monte Carlo simulations for heterogeneous catalysis: concepts, status and frontiers", in *Modeling and Simulation of Heterogeneous Catalytic Reactions, ed. by Olaf Deutschmann, Wiley-VCH, Weinheim, 2011.*

7. A.P.J. Jansen, *An Introduction to Kinetic Monte Carlo Simulations of Surface Reactions, Lectures Notes in Physics, Vol. 856, Springer-Verlag, Berlin Heidelberg, 2012.*

8. A.B. Bortz, M.H. Kalos and J.L. Lebowitz, *J. Comput. Phys.*, **17**, 10 (1975).

9. N.B. Luque and E.P.M. Leiva, *Electrochim. Acta*, **50**, 3161 (2005).

10. K. Pötting, N.B. Luque, P.M. Quaino, H. Ibach and W. Schmickler, *Electrochim. Acta*, **54**, 4494 (2009).

11. M. Mesgar, P. Kaghazchi, T. Jacob, E. Pichardo-Pedrero, M. Giesen, H. Ibach, N.B. Luque and W. Schmickler, *ChemPhysChem.*, DOI: 10.1002/cphc.201200621.

12. J.E. Müller and H. Ibach, *Phys. Rev. B*, **74**, 085408 (2006).

13. S.M. Foiles, M.I. Baskes and M.S. Daw, *Phys. Rev. B*, **33**, 7983 (1986).

14. N.B. Luque, H. Ibach, K. Poetting and W. Schmickler, *Electrochim. Acta.* **55**, 5411 (2010).

15. U. Kürpick, *Phys. Rev. B*, **64**, 075418 (2001).

16. M. Giesen, G. Beltramo, S. Dieluweit, J. Muller, H. Ibach and W. Schmickler, *Surf. Sci.*, **595**, 127 (2005).

17. K. Kleiner, A. Comas-Vives, M. Naderian, J.E. Mueller, D. Fantauzzi, M. Mesgar, J.A. Keith, J. Anton and T. Jacob, *Adv. Phys. Chem.*, **2011**, 252591 (2011).

18. C. Verdozzi, P.A. Schultz, R.Q. Wu, A.H. Edwards and N. Kioussis, *Phys. Rev. B*, **66**, 125408 (2002).

19. J.P. Perdew, K. Burke and M. Ernzerhof, *Phys. Rev. Lett.*, **77**, 3865 (1996).

20. B.D. Yu and M. Scheffler, *Phys. Rev. Lett.*, **77**, 1095 (1996).

21. K. Pötting, W. Schmickler and T. Jacob, *ChemPhysChem.*, **11**, 1395 (2010).

22. J. Gonzales-Velasco, *Chem. Phys. Lett.*, **313**, 7 (1999).

23. N. Hirai and K.-I. Watanabe, S. Hara, *Surf. Sci.*, **493**, 568 (2001).

24. R. Smoluchowski, *Phys. Rev.*, **60**, 661 (1941).

25. M. Giesen, *Prog. Surf. Sci.*, **68**, 1 (2001).

26. H. Ibach, *Physics of Surfaces and Interfaces, Springer, Berlin, 2006.*

27. N.C. Bartelt, J.L. Goldberg, T.L. Einstein and E.D. Williams, *Surf. Sci.*, **273**, 252 (1992).

28. S.V. Khare and T.L. Einstein, *Phys. Rev. B*, **57**, 4782 (1998).

29. S. Frank and W. Schmickler, *J. Electroanal. Chem.*, **564**, 239 (2004).

30. M. Giesen, C. Steimer and H. Ibach, *Surf. Sci.*, **471**, 80 (2001).

31. P. Stoltze, *J. Phys. Condens. Matter*, **6**, 9495 (1994).

32. R.C. Nelson, T.L. Einstein, S.V. Khare and P.J. Rous, *Surf. Sci.*, **295**, 462 (1993).

33. S.V. Khare and T.L. Einstein, *Surf. Sci.*, **314**, L857 (1994).

34. K. Morgenstern, *Phys. Status Solidi B*, **242**, 773 (2005).

35. P.A. Thiel, M. Shen, D.J. Liu and J.W. Evans, *J. Phys. Chem. C*, **113**, 5047 (2009).

36. E. Pichardo-Pedrero, G. Beltramo and M. Giesen, *Appl. Phys. A*, **87**, 461 (2007).

37. E. Pichardo-Pedrero and M. Giesen, *Electrochim. Acta*, **52**, 5659 (2007).

38. G. Schulze Icking-Konert, M. Giesen and H. Ibach, *Surf. Sci.*, **398**, 37 (1998).

39. C. Steimer, M. Giesen, L. Verheij and H. Ibach, *Phys. Rev. A*, **64**, 085416 (2001).

40. M. Giesen and D.M. Kolb, *Surf. Sci.*, **468**, 149 (2000).

41. R. Stumpf, *Phys. Rev. B*, **53**, R4253 (1996).

42. P.J. Feibelman, *Phys. Rev. Lett.*, **85**, 606 (2000).

43. W.L. Ling, N.C. Bartelt, K. Pohl, J. de la Figuera, R.Q. Hwang and K.F. McCarty, *Phys. Rev. Lett.*, **93**, 166101 (2004).

44. G.L. Kellogg, *Phys. Rev. Lett.*, **79**, 4417 (1997).

45. S. Horch, H.T. Lorensen, S. Helveg, E. Lagsgaard, I. Stensgaard, K.W. Jacobsen, J.K. Norskov and F. Besenbacher, *Nature*, 1999, 134 (**1999**).

46. N.D. Spencer and R.M. Lambert, *Surf. Sci.*, **107**, 237 (1981).

47. G.N. Kastanas and B.E. Koel, *Appl. Surf. Sci.*, **64**, 235 (1993).

48. M. Mesgar, P. Kaghazchi, T. Jacob, E. Pichardo-Pedrero, M. Giesen, H. Ibach, N.B. Luque and W. Schmickler, *ChemPhysChem*. **14**, 233–236 (2013).

49. R.A. Marcus, *J. Chem. Phys.*, **24**, 966 (1956).

50. N.S. Hush, *J. Chem. Phys.*, **28**, 962 (1958).

51. V.G. Levich, "Kinetics of reactions with charge transfer," in *Physical Chemistry, an Advanced Treatise, Vol. Xb, ed. by H. Eyring, D. Henderson and W. Jost, Academic Press, New York, 1970.*

52. J.P. Muscat and D.N. Newns, *Prog. Surf. Sci.*, **9**, 1 (1978).

53. P.W. Anderson, *Phys. Rev.*, **124**, 41 (1961).

54. W. Schmickler, *Chem. Phys. Lett.*, **237**, 152 (1995).

55. W. Schmickler, *Electrochim. Acta*, **41**, 2329 (1996).

56. E. Santos, A. Lundin, K. Pötting, P. Quaino and W. Schmickler, *Phys. Rev. B*, **79**, 235436 (2009).

57. E. Santos and W. Schmickler, *Ang. Chem. Int. Ed.*, **46**, 8262 (2007).

58. S. Davison and K. Sulston, *Green-Function Theory of Chemisorption, Springer, Dordrecht, 2006.*

59. B. Hammer and J.K. Nørskov, *Adv. Catal.*, **45**, 71 (2000).

60. S. Trasatti, *J. Electroanal. Chem.*, **39**, 163 (1972).

61. S. Trasatti, *Adv. Electrochem. Electrochem. Eng.* **10**, 213 (1977).

62. J.K. Norskov, T. Bligaard, A. Logadottir, J.R. Kitchin, J.G. Chen, S. Pandelov and U. Stimming, *J. Electrochem. Soc.*, **152**, J23 (2005).

63. B.E. Conway, E.M. Beatty and P.A.D. DeMaine, *Electrochim. Acta*, **7**, 39 (1962).

64. E. Santos, P. Quaino and W. Schmickler, *Phys. Chem. Chem. Phys.* **14**, 11224 (2012).

65. E. Santos, A. Lundin, K. Pötting, P. Quaino and W. Schmickler, *J. Solid State Electrochem.*, **13**, 1101 (2009).

66. E. Santos, K. Pötting and W. Schmickler, *Faraday Discuss.*, **140**, 209 (2009).

67. W. Schmickler and E. Santos, *Interfacial Electrochemistry, 2nd edition, Springer, Berlin, 2010.*

68. E. Santos, P. Hindelang, P. Quaino, E.N. Schulz, G. Soldano and W. Schmickler, *ChemPhysChem*, **12**, 2274 (2011).

69. J.K. Nørskov, T. Bligaard, A. Logadottir, J.R. Kitchin, J.G. Chen, S. Pandelov, U. Stimming, *J. Electrochem. Soc.*, **152**, J23 (2005); W. Schmickler and S. Trasatti, *J. Electrochem. Soc.*, **153**, L31 (2006).

70. E. Santos, P. Hindelang, P. Quaino and W. Schmickler, *Phys. Chem. Chem. Phys.* **13**, 6992 (2011).

71. S. Pandelov and U. Stimming, *Electrochim. Acta*, **52**, 5548 (2007).

72. L.A. Kibler, *ChemPhysChem*, **7**, 985 (2006).

73. J. Meier, J. Schiøtz, P. Liu, J.K. Nørskov and U. Stimming, *Chem. Phys. Lett.*, **90**, 440 (2004).

74. A. Roudger and A. Gross, *Chem. Phys. Lett.*, **409**, 157 (2005).

75. E. Santos, P. Quaino and W. Schmickler, *Electrochim. Acta*, **55**, 4346 (2010).

76. S. Pronkin, A. Bonnefont, P. Ruvinskiy and E. Savinova, *Electrochim. Acta*, **55**, 3312 (2010).

77. E. Santos, P. Quaino, P. Hindelang and W. Schmickler, *J. Electroanal. Chem.*, **649**, 149 (2010).

78. E. Gileadi, *J. Electroanal. Chem.*, **660**, 247 (2011).

79. N.M. Markovic, B.N. Grgur and P.N. Ross, *J. Phys. Chem. B*, **101**, 5405 (1997).

80. S. Chen and A. Kucernak, *J. Phys. Chem. B*, **108**, 13984 (2004).

81. O. Pecina and W. Schmickler, *Chem. Phys.*, **252**, 349 (2000).

4

ATOMISTIC MONTE-CARLO SIMULATIONS OF DISSOLUTION

STEVE POLICASTRO

Center for Corrosion Science and Engineering, Chemistry Division, Naval Research Laboratory, Washington, DC, USA

4.1 INTRODUCTION

One of the signature successes for atomistic Monte Carlo modeling was its description of binary dealloying and its prediction of morphological and compositional results [1] that had been observed in laboratory experiments and naturally occurring corrosion events. This successful model provided insight into the underlying physics of the selective dissolution process and helped settle understanding of the mechanism by which dealloying occurs in nature and in the lab. Because of this success, this article will use dealloying as a phenomenon to illustrate points about the use of atomistic Monte Carlo simulations of dissolution in general.

4.1.1 Dissolution and Dealloying

There are four environmental requirements that must be met for an electrochemical corrosion process to proceed: an electron path, an ion path, a cathode, and an anode [2]. The electron path can be as obvious as an external circuit connection between the anode and the cathode in an experimental setup or more subtly the transport of free electrons from an anodic to a cathodic site on a corroding metal surface. The ionic path refers to the medium that transports cations away from the anodic site and anionic or other reactive species transport to cathodic reaction sites. The cathode can be a separate material or a temporary reaction site on the corroding metal where a reduction reaction—such as the oxygen reduction reaction or hydrogen evolution reaction—occurs.

$$O_2 + 2H_2O + 4e^- \rightarrow 4OH^-$$ (4.1)

Molecular Modeling of Corrosion Processes: Scientific Development and Engineering Applications, First Edition.
Edited by Christopher D. Taylor and Philippe Marcus.
© 2015 John Wiley & Sons, Inc. Published 2015 by John Wiley & Sons, Inc.

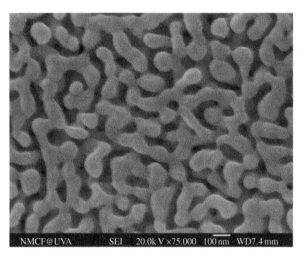

FIGURE 4.1   SEM image of nanoporous surface resulting from dealloying an $Au_{30}Ag_{70}$ film in 16 M HNO_3 for 30 min at 65°C.

$$2H^+ + 2e^- \rightarrow H_2 \tag{4.2}$$

The anode is the site where the oxidation reaction occurs. Schematically, this is shown in Equation (4.3) for the notional metal element, M.

$$M \rightarrow M^{n+} + ne^- \tag{4.3}$$

In corrosion processes, this reaction generally proceeds in the direction of ionizing metal atoms, followed by reaction with electrolyte species to form mineralized or oxidized corrosion products on the metal surface if the corrosion products are insoluble in water. This oxidation process that removes metal atoms from the sample is called dissolution.

Dealloying corrosion involves the preferential dissolution of one or more components of a metallic alloy. This selective oxidation eliminates the more reactive components from the alloy and alters the residual material so that it is less dense and frequently has lost much of its strength, hardness, and macroscopic ductility. During the dealloying process, the rearrangement of the residual atoms of the less reactive components can result in the development of a filamentary layer called a nanoporous structure as shown in Figure 4.1.

4.1.2   A First Description of Dissolution

To a first approximation, and considering only the metal alloy, an atomistic description of dissolution can be found in Figure 4.2. The least tightly bound atoms—those found at kink sites along step edges, for example—would be the first to undergo dissolution. And, for reactive metal samples of high purity, dissolution can unspool the sample along step edges. However, once the atoms at these sites became exhausted, dissolution can slow if the potential is not sufficiently high enough to oxidize atoms from within the step edge face or from the terrace level itself.

(a)

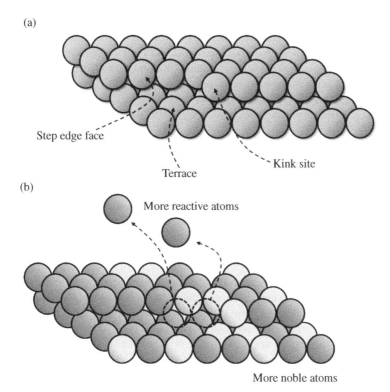

Step edge face

Terrace

Kink site

(b)

More reactive atoms

More noble atoms

FIGURE 4.2   (a) Schematic illustration of atoms located in the different sites on a metal surface. (b) Schematic illustration of dissolution of more reactive atoms from a kink site, eventually uncovering a more noble atom.

The situation becomes more complex in the case of an alloy in which the applied potential is above the reversible potential for the most reactive component but below the reversible potential of the others. In that case, selective dissolution of the more reactive component occurs, but problems arise with the dissolution description provided in Figure 4.2a. As seen in Figure 4.2b, once the dissolution of the more reactive A atoms at kink sites uncovers a noble B atom, that reaction site is shut down. In the same manner, even if reactive A atoms are dissolved from the step edge faces or terraces occurs, eventually, the surface becomes enriched in B atoms and dissolution stops. As this does not occur—as seen in Figure 4.1—the dissolution description provided in Figure 4.2 is lacking.

4.1.3   Evolution of Dissolution and Selective Dissolution Mechanisms

Oxidation has been known since Sir Humphrey Davy electrolytically decomposed potash in 1806 and dissolution as a form of corrosion was investigated by the British Navy in 1823 to develop a method preventing the degradation of their copper-hulled ships. Dealloying has been known by other names such as leaching, dezincification or parting, graphitization, and so on which refer to the removal of a reactive metal component from an alloy or the purification of a noble metal from an alloy, respectively.

The first modern investigations into selectivity in the dissolution of alloy components were performed by Calvert and Johnson [3] in their experiments on the behavior of brass and bronze alloys under the action of various types of acids. Their experiments suggested that for alloy compositions in which the noble component was above a certain concentration, it interfered in the dissolution of the more reactive component, thereby foreshadowing the concept of the parting limit. They did not, however, propose a mechanism to explain the differences in the dissolution behavior of their alloys in response to acids of different strengths.

However, since then, several mechanisms had been advanced to explain the dealloying process. They are volume diffusion, vacancy diffusion, surface diffusion, oxide formation, percolation, and ionization and re-deposition [4].

In the volume diffusion model, developed by Wagner [5], it was proposed that bulk diffusion of the less noble (LN) component to the dealloying front was the mechanism by which the selective dissolution reaction proceeded into the alloy. It was shown later by Cook and Hilliard [6] that bulk diffusion at room temperature was too slow to account for the rates of dealloying that were observed.

In the enhanced diffusion model proposed by Pickering [7], he suggested that with a sufficiently high concentration of divacancies to facilitate the diffusion process bulk diffusion of the more reactive component to the interface could account for dealloying. However, this proposal did not provide a mechanism for overcoming the higher energy of formation for the divacancies in order to reach the higher concentrations needed [4].

Tischer and Gerischer suggested surface diffusion as a mechanism that could account for the replenishment of the more reactive component at the dealloying interface in order to prevent the dealloying process from extinguishing itself [8]. Forty proposed that the surface rearrangement of the more noble (MN) atoms created islands enriched in MN atoms [9], and Young extended that proposal by suggesting that the surface rearrangement in response to the loss of the LN atoms was a way to reduce the surface energy of the residual material [10].

In the percolation model, it is the interconnectedness of atoms of the LN component that allows the dealloying front to proceed into the alloy. However, without allowing for surface diffusion, the dissolution process stops after penetrating a short distance into the crystal as the MN component covers the surface and prevents further dissolution [11].

In the dissolution and precipitation mechanism, proposed by various authors [12–14] working with brasses, both the LN and MN components dissolve, but the MN component is deposited. Several difficulties arose with this mechanism outside of the brass system in that ions of the MN component were not detected in the solution during dealloying, but instead alloys of intermediate compositions were found on the dealloyed surface.

4.2 METROPOLIS MONTE CARLO AND KINETIC MONTE CARLO SIMULATIONS

4.2.1 Overview and Background of Monte Carlo Model

Classical Monte Carlo methods have been used to study a wide variety of different systems since they were first proposed. They were first developed by Stanislaw Ulam Nicholas Metropolis, and John von Neumann at the Los Alamos National Laboratory in the late 1940s [15, 16] and are defined by algorithms with the characteristic of using random sampling or number generation to obtain numerical solutions to equations. They enjoy wide application in the solution of physical problems in which it is impossible to obtain an analytical solution or in which a deterministic method isn't feasible [17]. For example, Monte Carlo methods can be used for problems in areas such as numerical integration of statistical mechanics integrals and generation of samples from a probability distribution. In other applications, a number of Monte Carlo algorithms have been developed for calculating static behaviors, such as solving multidimensional integrals or calculating static critical exponents. These types of static calculations permit a wide variety of trial moves because only the final sum or integral is of interest, and the individual attempts do not necessarily have any physical meaning [18].

As an example for using a Monte Carlo method to solve an integration problem in order to obtain an estimate for π, consider a circle of radius, r, inscribed by a square, of length $2r$, as shown in Figure 4.3. The area of the circle, $A_c = \pi r^2$ and the area of the square, $A_s = (2r)^2 = 4r^2$. The ratio of the area of the circle to the area of the square is $A_c/A_s = \pi r^2/4r^2 = \pi/4$. Randomly selecting two numbers from $-r$ to r, N times, and letting them be the x- and y coordinates for a location inside the area of the square, then the ratio of locations that fall inside the circle area, n_c, to the total locations, N, inside the area of the square is $n_c/N = \pi/4$ so $\pi = 4n_c/N$. The larger the number N becomes, the more accurate the determination of π becomes.

Monte Carlo methods are useful for simulating systems with many coupled degrees of freedom, such as in evaluating a multivariable statistical mechanics integral. That is, they can be used to obtain the expectation value for a macroscopic variable, A, for a system of N particles in which [1] the Hamiltonian, $U(r)$, is known [2], the system is at some temperature, T, and [3] Boltzmann statistics apply:

$$A\left(r^N\right) = \frac{\int A(r^N)e^{-\frac{U(r^N)}{kT}}}{Z} \, dr^N$$

$$Z = \int P(r)dr \qquad (4.4)$$

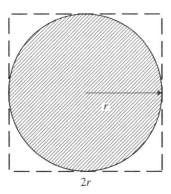

FIGURE 4.3 Illustration of a circle of radius r, inscribed by a square of side, $2r$.

Where $P(r)$ is the particular partition needed for the statistical ensemble and the integrals are over all configuration space.

An important point that needs to be mentioned here is that random sampling from all of the configuration space is inefficient if there are no contributions to the integral coming from large fractions of the available space. For example, if the particles in the configuration are treated as hard spheres, then configurations in which the particles overlap are nonsensical and should not be sampled. The technique of importance sampling can be used to restrict sampling to areas of configuration space that are important to the integral. Importance sampling weights the random sampling using a probability distribution, $P(r)$.

Last, for an "evolving" system, that is, a system, in some configuration, A, which is randomly walking though configuration space, a criterion for accepting or rejecting the next configuration is needed. The approach used in the Metropolis Monte Carlo method is to automatically accept a transition that leads to a lower overall system energy is accepted, while a transition that raises the system energy is accepted based on a probability defined by the ratio of probabilities of the initial and final states.

An example algorithm [19] for performing a Metropolis Monte Carlo simulation is as follows:

1. Choose an initial system configuration, n, and determine its energy, U_n.
2. Choose a new configuration, m, for the system and determine its energy, U_m.
3. Decide whether to allow the transition to occur.
    a. If $U_m-U_n < 0$, then transition occurs and the overall system energy is lowered.
    b. If $U_m-U_n > 0$, then
        i. Calculate a transition probability, W, of the transition from n to m:
        $W(n \rightarrow m) = e^{-\left(\frac{U_{mn}}{kT}\right)}$.
        ii. Select a random number, R, from 0 to 1.
        iii. If $W(n \rightarrow m) > R$, allow the transition to occur.
        iv. If not, reject the transition.
4. Repeat step 2. Accumulate counts for averages. If system configuration, n, is maintained, it is counted as a "new" configuration for the random walk.

However, if the Monte Carlo algorithm is applied to a dynamic problem in which the physical meaning of the trial moves is an essential part of the model—the Ising model of ferromagnetism, for example—then calculating the behavior of the system can become very computationally expensive for systems of even modest size. Advanced dynamic Monte Carlo algorithms have been developed to address larger systems that evolve over long periods of time—including the kinetic Monte Carlo (KMC) algorithm, which was developed from the n-fold way algorithm [20].

KMC models are Monte Carlo algorithms that use a simplified transition state theory [21] to model the kinetics of slow processes—as opposed to the fast kinetics (on the order of picoseconds) modeled by molecular dynamics. KMC models reduce the computational load of determining the individual motions of every particle to the less intensive task of tracking only successful outcomes to transitions. Thus, before beginning a KMC simulation, all transitions that can occur must be listed and rates assigned to each—an example of which is shown in Table 4.1.

TABLE 4.1 Example table of rates and probabilities for three transitions in a KMC model of selective dissolution[a]

Transition	Rate	Probability
Au_{ads} Au_{ads}	R_1	$P_1 = \frac{R_1}{R_1 + R_2 + R_3}$
Ag_{ads} Ag_{ads}	R_2	$P_2 = \frac{R_2}{R_1 + R_2 + R_3}$
Ag_{ads} $Ag^+ + e^-$	R_3	$P_3 = \frac{R_3}{R_1 + R_2 + R_3}$

[a]Reproduced with permission from the *Journal of the Electrochemical Society* 2012, 157(10) C328, © The Electrochemical Society, 2012.

4.2.2 Application of the KMC Algorithm to Simulate Dissolution and Selective Dissolution

This section discusses, in general terms, the development of a KMC model of dissolution and dealloying that can incorporate an alloy and electrolyte and develops rules for determining energy barriers and interactions. The beginning of the section provides a brief description of the metal–electrolyte interface in order to set the stage for using the KMC technique in modeling the dissolution processes. The discussion then moves to providing the details of implementing a KMC model and to points out some of the choices that can be made in how the model is implemented. There is a brief discussion of the procedures used to obtain dealloyed charge and current densities from dissolution reactions and the resultant porosity of dealloyed samples because they are simulation results that can be easily compared to experimental measurements. The final part of the section delves into a derivation of the diffusion equation from the random-walk model and links the diffusivity parameter obtained from the random-walk calculation to the macroscopic diffusivity. A brief derivation of the general Einstein relation is provided that will be used to determine metal atom diffusivities. These derivations provide the justification for the calculation of surface diffusivities of atoms from the KMC model. Because surface and volume diffusivity values are available in the literature, they are a useful parameter that can be compared to simulation outputs and thus can be used to "calibrate" model parameters, such as bonding energies, by comparing surface diffusion in the model to the experimentally determined surface diffusivities. However, diffusivities are only one of many possible options for calibrating the simulation. Other options could include using porosity, dissolution current, critical potential, and so forth.

4.2.3 The Alloy–Electrolyte Interface

The Stern model is used as the basis for describing the interface between the charged metal surface and the electrolyte. Briefly, the Stern model synthesizes the Helmholtz–Perrin and the Gouy–Chapman interface models by assuming that electrolyte ions are not allowed to come into direct contact with the metal surface—thereby forming the outer Helmholtz plane (OHP)—while the electrolyte charge that is not fixed in the OHP is diffusely spread out into the solution, decaying away exponentially as distance from the metal surface increases. Figure 4.4 illustrates a Stern model interface [22].

During the discussion of the development of a KMC model for dissolution, various methods for treating the electrolyte are discussed, and one approach is to differentiate them with respect to the level of interaction of the electrolyte anions with the metal surface. In that case, the noninteracting electrolyte most closely approximates the Stern interface model. The

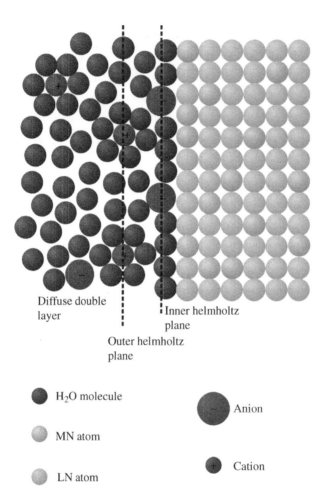

FIGURE 4.4 Schematic representation of the Stern model of the structure of the double layer at the metal–electrolyte interface showing the ions and water molecules. The inner and outer Helmholtz planes are labeled, along with the diffuse double layer. In the figure, the metal has been positively polarized.

other electrolytes require specific adsorption to the metal surface in order to interact with the metal.

Sites where dissolution is occurring are assumed to be generally anodic to the equilibrium potential and sites where oxygen reduction or hydrogen evolution is occurring are assumed to be cathodic. At the anodic sites, because of local interactions between the metal atoms and the electrolyte, the electron cloud is assumed to be oriented toward the electrode. The capacitance of the double layer is thereby decreased and makes the metal atom cores more susceptible to ionization and dissolution.

In a physical electrode–electrolyte interface, the density of the water molecules in the layer closest to the metal surface can be much greater than the bulk water density due to electrostriction [23] and, in fact, can exceed the availability of adsorption sites.

Equation (4.5) presents the starting points for obtaining a numerical solution to the areal water density [24]:

$$\frac{\sigma(\varepsilon - n^2)}{\varepsilon NB} = \tanh\left(\frac{A\sigma}{\varepsilon + \frac{n^2}{2}}\right)$$

$$\frac{1}{v^o}\int_{P^0}^{P} v(P)dP = \frac{N^o}{\varepsilon_0}\left[\left(\frac{\partial f}{\partial y}\right)\left(\frac{\partial y}{\partial \varepsilon}\right)\right]_N\left(\frac{\partial \varepsilon}{\partial N}\right)_\sigma \tag{4.5}$$

Where σ is the surface charge density, ε is the permittivity, n is the refractive index (1.333), N the number of water molecules, $A = \mu(n^2 + 2)/(2kT\varepsilon_0)$ and $B = \mu(n^2 + 2)/(3V)$, μ is the dipole moment of a water molecule (6.023×10^{-30} cm), $T = 293$ K, k = Boltzmann constant (1.3807×10^{-23} V C K^{-1}), the superscript o refers to quantities outside the interfacial electric field, f is the free energy density $= \varepsilon_0\Delta F/V$, y is the distance from the interface, P is the striction pressure, and v is the volume per water molecule. Results of this calculation are presented elsewhere [24], but the important point here is that in developing a KMC model of the electrolyte–metal interface, some care must be taken in how the adsorbed concentration of water molecules is to be considered—if the electrolyte is to be explicitly considered in order to predetermine the various rates for transitions, especially if they are affected by the number of nearest neighbors.

Last, in order for an anion to adsorb onto a metal surface from the solution, three processes must occur. The first process that occurs removes the water molecule from the adsorption site—including adjacent sites if the anion is larger than a water molecule. The second process occurs near the OHP as the anion must partially or fully dehydrate its solvation sheath and, finally, the anion must adsorb into the site created by the vacated water molecules. One approach is to assume the adsorption of the anions to be approximately governed by the Langmuir isotherm in that an anion adsorbed near the metal surface has occupied that adsorption site, and so another anion in the electrolyte cannot occupy that site and must diffuse to a vacant site in order to adsorb.

4.2.4 Effect of the Electrolyte on Dissolution and Dealloying

There were a series of papers in the early 1990s [25–28] that began looking at the processes occurring at the interface between a metal sample and the electrolyte. First, Schott and White looked at the effect of electric fields on the surface reconstruction of Au(111) surfaces and then Gao et al. examined the reconstruction of Au(100)-aqueous electrolyte (0.1 M HClO$_4$) interfaces *in situ*. They observed that various applied overpotentials could alter the surface reconstruction of the gold in contact with the electrolyte, thus paving the way to considering the role of the electrolyte in the dealloying process.

Doña and González-Velasco [26] measured the surface diffusivities of gold atoms as a function of the applied potential for surfaces in contact with a 1 M HClO$_4$ electrolyte and found that as the potential increased so too did the gold atom diffusivity. They theorized this increase in diffusivity was due to the gold atoms' increasing interaction with the electrolyte and a corresponding reduction in the bonding of the surface atoms to the bulk gold atoms.

They [27] then turned to experimentally measuring the activation enthalpies and entropies for the surface diffusion process for gold atoms in contact with a 0.5 M HClO$_4$ electrolyte at increasing temperatures and at various potentials. They observed negative activation entropies at higher potentials and suggested that this was due to increased ordering in the double layer as the H_2O molecules became oriented in the double layer as the result of a partial charge transfer from the gold adatom to the electrode surface prior to the onset of surface diffusion.

Last, Dursun et al. [29] experimented with the effect of halide-containing electrolytes on dealloying silver–gold alloys. They used $Ag_{0.7}Au_{0.3}$ and $Ag_{0.65}Au_{0.35}$ samples in 0.1 M $HClO_4$ with the addition of 0.1 M KCl, 0.1M KBr, or 0.1M KI and observed not only a decrease in the critical overpotential for dealloying with the addition of the halides but also an increase in the pore sizes and a corresponding increase in the gold atom diffusivity.

4.2.5   An Algorithm for Implementing a Dissolution and Dealloying KMC Model

The metal sample on which the KMC algorithm operates can comprise a single or multiple components with one of the components having a substantially lower reversible potential than the others. Table 4.2 provides a list of reversible potentials for some oxidation–reduction reactions focused but are available in the literature. In general, while dissolution will proceed if an externally applied potential or local galvanic couple drives the potential of the sample above its reversible potential, selective dissolution will in general occur only for potentials that fall within the gap between the reversible potentials for the components of the alloy.

Furthermore, for concentrations of the LN component below a certain level, a type of passivation behavior can be observed. That is, if the MN component is present in sufficient concentration that it completely covers the dealloying surface after a limited amount of dissolution of the LN component, then the selective dissolution reaction will stop.

If an aqueous electrolyte is to be considered in the model, it consists of water molecules and dissolved anions and cations. The interaction between the electrolyte and the electrode can be used to establish the conditions for the enhanced surface diffusivity of the alloy components and facilitates the oxidation of the LN component. The electrolyte also provides the medium for the diffusion and electromigration of anions and cations to and from the interface, respectively.

The algorithm, discussed in the following text, adopts the Bortz–Kalos–Lebowitz [20] model for a KMC implementation. The algorithm to be used is as follows:

1. Initialize the system at time $t = 0$. That is, generate the crystal structure with the specified alloy composition. There are a couple of different approaches that can be taken here. One approach [1]—which uses fewer memory resources, generally runs more quickly depending on the computer architecture, and can simulate larger

TABLE 4.2   Excerpted listing of standard electromotive force potentials[a]

Reaction	Standard potential, E_0 (V_{NHE})
$Au^{3+} + 3e^- \leftrightarrow Au$	+1.498
$Pt^{2+} + 2e^- \leftrightarrow Pt$	+1.118
$Ag^+ + e^- \leftrightarrow Ag$	+0.799
$Cu^{2+} + 2e^- \leftrightarrow Cu$	+0.342
$Ni^{2+} + 2e^- \leftrightarrow Ni$	−0.250
$Fe^{2+} + 2e^- \leftrightarrow Fe$	−0.447
$Cr^{3+} + 3e^- \leftrightarrow Cr$	−0.744

[a]Haynes et al. [30].

structures—is to generate only the layer that is in contact with the electrolyte. That is, the bulk atoms that occupy the ligaments and such are not generated or tracked. Another approach, used here, occupies more memory resources though it ensures that the alloy composition is accurate to generate the entire crystal sample at the outset. Populate the electrolyte with the desired concentration of ions if the electrolyte is to be included in this model.

2. Create a look-up table of atom hop positions for the metal atoms. One approach is to extend the crystal lattice throughout the space of the alloy sample and assume that atom hops are to vacant nearest-neighbor positions in the crystal structure. Look-up tables speed up the main program because they do not have to be recalculated for each iteration.

3. Create a look-up table of energy barriers for all of the possible energy barriers that could occur in the system. One approach is to determine the energy barrier for a diffusion or dissolution event based on the number and type of metallic or electrolyte neighbors, but other approaches can be taken based on quantum mechanical calculations or experiments. Again, the look-up table speeds the main program execution because it does not need to be recalculated for each loop iteration.

4. Determine the simulation cut-off time. Because each iteration of the main execution loop can advance the simulation by different amounts of time—as discussed in step 10—keeping track of the amount of time that has passed assists in comparing simulation results to experimental measurements. A cut-off time of approximately 1×10^5 s could be chosen, for example, in order to compare with 24-h experimental measurements.

5. Begin iterating, the main execution loop based on the number of iterations desired for that simulation.

6. Enumerate all of the metal atoms and electrolyte components in the alloy–electrolyte interface and store their dissolution (if appropriate) and diffusion rates along with potential exchange sites.

7. Calculate the cumulative rate function: $R_T = \sum_i R_i$, where R_i is the rate for a given process and i is an index that loops over the dissolution and diffusion processes for all of the interfacial atoms and electrolyte components. R_T is the total rate. Each rate, R_i, is calculated using an Arrhenius expression that is discussed further.

8. Determine the probability for each event, i, to occur by obtaining $P_i = \frac{R_i}{R_T}$.

9. Obtain a uniform random number, u_1, between 0 and 1 and determine the event that will be carried out by finding the event, i, for which $P_{i-1} < u_1 < P_{i+1}$. It is very important—in this type of implementation—that the random number generator provide a uniform distribution of random numbers between 0 and 1—and not a Gaussian distribution, for example. Additionally, the seed value that initiates the generator should be tied to a system event such as the clock time when the loop is initiated or another randomly chosen number to ensure that useful statistical results are obtained from the simulation.

10. Carry out event i by exchanging an atom with an electrolyte component, performing the oxidation reaction, or transferring electrolyte components. A relatively straightforward way for handling time is to let the increase in time for each iteration be inversely proportional to the total rate, R_T, of all the processes available in that time

step. Another approach is to allow the increase in time for each time step to be the same, but that Δt has to be smaller than the inverse of the fastest rate and thus there may be many iterations in which nothing occurs in the simulation.

11. Obtain a new uniform random number, u_2, between $0 \rightarrow 1$ and update the time step such that $t = t + \Delta t = t + \frac{-\ln(1/u_2)}{R_T}$. The additional random number generation in this step allows the delay associated with the transition to be selected from the distribution of r_i.

12. Check to see if the simulation stop count or total time was exceeded. If not, return to step 5.

As an example of the complexity of nearest-neighbor types that could influence the energy barrier calculations in step 3 of the algorithm, there would be two metal atom types for a binary alloy (three in the case of a ternary alloy) and up to four electrolyte components. For example, for a simple NaCl solution, there would be H_2O molecules, Na^+ and Cl^- ions, and cations of the most reactive alloy component—assuming that the noble alloy components do not dissolve. The alloy components are designated as MN or LN in the case of binary alloys. See Figure 4.5 for an example simulation cell.

The two types of transitions allowed for metal atoms in the algorithm are hop transitions among vacant lattice sites and dissolution events in which the atom lost electrons to become

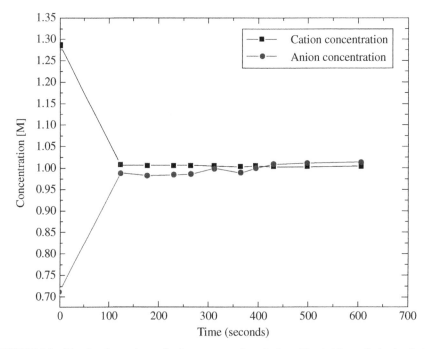

FIGURE 4.5 Plot showing cation and anion concentrations during a kinetic Monte Carlo simulation of silver dissolution with the initial electrolyte concentrations for the anions and cations set at ~1 M. Reproduced with permission from Policastro et al. [31]. © The Electrochemical Society, 2012.

an ion in the electrolyte. The form of the rate equations, as mentioned earlier, is of an Arrhenius-type equation and is the same form for both diffusion and dissolution:

$$R_i = v_i e^{-\frac{E_b{}^i}{k_b T}} \tag{4.6}$$

where R_i is the rate of process i, v_i is an oscillation frequency, E_b^i is an energy barrier, k_b is the Boltzmann constant, and T is the simulation temperature in Kelvin. See Figure 4.6 for a schematic diagram of this process.

The oscillation frequency, v, for diffusion can be considered to be on the same order as the Debye frequency of the lattice: $1 \times 10^{13}/s$, after Erlebacher [32], for both alloy components. For dissolution, v can be considered to be related to the exchange current density. As an example, for Ag, $v_{Ag} \approx 1 \times 10^4/s$ and for Au, $v_{Au} \approx 1 \times 10^{-3}/s$.

As mentioned earlier, the energy barriers, E_b, to diffusion and dissolution can be determined using a local bond-breaking approximation—that is, the nearest-neighbor bonds for the atom of interest establish its barrier to motion or dissolution [33]—along with other modifiers. As an example, Figure 4.7 shows the nearest-neighbor configuration for atoms in an FCC lattice oriented in the (001) direction.

For diffusion in a binary system, the energy barrier for lattice hopping is given by Equation (4.7) for an atom, j, with a given configuration of A and B metal and electrolyte neighbors.

$$E_b{}^j = mE_{A-A} + nE_{A-B} + pE_{A-e} \tag{4.7}$$

E_{A-A} is the bond energy between A–A atoms, E_{A-B} is the bond energy between A–B atoms, and E_{A-e} is the interaction energy between an A atom and a neighboring electrolyte component, m is the number of A–A bonds, n is the number of A–B bonds, and p is the number of A–e interactions.

Likewise, for dissolution, the energy barrier ionization is given by Equation (4.8) for a given configuration of A and B atom neighbors and electrolyte neighbors for an atom, j, while an applied potential can bias the energy of the initial energy level in relation to the energy barrier for dissolution

$$E_j^B = mE_{A-A} + nE_{A-B} + pE_{A-e} \tag{4.8}$$

E_{bias} accounts for the externally applied potentials, V_{app}, and E_r is the energy of the initial state with respect to the energy barrier and related to the Nernst potential given by Equation (4.9) for the electrochemical reaction $R + ze^- \Leftrightarrow P$, where R stands for the reacting species and P for the product species. E_0 is the standard potential in V_{NHE}, z is the number of electrons exchanged, F is Faraday's constant, and a_p and a_r are the activities of the products and reactants—using ideal solution thermodynamics so that their activities are based on their solution concentrations—respectively.

$$E_{bias} = V_{app} - E_r = V_{app} - \left[E_0 - \frac{RT}{zF} \ln \left(\frac{\prod_{p=1}^{n} a_p}{\prod_{r=1}^{n} a_r} \right) \right] \tag{4.9}$$

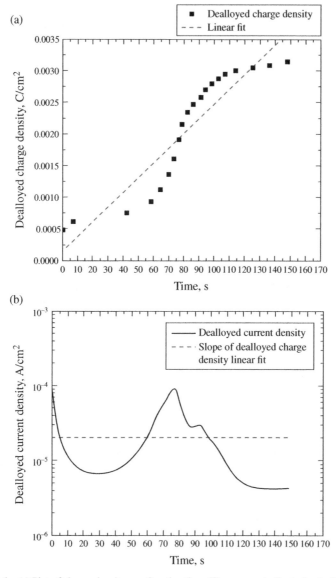

FIGURE 4.6 (a) Plot of charge density as a function time. The squares indicate the dealloyed charge density obtained from the simulation, and the dashed line indicates the best linear regression fit. (b) Plot of the log of the current density as a function of time where the dashed line indicates the slope of the linear regression and the solid line indicates the log of the instantaneous current density.

As an example, Table 4.3 provides data on the bonding energies for gold, silver, and copper. The interspecies bonding energies listed in the table were obtained using the assumption that the metal components formed an ideal solution. $E_{A-B} = \frac{1}{2}(E_{A-A} + E_{B-B})$, but regular solution models or other approximations could be used for more complex alloying behavior.

(a)   (b)   (c)

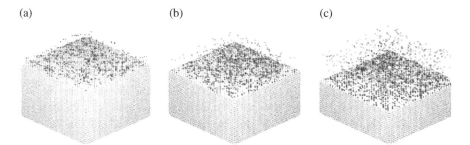

FIGURE 4.7   (a–c) Images from the progression of the microstructural evolution of a single-component alloy undergoing dissolution. Water molecules have been blanked out to aid in the visualization.

TABLE 4.3   Binding energies for atoms in the simulation alloys

Bond type	Energy per bond (eV)
Au–Au[a]	−0.64
Au–Ag	−0.56
Au–Cu	−0.61
Ag–Ag[a]	−0.48
Ag–Cu	−0.53
Cu–Cu∗	−0.58

[a]These values were obtained from the literature Howe [34].

There were three ways in which the behavior of the MN and LN atoms can be distinguished in the algorithm discussed earlier: the first is through the value used for the dissolution hop frequency—which is related to the exchange current density, the second is through its bonding energy with like and unlike neighbors, and the third is in its reversible potential, based on the concentration of its cations assumed to be in the electrolyte and the standard potentials as given in the literature.

Three considerations are to be addressed in modeling the electrolyte: account for ionic transport through the bulk solution, maintain charge neutrality and concentration, and differentiate among the interaction of the electrolyte components with the alloy. The ionic motion should not governed by the KMC algorithm used for the metal atoms because ionic diffusivities in solution are, as a rule, much higher than surface diffusivities for atoms—in which case many computational cycles would be spent moving ions around in the solution. Instead, the ion motion can be incorporated into a separate molecular dynamics simulation or treated in a deterministic manner such as having all cations biased to drift away from anodic dissolution sites and anions generally biased to move toward local concentrations of cations.

In order to account for system charge neutrality and maintenance of the ionic concentration in the electrolyte, the following deterministic rules can be used to constrain anion and cation motion without needing to set up another simulation to account for ionic motion. However, the simulation will not accurately replicate ionic behavior in solution far from the interface using these rules.

1. Electrolyte cations—Forced to move away from anodic dissolution sites because of charge repulsion. A small fraction of hops can be allowed in the "wrong" direction in order to incorporate more random motion into the motion of the cations, and to allow anions the opportunity of escaping out of small pits on the crystal surface.
2. Electrolyte anions—Constrained to move, in general, toward anodic sites where there is a higher concentration of metal cations with the exception of the occasional hop in the "wrong" direction to allow it to escape becoming trapped on the alloy surface.
3. Reactive alloy component cations—Can follow similar rules as to the electrolyte cation.

Some care must be taken as metal and electrolyte cations diffuse to the limits of the computational cell in order to maintain system charge neutrality because, while the system can be allowed to be periodic along the x- and y axes, that is, parallel to the surface of the metal alloy, it can't be periodic in the z-direction, that is, perpendicular to the alloy surface. One approach is to remove cations from the system as they reach the top of the computational cell and replace them using the following rules:

- If, within the electrolyte region modeled, the total number of cation charges equals the total number of anion charges and the concentration of the cations is approximately equal to the concentration set at the beginning of the simulation, then the outgoing cation can be replaced with a water molecule.
- If the previous conditions are not true, then the outgoing cation has to be replaced with an electrolyte cation in order to maintain charge neutrality.

In this way, outgoing metal cations are replaced with electrolyte cations and charge neutrality and concentrations are maintained. Figure 4.5 shows a plot of ion concentrations during an example kinetic Monte Carlo simulation using the rules outlined earlier.

As mentioned previously, electrolytes can be differentiated in the manner in which their ionic components interact with the alloy components. Table 4.4 shows interaction energies that were determined for various notional electrolytes interacting with a binary MN_xLN_{1-x} alloy. For every alloy atom that was on an exposed surface, electrolyte components within

TABLE 4.4 Initial interaction energies for atoms and electrolyte components

Interaction type	Energy per interaction (eV)			
	Non-int	MN-int	LN-int	Int
MN–H$_2$O	0.160	0.160	0.160	0.160
MN–cation	0.000	0.000	0.000	0.000
MN–anion	0.000	0.297	0.000	0.297
LN–H$_2$O	0.140	0.140	0.140	0.140
LN–cation	0.000	0.000	0.000	0.000
LN–anion	0.000	0.000	0.297	0.297

[a]Reproduced with permission from the *Journal of the Electrochemical Society* 2012, 157(10) C328, © The Electrochemical Society, 2012.

The values listed for the MN and LN interactions with anions reflect the various types of anions: those that did not interact at all, those that would interact only with the MN atoms, and those that would interact with both types of atoms.

a nearest-neighbor distance, the energy well that kept the atom bound in the alloy was altered by the amount given in Table 4.4.

4.2.6 Obtaining Current Densities from Dissolution Simulations

Provided that the KMC simulation "counts" electrons lost during ionization events, the dissolution charge density for the simulation can be obtained from Equation (4.10). Some caution should be used in defining the value of the area because, as the dissolution reaction proceeds—especially in the case of selective dissolution in which the noble alloy components remain behind—the exposed surface area will change as the dissolution reaction interface proceeds into the alloy. However, provided that the explicit path of the electrons is not needed, the electrons lost by the oxidized metal components

$$Q = \frac{z(q)}{A} \tag{4.10}$$

Q is the charge density in C/cm^2, z is the number of electrons, q is the charge on an electron 1.6×10^{-19} C, and A is the cross-sectional area of the alloy sample. The current density can then be obtained by calculating the slope of the linear regression fit to the charge density as a function of time. The slope of the linear regression fit gives a constant value for the steady-state dissolution current density versus what was obtained from the instantaneous current density calculation. The slope of the linear regression fit provides a constant value that can be used to compare with the dissolution current obtained from experimental results or other simulations using other approached. Figure 4.6 provides an example of the process used for obtaining the steady-state dissolution current density.

4.2.7 Morphology and Porosity from Simulation Results

The output of a KMC simulation incorporated with visualization software can lend itself to insight into the behavior of the dissolving system. For example, in Figure 4.7, a metal sample is undergoing dissolution. Because the sample is composed entirely of a single element that can oxidize, the dissolution interface proceeds in a stable manner into the sample. Atoms in kink sites or step edge faces are more susceptible to dissolution and so pits greater than one monolayer deep are difficult to form. However, for a multicomponent alloy in which at least one of the components is noble enough that it does not dissolve while in galvanic contact with the more reactive component, then more complex behavior can occur. Figure 4.8 illustrates selective dissolution in a binary alloy in which the sample starts out with a (100) crystalline orientation but as dissolution proceeds, (111) facets are observed where the dissolution interface has penetrated into the structure.

Determining the porosity or void fraction of the alloy occupied by electrolyte following a dissolution simulation are more relevant to selective dissolution simulations in which at least one noble component of the alloy does not dissolve so that a coarsened filamentary structure results. As noted earlier, in the case of dissolution of a single component sample, there is no porous structure at the completion of the simulation. Porosity as a function of depth can be used to characterize different types of attack and to determine how uniform the dissolution was throughout the sample. One approach to determine void fraction is to extract metal atom position information from cross sections through the sample and color code the cross

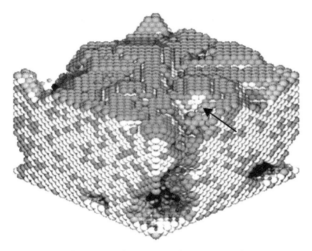

FIGURE 4.8 Image taken from the microstructural evolution of a binary alloy with a (100) crystalline orientation undergoing dissolution. The arrow indicates a (111) facet.

sections to indicate metal versus electrolyte or void spaces, if the electrolyte is not explicitly treated. Figure 4.9 provides an overview of this process used to obtain void fraction in a selectively dissolved sample using the contrast-determination algorithms in WCIF ImageJ to analyze the sample cross sections.

4.2.8 Obtaining the Expectation Value for Atomic Positions from a Random Walk Model

The purpose of this subsection is to obtain an expression for the expectation value of a particle's position. The focus is on a particle undergoing random-walk motion on a one-dimensional lattice in order to obtain an expression that can be associated with the macroscopic surface diffusivity parameter in the following subsection. The results from one-dimension are then extended to two and three dimensions.

Figure 4.10 depicts a one-dimensional chain with lattice sites placed at a distance, λ, along the line. A random-walk sequence is performed by a particle that starts at the origin. An example of such a sequence, described by a Markov chain, is also shown in Figure 4.9.

Let n be the number of steps or jumps that the particle makes. The number of possible sequences for n jumps is 2^n and each sequence is equally probable, $P_n = \frac{1}{2^n}$. Assuming that the random walk produces p steps to the right and $n-p$ steps to the left; then the net displacement is $p\lambda - (n-p)\lambda = (2p-n)\lambda = \lambda_p$. The *a priori* probability—the probability before the event occurs—of a net displacement of λ_p is

$$\frac{1}{2^n} \frac{n!}{(n-p)!p!} \tag{4.11}$$

$\langle \lambda_p \rangle$ is defined as the mean displacement which can be written as follows:

$$\lambda_p = \sum_{p=0}^{n} \frac{\lambda_p}{2^n} \frac{n!}{(n-p)!p!} = \frac{\lambda}{2^n} \sum_{p=0}^{n} \frac{n!}{(n-p)!p!} (2p-n) \tag{4.12}$$

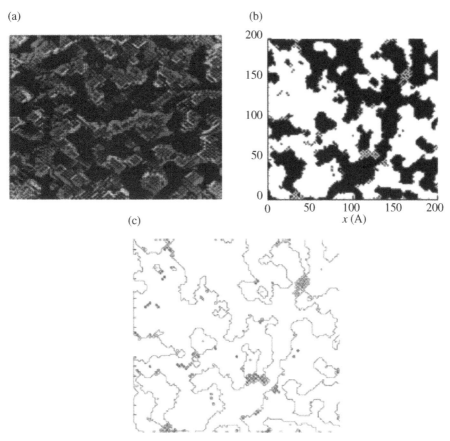

FIGURE 4.9 (a) Plan views of a $20\,\mathrm{nm} \times 20\,\mathrm{nm} \times 20\,\mathrm{nm}$ MN_{30}–LN_{70} sample that underwent a selective dissolution simulation in a noninteracting electrolyte. The electrolyte has been blanked out to aid in visualizing the morphology. (b) Cross-sectional cut at a depth of $6.4\,\mathrm{nm}$ below the sample surface. Dark areas indicate the presence of the electrolyte and light areas indicate the presence of metal atoms. (c) Outline of electrolyte regions calculated by the WCIF ImageJ software. This area calculation performed on this image resulted in a determination that, at this level, the cross section contained 46.8% of electrolyte.

For n odd,

$$\lambda_p = \frac{\lambda_p}{2^n}\left\{\sum_{p=0}^{(n-1)/2}\frac{n!(2p-n)}{(n-p)!p!} + \sum_{(n-1)/2}^{n}\frac{n!(2p-n)}{(n-p)!p!}\right\} \tag{4.13}$$

Let $j = n - p \rightarrow p = n - j$, $p = (n+1)/2 \rightarrow n - p = n - (n+1)/2 = n/2 - 1/2$
$p = n$, and therefore $j = 0$

$$2p - n = 2n - 2j - n = n - 2j$$

$$\lambda_p = \frac{\lambda_p}{2^n}\left\{\sum_{p=0}^{(n-1)/2}\frac{n!(2p-n)}{(n-p)!p!} + \sum_{(n-1)/2}^{n}\frac{n!(n-2j)}{(n-j)!j!}\right\} \tag{4.14}$$

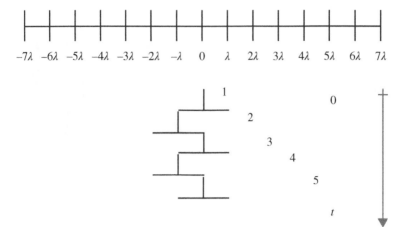

FIGURE 4.10   The top row shows a one-dimensional chain along which sites are spaced a distance, λ, apart. The bottom row shows an example of a Markov chain for a sample random walk sequence. The arrow to the right indicates the passage of time. The numbers indicate the step # and the lines indicate the directions available to hop in each time step.

But, for n odd, $\lambda_p = 0$. So using the mean square displacement,

$$\lambda_p^2 = \sum_{p=0}^{n} \frac{\lambda_p^2}{2^n} \frac{n!}{(n-p)!p!} = \frac{\lambda^2}{2^n} \sum_{p=0}^{n} \frac{n!}{(n-p)!p!} (2p-n)^2 \tag{4.15}$$

The binomial theorem is

$$(a+b)^n = \sum_{p=0}^{n} \frac{n!}{(n-p)!p!} a^p b^{n-p} \tag{4.16}$$

$$(2p-n)^2 = 4p^2 - 4pn + n^2 = -4(n+p)p + n^2$$

$$\lambda_p^2 = \frac{\lambda^2}{2^n} \left\{ -4\sum_{p=0}^{n} \frac{n!(n-p)p}{(n-p)!p!} + n^2 \sum_{p=0}^{n} \frac{n!}{(n-p)!p!} \right\} \tag{4.17}$$

With the second summation in Equation (4.17) equal to 2^n:

$$\lambda_p^2 = \frac{\lambda^2}{2^n} \left\{ -4\sum_{p=1}^{n-1} \frac{n!}{(n-p-1)!(p-1)!} + 2^n n^2 \right\} \tag{4.18}$$

$$\lambda_p^2 = \frac{\lambda^2}{2^n} \left\{ -4n(n-1)\sum_{p=1}^{n-1} \frac{(n-2)!}{(n-p-1)!(p-1)!} + 2^n n^2 \right\} \tag{4.19}$$

Now, let $j = p-1$, so if $p=1$ then $j=0$ and $p=n-1$ means $j=n-2$.

$$\lambda_p^2 = \frac{\lambda^2}{2^n}\left\{-4n(n-1)\sum_{j=0}^{n-2}\frac{(n-2)!}{(n-j-2)!j!} + 2^n n^2\right\} \tag{4.20}$$

Let $N = n - 2$, then

$$\lambda_p^2 = \frac{\lambda^2}{2^n}\left\{-4n(n-1)\sum_{j=0}^{N}\frac{(N)!}{(N-j)!j!} + 2^n n^2\right\} \tag{4.21}$$

And from the binomial theorem in Equation (4.16),

$$\lambda_p^2 = \frac{\lambda^2}{2^n}\left\{-4n(n-1)2^{n-2} + 2^n n^2\right\} \tag{4.22}$$

Factor out the 2^n and cancel and $<\lambda_p^2>$ is

$$\lambda_p^2 = \lambda^2\left\{-4n(n-1)2^{-2} + n^2\right\} = n\lambda^2 \tag{4.23}$$

Thus, the expectation value for a given position, λ^p, of a particle $<\lambda_p^2>$ is given by $n\lambda^2$.

4.2.9 Derivation of the Relation between the Random Walk Model and Macroscopic Diffusivity

The next step is to show that Equation (4.23), obtained earlier, which expressed the expectation value of an atom's position as a function of the number of hops it made on a one-dimensional lattice of lattice-spacing, λ, can be related to the continuum quantities of diffusivity and composition fields.

Starting from the diffusion equation as expressed in Fick's second law,

$$\frac{\partial c(x,t)}{\partial t} = D\frac{\partial^2 c(x,t)}{\partial x^2} \tag{4.24}$$

And assuming that the diffusion is confined to one dimension and that the initial condition on the composition field is $c(x,0) = \delta(x)$, where $\delta(x)$ is the Dirac delta function and D is the diffusivity of the material in the space.

Then the solution to the diffusion equation is

$$c(x,t) = \frac{1}{(4\pi Dt)^{1/2}}e^{-x^2/4Dt} \tag{4.25}$$

Equation (4.24) can be considered the linear probability distribution of the atoms in the initial concentration peak. Multiply Equation (4.24) by r^2 and then integrate over all space.

$$\lambda_p^2 = \int_{-\infty}^{\infty}\frac{r^2}{(4\pi Dt)^{\frac{1}{2}}}e^{-r^2/4Dt}dx \tag{4.26}$$

Perform a change of variables. Let $q = \frac{r}{(4Dt)^{1/2}}$, then $dq = \frac{1}{(4Dt)^{1/2}} dx$ and, upon substituting these expressions into Equation (4.25)

$$\lambda_p^2 = \frac{(4Dt)^{1/2}}{(4\pi Dt)^{\frac{1}{2}}} \int_{-\infty}^{\infty} q^2 e^{-q^2} dq \tag{4.27}$$

Using integration by parts gives the following: $u = q$, $du = -dq$, $dv = qe^{-q^2}$, and $v = \frac{1}{2} e^{-q^2}$. Thus,

$$\lambda_p^2 = \frac{4Dt}{(\pi)^{\frac{1}{2}}} \left\{ \frac{-q}{2} e^{-q^2} \Big|_{-\infty}^{\infty} + \frac{1}{2} \int_{-\infty}^{\infty} e^{-q^2} dq \right\} \tag{4.28}$$

The first term inside the braces in Equation (4.27) equals zero because it is an odd function that leaves

$$\lambda_p^2 = \frac{4Dt}{2(\pi)^{\frac{1}{2}}} \left\{ \int_{-\infty}^{\infty} e^{-q^2} dq \right\} = \frac{2Dt}{(\pi)^{\frac{1}{2}}} \pi^{\frac{1}{2}} = 2Dt \tag{4.29}$$

And, last,

$$\lambda_p^2 = n\lambda^2 = 2Dt\ 2Dt \tag{4.30}$$

Or, in two and three dimensions, respectively:

$$\lambda_p^2 = n\lambda^2 = \overline{r^2} = 4Dt \tag{4.31}$$

$$\lambda_p^2 = n\lambda^2 = \overline{r^2} = 6Dt. \tag{4.32}$$

The distance, λ, is obtained from the nearest-neighbor positions to which the particle or atom can move. In the case of an FCC lattice, where a_0 is the lattice parameter,

$$\lambda = \sqrt{(a_0)^2 + (a_0)^2} \tag{4.33}$$

And, thus, the macroscopic diffusivity can be obtained from a consideration of the random atomic motions. The importance of this and the previous derivation is that a_0 and t are both parameters that can be easily extracted from a KMC simulation and thus diffusivities can be obtained that can be used to compare with experimentally determined values or that can be used to calibrate less easily measured parameters, such as atom–electrolyte interactions or interspecies bonding, that are used to determine the energy barriers in the local bond-breaking model for diffusion and dissolution.

4.3 DISCUSSION

Newman, Sieradzki et al. [35–37] were among the first to use simulations to investigate dealloying in 2D and 3D simple cubic structures—capturing parting limits, critical potentials, and percolation thresholds—in their models. They were investigating the role dealloying played in the onset of stress corrosion cracking in ductile alloys. There had been strong, but indirect, evidence that dealloying was the corrosion reaction occurring in such alloys as brass and austenitic steels that suffered from transgranular fracture but were failing in environments under which the noble alloying element would not dissolve. The results of their modeling and investigations indicated that thin layers of dealloyed material that remained coherent with the undealloyed substrate could initiate film-induced cleavage cracks into the substrate.

Of the aforementioned theoretical descriptions of dealloying, the most successful to date [1] has been a model that relies on surface diffusion of metal atoms in the alloy–electrolyte interface as the principal reason for observed behavior. And the crucial evidence for the surface diffusion explanation is found in KMC simulations of the dealloying process. They are able to replicate—with a minimum number of assumptions—the microscopic features observed in dealloying. In these simulations, the disordering of the surface by the dissolution of the LN component is relaxed by the surface diffusion of the remaining MN component. Alloy compositions with the LN component below a percolation threshold—which deals with the degree of connectedness of the LN component in the alloy [38]—either will not dealloy or will passivate shortly after dealloying commences. Several papers [1, 35, 39] have described models that incorporate these aspects of the theoretical description of dealloying. The underlying assumption of the models is that both the dissolution rate of the LN member of the alloy and the surface diffusion rates of both components are dependent on the coordination number of the bonds of the atoms at the surface.

Many explanations of dealloying [32] focus on the roles of the alloy components in the process, with the surface diffusivity of the MN component seen as the dominant parameter in controlling both the morphology of the dealloyed layer and the rate of dealloying. Because the surface diffusivities of both the MN and LN components of the alloy are required to be higher than they would be in vacuum [40], most models provide for weaker bonding energies of the alloy components at the surface where dealloying is occurring in order to capture the higher diffusivities and account for the morphologies and dealloying rates that are observed experimentally. However, they do not take into account differences in bonding energies between atoms of different types, nor do they account for the interaction of the electrolyte constituents with the alloy components or for the transport of ions to and from the electrolyte–alloy interface; but as was seen in the work by Dursun and others, the choice of the electrolyte affects the morphology of the remaining MN component and should be considered in models for selective dissolution. A method for incorporating the electrolyte and its effects was suggested in the KMC algorithm outlined earlier.

From a practical standpoint, the length scales of the pores and ligaments that make up the residual filamentary structure left behind by dealloying can be on the order of hundreds of nanometers. The ligament diameters and spacing are influenced by the conditions under which the dissolution occurs. In some binary alloy systems, such as silver–gold or copper–gold in which the dealloying process has been well-studied in the laboratory, engineering applications for the porous structures have been developed. Applications for dealloyed materials, such as flexible electrodes, high surface area catalysts, and storage cells for providing controlled drug elution take advantage of the material properties of the nanoporous structures. Nanoporous gold has been investigated as a basis for chemical sensors [41]. The complete solid-solubility of gold in silver and the wide gap between their reversible potentials make the Ag–Au system attractive for dealloying studies. The porous gold structure left behind by the dissolution of silver not only has the high surface area characteristic of porous materials, but it is also chemically stable and biocompatible. Nanoporous platinum has been suggested as a high-surface area catalyst to take advantage of its high exchange current density for hydrogen evolution. Nanoporous platinum has been made by dealloying $Pt_{0.25}Cu_{0.75}$ [42]. Additionally, selective dissolution of one of the components of a semiconductor alloy has been found to modify the semiconductor band gap and dealloying of aluminum alloys has led to the creation of porous aluminum templates that can be used as patterns to create other nanoscale materials [38]. Predicting the nanoporous structure that will result from a combination of factors such as applied overpotential, sample composition, temperature, electrolyte type, and duration of dealloying is of prime importance in obtaining the desired nanostructure with the goal of many simulations being to make the search for those parameters more efficient.

To this point, the discussion of modeling dissolution and dealloying has focused on simulating oxidation reactions in which the modeled metal atoms undergo instantaneous electron transfers, become ions, and then transition either out of the system or across the metal–electrolyte interface into the electrolyte. However, for many alloys of engineering interest, the oxidized metal ions undergo multistep electron transfers and react to form oxide films, which can passivate a dissolving surface. Legrand et al. [43] used KMC simulations to study how changes in chromium concentration affected passivation in Fe–Cr alloys which allowed them to probe the effects of changing different parameters, such as chromium diffusivity and the critical size of Cr clusters needed to passivate a region.

Last, in order to overcome some shortcomings in the KMC approach—such as the requirement for defining all transitions prior to the start of the simulation—linking KMC with molecular dynamics or with inputs from atomistic calculations would provide more quantitative support for the energy barrier calculations and allow for less rigid constraints on the simulated structure [44–46].

4.4   SUMMARY

The article has briefly considered the role of Monte Carlo and kinetic Monte Carlo simulations in understanding dissolution and selective dissolution processes that can occur spontaneously in the natural environment and under directed control in laboratories. Algorithms for both Metropolis Monte Carlo and KMC models were discussed, and some results from an implementation of the KMC algorithm were shown as examples. Last, the article surveyed several areas where KMC models have been used to study corrosion processes and where they can contribute in engineering applications.

ACKNOWLEDGMENTS

The authors gratefully acknowledge the financial support of the Naval Research Laboratory and the Office of Naval Research.

REFERENCES

1. J. Erlebacher, M. J. Aziz, A. Karma, N. Dimitrov, K. Sieradzki, "Evolution of nanoporosity in dealloying," *Nature*, vol. 410, pp. 450–453, 2001.

2. J. R. Davis, *Corrosion: Understanding the Basics*. Baltimore, MD: ASM International, 2000.

3. C. A. Calvert, "Action of acids on meals and alloys," *Journal of the Chemical Society*, vol. 19, pp. 434–455, 1866.

4. A. Chidester Van Orden, *Dealloying. Corrosion: Understanding the Basics*. Baltimore, MD: ASM International, 2005.

5. C. Wagner, "Theoretical analysis of the diffusion processes determining the oxidation rate of alloys," *Journal of the Electrochemical Society*, vol. 99, pp. 369–380, 1952.

6. H. E. Cook and J. E. Hilliard, "Interdiffusion in Au–Ag alloys at low temperatures," *Applied Physics Letters*, vol. 8, pp. 24–26, 1966.

7. H. W. Pickering, "Formation of new phases during anodic dissolution of Zn-rich Cu–Zn alloys," *Journal of the Electrochemical Society*, vol. 117, pp. 8–15, 1970.

8. V. Tisher and H. Gerischer, "Elektroltische auflosung von gold = silber-legierungen und die frange der reistenzgrenzen," *Zeitscrift für Elektrochemie*, vol. 62, pp. 50–64, 1958.

9. A. J. Forty, "Micromorphological studies of the corrosion of gold alloys," *Gold Bulletin*, vol. 14, pp. 25–35, 1981.

10. D. J. Young, "Dealloying reactions as cellular phase transformations," in *Advances in Phase Transformations*. New York: Pergamon Press, 1988.

11. K. Sieradzki, R. R. Corderman, and K. Shukla, "Computer simulation of corrosion: selective dissolution of binary alloys," *Philosophical Magazine A*, vol. 59, pp. 713–746, 1989.

12. D. B. Bird and K. L. Moore, "Dezincification of brasses in seawater," *Materials Protection*, vol. 1, pp. 65–137, 1962.

13. H. T. Storey, "Corrosion of metallic brasses," *Metallurgical and Chemical Engineering*, vol. 17, pp. 650–654, 1917.

14. G. D. Bengough, G. Jones, and R. Pirret, "Diagnosis of brass condenser tube corrosion," *Journal of the Institute of Metals*, vol. 23, pp. 65–137, 1920.

15. N. Metropolis and S. Ulam, "The Monte Carlo Method," *Journal of the American Statistical Association*, vol. 44, no. 247, pp. 335–341, 1949.

16. N. Metropolis, A. W. Rosenbluth, M. N. Rosenbluth, A. H. Teller, and E. Teller, "Equation of state calculations by fast computing machines," *Journal of Chemical Physics*, vol. 21, pp. 1087–1092, 1953.

17. Wikipedia.org (2012, December). Online. http://www.wikipedia.org/

18. M. A. Novotny, "A tutorial on advanced dynamic Monte Carlo methods for systems with discrete state spaces," *Annual Reviews of Computational Physics*, vol. 9, pp. 153–210, 2001.

19. L. Zhigilei, Modeling in Materials Science. Course lecture notes, Materials Science and Engineering, University of Virginia, 2003.

20. A. Bortz, M. H. Kalos, and J. L. Lebowitz, "A New algorithm for Monte Carlo simulation of Ising spin systems," *Journal of Computational Physics*, vol. 17, pp. 10–18, 1975.

21. A. Voter, "Classically exact overlayer dynamics: diffusion of rhodium clusters on Rh(100)," *Physical Review B*, vol. 34, pp. 6819–6829, 1986.

22. J. O. Bockris, A. K. Reddy, and M. Gamboa-Aldeco, *Modern Electrochemistry*. New York: Kluwer Academic/Plenum Publishers, 2000, vol. 2A.

23. M. F. Toney, J. N. Howard, J. Richer, "Distribution of water molecules at Ag{111}/electrolyte interface as studied with surface X-ray scattering," *Surface Science*, vol. 335, pp. 326–332, 1995.

24. I. Danielwicz-Ferchmin and A. R. Ferchmin, *The Journal of Physical Chemistry*, vol. 100, pp. 17281–17286, 1996.

25. X. Gao, A. Hamelin, and M. J. Weaver, "Potential-dependent reconstruction at ordered Au(100)-aqueous interfaces as probed by atomic-resolution scanning tunneling microscopy," *Physical Review Letters*, vol. 67, pp. 618–621, 1991.

26. J. M. Doña and J. González-Velasco, "The dependence of the surface diffusion coefficients of gold atoms on the potential: its influence on reconstruction of metal lattices," *Surface Science*, vol. 274, pp. 205–214, 1992.

27. J. M. Dona and J. Gonzalez-Velasco, *The Journal of Physical Chemistry*, vol. 97, pp. 4714–4719, 1993.

28. J. H. Schott and H. S. White, "Electric field induced phase transitions in scanning tunneling microscopy experiments on gold (111) surfaces," *Langmuir*, vol. 8, pp. 1955–1960, 1992.

29. A. Dursun, D. V. Pugh, and S. G. Corcoran, "Dealloying of Ag-Au alloys in halide-containing electrolytes: affect on critical potential and pore size," *Journal of the Electrochemical Society*, vol. 150, pp. B355–B260, 2003.

30. W. M. Haynes, T. J. Bruno, and D. R. Lide (eds), CRC Handbook of Chemistry and Physics. 95th ed. CRC Press: Boca Raton, 2014–2015, pp. 5–80, -5-84.

31. S. A. Policastro, J. C. Carnahan, G. Zangari, H. Bart-Smith, E. Seker, M. R. Begley, M. L. Reed, P. F. Reynolds and R. G. Kelly, Journal of Electrochemical Society, vol. 157(10), pp. C328–C337, 2010.

32. J. Erlebacher, "An atomistic description of dealloying," *Journal of the Electrochemical Society*, vol. 151, pp. C614–C626, 2004.

33. G. S. Bales and D. C. Chrzan, "Dynamics of irreversible island growth during submonolayer epitaxy," *Physical Review B*, vol. 50, pp. 6057–6067, 1994.

34. J. M. Howe, Interfaces in Materials. *Wiley: New York*, 1997; Appendix B.

35. K. Sieradzki, R. R. Corderman, K. Shukla, and R. C. Newman, "Computer simulations of corrosion: selective dissolution of binary alloys," *Philosophical Magazine A*, vol. 59, pp. 713–746, 1989.

36. R. C. Newman, F. T. Meng, and K. Sieradzki, "Validation of a percolation model for passivation of Fe-Cr alloys: I current efficiency in the incompletely passivated state," *Corrosion Science*, vol. 28, pp. 523–527, 1988.

37. K. Sieradzki, J. S. Kim, A. T. Cole, and R. C. Newman, "The relationship between dealloying and transgranular stress-corrosion cracking of Cu-Zn and Cu-Al alloys," *Journal of the Electrochemical Society*, vol. 134, pp. 1635–1639, 1987.

38. E. Schofield, "Anodic routes to nanoporous materials," *Transactions of the Institute of Metal Finishing*, vol. 83, no. 1, pp. 35–42, 2005.

39. J. Erlebacher, "An atomistic description of dealloying: porosity evolution, the critical potential, and rate-limiting behavior," *Journal of the Electrochemical Society*, vol. 151, pp. C614–C626, 2004.

40. E. G. Seebauer and C. E. Allen, "Estimating surface diffusion coefficients," *Progress in Surface Science*, vol. 49, pp. 265–330, 1995.

41. Z. Liu and P. C. Searson, "Single nanoporous gold nanowire sensors," *Journal of Physical Chemistry B*, vol. 110, pp. 4318–4322, 2006.

42. D. V. Pugh, A. Dursun, and S. G. Corcoran, "Formation of nanoporous platinum by selective dissolution of Cu from $Cu_{0.75}Pt_{0.25}$," *Journal of Materials Research*, vol. 18, pp. 216–221, 2003.

43. M. Legrand, B. Diawara, J. J. Legendre, and P. Marcus, "Three-dimensional modeling of selective dissolution and passivation of iron–chromium alloys," *Corrosion Science*, vol. 44, pp. 773–790, 2002.

44. A. Heyden, A. T. Bell, and F. J. Keil, "Efficient methods for finding transition states in chemical reactions: comparison of improved dimer method and partitioned rational function optimization method," *Journal of Chemical Physics*, vol. 123, p. 224101, 2005.

45. L. J. Xu and G. Henkelman, "Adaptive kinetic Monte Carlo for first-principles accelerated dynamics," *Journal of Chemical Physics*, vol. 129, p. 114104, 2008.

46. H. Xu, Y. N. Osetsky, and R. E. Stoller, "Simulating complex atomistic processes: on-the-fly kinetic Monte Carlo scheme with selective active volumes," *Physical Review B*, vol. 84, p. 132103, 2011.

5

ADSORPTION OF ORGANIC INHIBITOR MOLECULES ON METAL AND OXIDIZED SURFACES STUDIED BY ATOMISTIC THEORETICAL METHODS

DOMINIQUE COSTA AND PHILIPPE MARCUS

Institut de Recherche de Chimie Paris/Physical Chemistry of Surfaces, Chimie ParisTech-CNRS, Ecole Nationale Supérieure de Chimie de Paris, Paris, France

5.1 INTRODUCTION

The corrosion rate of reactive metals such as Fe, Cu, Zn, and Al can be reduced significantly by a modification of their surface by organic molecules or polymer coatings. Corrosion inhibition by adsorbed organic molecules is thus often used as a way to protect metals and alloys against corrosion [1]. Inhibition to the extent of 98% efficiency may be achieved. It is very likely that natural compounds will become more important in the future as effective corrosion inhibitors due to their biodegradability, easy availability, and nontoxic nature. Careful perusal of the literature clearly reveals that the era of green inhibitors has already begun.

Numerous experimental works have been devoted to corrosion inhibition, whereas atomistic theoretical studies appeared only recently. A detailed understanding of the chemical interaction between the organic molecule and the substrate (metal without or with a surface oxide film) is the key to understanding and control of corrosion inhibition [2].

The resistance of the chemical bond to the possible attack by water is a major issue, as is the inhibition, by the adsorbed organic molecule, of the electrochemical reactions (i.e., the reduction of oxygen) [3].

Molecular Modeling of Corrosion Processes: Scientific Development and Engineering Applications, First Edition.
Edited by Christopher D. Taylor and Philippe Marcus.
© 2015 John Wiley & Sons, Inc. Published 2015 by John Wiley & Sons, Inc.

Reviews proposing an exhaustive research of all possible natural or drugs corrosion inhibition properties are reported [4–6]. Seeing the enormous number of natural molecules capable of corrosion inhibition properties, it is mandatory to extract guidelines from fundamental studies to adequately select the best potential candidates and avoid costly and time-consuming experimental screening tests. Thus, in the recent years, a great number of works devoted to the quantitative structure–activity relationship (QSAR) approach have been done (see, e.g., Refs. [7–12]). To our knowledge, only one review already exists trying to synthesize this topic: Gece [6] has reported all molecules that were investigated through atomistic methods in relationship with their corrosion inhibition properties.

As will be demonstrated in this chapter, there has been a strong evolution of fundamental studies recently, giving rise to new concepts that will be explored. In the present review, we summarize fundamental approaches used to characterize organic molecule adsorption on model metal (and oxidized) surfaces, in relation to their corrosion inhibition properties, by means of atomistic simulations.

With respect to the broad spectrum of studies available in the literature, we focus in this chapter on the following metals Al, Cu, Fe, and Zn and their oxidized surfaces. We present results of increasing degree of system complexity (from the isolated molecule to the molecule adsorbed on a surface in the presence of solvent) and calculation accuracy (from classical MD to quantum methods).

5.2 STATE OF THE ART IN MODELING INHIBITION PROPERTIES THROUGH ATOMISTIC METHODS

In 1981, Sanyal [13] gave an account of organic corrosion inhibitors including a classification and mechanisms of action. The corrosion inhibition potential was assigned to the donation of a lone pair of electrons to metal atoms. Since then, a great number of publications have been devoted to the study of the intrinsic properties of the inhibitor molecule by means of QM approaches. This is also called the QSAR approach. Organic compounds that can donate electrons to unoccupied d orbitals of metal surfaces to form coordinate covalent bonds and can also accept free electrons from the metal surface by using their antibonding orbitals to form retrodonating bonds are good candidates for corrosion inhibition purposes. The most effective corrosion inhibitors are compounds containing heteroatoms like nitrogen, oxygen, sulfur, and phosphorus as well as aromatic rings. The inhibitory activity of these molecules is due to the formation of a monolayer on the metal surface or even to the formation of a mixed compound, as for example, Cu-benzotriazole on the Cu surface, *vide infra*. Free electron pairs on heteroatoms or p electrons are readily available for sharing to form a bond and act as nucleophile centers of inhibitor molecules and greatly facilitate the adsorption process over the metal surface, whose atoms act as electrophiles. The effectiveness of a corrosion inhibitor molecule has been related to its electronic structure. Correlations between corrosion inhibition efficiency and a number of molecular properties such as dipole moment (μ), highest occupied (HOMO) and lowest unoccupied (LUMO) molecular orbitals, and the gap between them (HOMO–LUMO), Fukui indices [14, 15], charge density, polarizability, and molecular volume have been established: [7, 10, 11, 16–27]. Quantum chemical methods are ideal tools for investigating these parameters and are able to provide an insight into the inhibitor–surface interaction. Density functional theory (DFT) is very reliable in explaining the hard and soft acid–base behavior of inhibitor molecules introduced by Pearson, including, HOMO–LUMO gap, hardness, softness and Fukui indices. More explanations

TABLE 5.1   Molecular functions that have been considered through QSAR approaches, and their combinations

Molecules families	Active atom(s)
Phenols, naphthol	OH
Azoles, imidazoline	N
Aminoacids, aniline, methionine, phenylalanine	N, COO
Amines-, pyridine	N
Sugars	O, OH
2-butyne-1,4-diols	OH
Vitamins, riboflavin, niacin	N, O, OH, COOH
Urea	N, O
Amides	N
Carbazides	N, O
Carbazones	N, O
Porphyrins	N
Benzopyrone, pyrophthalone, cycloheptendione	O
-Diene	C=C
Guanine, adenine, purine	N

Data from Gece [6].

can be found in Refs. [6, 27]. A nonexhaustive list of molecules that have been studied is reported in Table 5.1, mainly from Ref. [6].

However, a strong weakness of the QSAR approach is the lack of molecule–surface interaction description. Recently, attempts to model the molecule–surface interaction have been made. The surface was modeled by considering one single metal atom, a cluster (typically 10–20 atoms), or a periodic infinite surface in vacuum or in interaction with a solvent. Also the molecular coverage was considered, from one isolated molecule to one adsorbed layer.

The simplest way indeed consists in representing separately the metal surface and the organic molecule [28–34]. The intrinsic properties of the molecule, dipole moment, HOMO, and LUMO, are calculated taking into consideration using the technique of continuum solvation (generally through the SCRF theory with PCM approach [35] method). The surface is represented by a cluster of a few atoms, and a thermodynamic approach allows evaluating the solvation free energies. Within this approach, molecular properties such as deprotonation, pK_A near the interface inside the electrical double layer may be calculated and correlated with inhibition properties [32]. However, even if the properties of the two constituents are considered, the surface/molecule interaction was not directly modeled.

The molecule–surface interaction was considered with classical molecular dynamics (MD) using a force field (FF) without *ab initio* preliminary study. Section 5.2.1 is devoted to the description of these FF-based studies of the inhibitor–surface interaction. Another strategy consists in intuiting the fragment of the molecule that might be the most reactive toward the metallic surface from the *ab initio* study of the isolated molecule, for example, by characterizing the fragment of the HOMO/LUMO, then to perform classical MD approach to explore the molecule–surface interaction, molecule organization and orientation toward the metal surface.

In recent years, the molecule/surface interaction could be modeled with ab initio tools. This will be developed in Section 5.2.2. Free energy calculations including solvent effects through atomistics thermodynamics, and "real" solid/liquid interfaces, are considered in the last section.

5.2.1 Organic Inhibitor/Surface Interaction Studied by Classical MD

Molecular dynamics simulation is an atomistic treatment that aims at exploring the potential energy surface, that is, solving the electronic Schrödinger equation at different nuclei positions. It allows determining thermodynamically stable structures (minima) and transition structures (first-order saddle points). Realistic applications of MD require a huge computational effort to explore the system behavior even for a very short-time window. For that reason, MD has developed using a description of the internal forces resulting from classical force fields, which ensures a fast evaluation of energy and forces at the expense of accuracy. Classical force fields, derived either from empirical data or from quantum chemical calculations, can extend up to the microsecond time scale. The reliability of these force fields, however, is a critical aspect that should be carefully analyzed, particularly when entering into large time scales.

In classical force fields, the molecular structure is represented by means of "mechanical springs" whose features (length, strength, etc.) mimic the different interactions occurring within the molecule and between different molecules. The set of parameters associated to the springs is known as "force field" and the relaxation of the mechanical energy is the result of a "molecular mechanics calculation." Force fields exist for organic molecules, water, and inorganic solids. The complexity of inorganic materials and the paucity of experimental data needed to derive the force field parameters, forced the adoption of simple two-body potentials for the description of each spring, compared to the most sophisticated terms (three-body and four-body terms) used for molecular cases. The challenge still is to model properly the interfaces. Corrosion inhibition modeling presents one of the most intricate case, exhibiting complex interfaces: metal or oxide/water, molecule/surface, and metal/oxide interface. The advantage of classical MD is to bring information on the entropic factors of the system by accumulating statistics over times that reach now the microsecond. However, important chemical events cannot yet be described with force fields. In particular, change in oxidation state, dissolution, dissociation, and deprotonation is still difficult to evaluate, even if some aspects are achieved with promising tools such as ReaxFF [36, 37]. However, as it will be demonstrated later, molecular aspects of interfaces are just now being discovered in the present time at the atomic scale using Born–Openheimer MD.

As can be noted from Table 5.2, there are to our knowledge only a few works reporting molecule/surface interactions with classical force fields. Most of the time, the calculations are presented to complement experimental data that consist of electrochemical studies to estimate the corrosion inhibition efficiency of some molecules. The calculations are performed to better understand the molecule/surface interaction. Force fields used are universal force field (UFF) [51–53], consistent-valence forcefield (CVFF) [39, 40, 55], and optimized molecular potentials for atomistic simulation studies (COMPASS) [47] that have been developed in the scope of treating a broad range of systems. Universal force field (UFF) is an -all atom potential containing parameters for every atom. The force field parameters are estimated using general rules based only on the element, its hybridization, and its connectivity. The CVFF is a generalized valence forcefield. Condensed-phase COMPASS is an *ab initio* force field because most of the parameters are derived from *ab initio* data. The three force fields allow treating organic elements as well as metals; however, no specific validation of the inorganic/organic interface is presented in the papers mentioned here.

MD simulations were mostly performed on iron as a model of mild steel. Phenol, methionine, and azoline were studied. Kong et al. [38] performed an experimental STM study to identify the overlayer structure adopted by the organic molecules adsorbed on the Fe surface.

TABLE 5.2 Molecule–surface systems of interest in corrosion inhibition, studied by FF

Metal/oxide	Molecule	Experimental data	Coverage	Methods	References
Fe(110)	o-aminothiophenol (OATP)	[38]	0.25–1 ML	Consistent valence force field (cvff) [39–43], PBC	[38]
Fe(100), Fe$_2$O$_3$(110) Fe$_3$O$_4$(100)	Imidazoline	[44]	Low	Ab initio preliminary study FF PBC study of the molecule/surface interaction Interaction with the metal surface and the oxide surfaces	[45]
Fe	Methionine	[46]	Low	MD PBC in the presence of solvent (water), with COMPASS [47] classical FF	[46]
Fe	Azoles	[48]	Low	MD PBC in the presence of solvent (water), with COMPASS [47] classical FF	[48]
Al(110)	Thiamine	[49, 50]	Low	UFF [51–53] and COMPASS	[50]
Al(110)	L-lysine (LY) and cystein (CY) from CC; emodine (EMD), and glycoside (GLY)	[54]	Low	COMPASS, MD, PBC	[54]

Related experimental data are also reported. In some cases a joint experimental and theoretical work was conducted. Experimental studies consist in electrochemical measurements to estimate the corrosion inhibition efficiency of the studied molecules.

In this work, several molecular coverages are considered, and a dense, self-assembled monolayer (SAM) of inhibitor is modeled. The other studies are also interesting for some aspects: In Ref. [45], both the metal and oxidized surfaces are considered when studying the molecule/surface interaction. This is interesting because for metals protected by a passive film corrosion (e.g., localized corrosion) may occur also in electrochemical conditions where the passive oxide film is present on the metal surface. We also note that in Ref. [46], the solvent was included in the study of the interaction.

The interaction of imidazoline with Fe and Fe oxides has been studied. Indeed, corrosion protection is due to the adsorption of the inhibitor molecule on the metal surface, but interaction with the oxidized surface is also interesting, as iron oxide was detected on the surface in the presence of the inhibitor. Preliminary DFT calculations suggest that the $N = C - N$ region in imidazoline ring is the most active reaction site for the inhibitor adsorption on the metal surface. This should occur via donor–acceptor interactions between the lone–electron pair on nitrogen atoms together with the π-electrons of heterocyclic and the vacant d orbital of iron atoms [45]. Then the adsorption of inhibitor on three typical surfaces (Fe(100), Fe_2O_3(110), and Fe_3O_4(100)) was studied by MD simulations. The adsorbed molecule is nearly parallel to the surface so as to maximize its contact with the surface (Fig. 5.1). Large adsorption energies are found, $-284\,kJ\,mol^{-1}$ on Fe, $-226\,kJ\,mol^{-1}$ on Fe_3O_4, and $-157\,kJ\,mol^{-1}$ on Fe_2O_3 [45]. Unfortunately, no experimental data are available to check those values.

The corrosion inhibition effect and structure of aminothiophenol (OATP) monolayers on Fe(110) surface were investigated by combining electrochemical, STM, and a FF approach [38]. It was shown that the OATP adlayers formed by self-assembling on Fe(110) surface in solution are highly ordered, densely packed, firmly attached and well protective. STM images revealed long-range ordering in the SAM of OATP with a p(2×2) commensurate packing structure, which yields a surface coverage of 0.25. Classical MD simulations were in good agreement with this result, since for OATP monolayers formed on Fe(110) the p(2×2) pattern is the lowest energetic structure among the studied typical packing patterns for SAM formation (Fig. 5.2). OATP molecules were found taking a tilt orientation with a herringbone configuration on the Fe(110) surface, yielding a film thickness of 0.5 nm. This structure is more stable than a structure in which all molecules are oriented parallel to the surface.

The adsorption of triazoles on the dense Fe(110) surface was investigated by performing molecular mechanics (MM) and the COMPASS force field in PBC approach [46, 48]. The MD study suggested that the close contact between Fe(110) and aminotriazole is more efficient than with triazole and benzotriazole. Complementary *ab initio* calculations of the isolated molecules show that the charge density distribution on aminotriazole is larger than on triazole and benzotriazole, which enhances the possibility of aminotriazole to adsorb more strongly on iron surface than triazole and benzotriazole. It is thus confirmed that the more negative the atomic charges of the adsorbed center, the more easily the atom donates its electrons to the unoccupied orbital of the metal.

Oguzie et al. [54] performed a joint experimental and theoretical study of the inhibition of Al corrosion by plant extracts. The experimental approach consists in hydrogen gas evolution measurements during the corrosion of aluminum in alkaline medium. *Bucolzia coriacca* (BC) and *Cninodoscolus chayansa* (CC) plant extracts inhibit Al corrosion with an efficiency of 95 and 70%, respectively. The effective energy of activation which is related to the strength of molecule/surface interaction was calculated by application of the Arrhenius relation to 56.4 (CC) and 71.4 kJ mol^{-1} (BC). The plant extracts were modeled with L-lysine

(a)   (a')

(b)   (b')

(c)   (c')

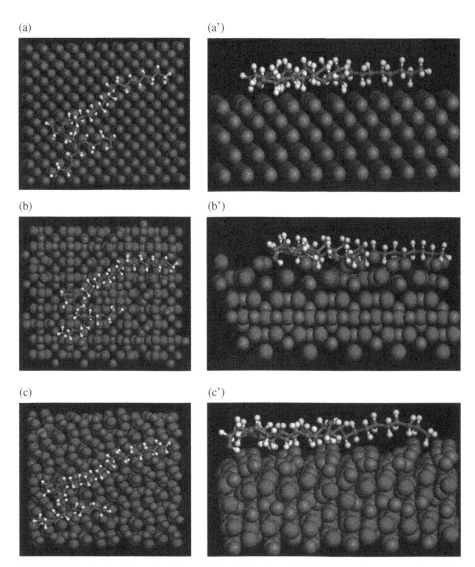

FIGURE 5.1 Imidazoline adsorbed on (a) Fe, (b) Fe_3O_4, and (c) Fe_2O_3 surfaces, respectively. The left is a top view, and the right is a side view. With permission from Feng et al. [45].

and cysteine (CC) and emodine and glycoside (BC). Strong binding energies were calculated: -56.2, -50.6, -59.9, and -57.7 kcal/mol on Al(110). Note that those energies are four times higher than the values derived from experimental data. Whereas it is difficult to understand the origin of the discrepancy between experience and theory, being certainly multifactorial (complex versus simplified system, the absence of solvent in the MD approach, and the use of a force field method that was not quantitatively benchmarked with experience for adsorption values), we may comment that the calculated adsorption energy values enter in the range of calculated *ab initio* adsorption energies of molecules on metal surfaces (see next paragraph). The authors mention a higher adsorption energy on Al_2O_3, but

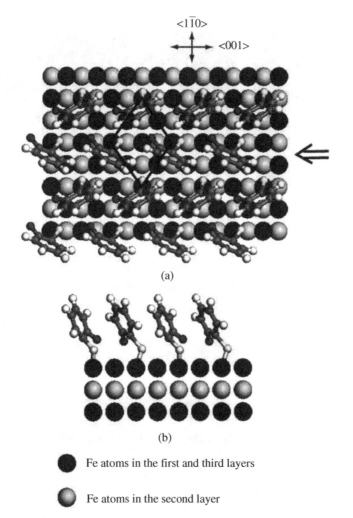

FIGURE 5.2 (a) Top and (b) side views of the detailed molecular packing structure of OATP monolayer with the p(2×2) pattern on Fe(110) from classical MD simulations. With permission from Kong et al. [38].

the values are not reported. Unfortunately, the geometries obtained are not reported in the paper, and the binding energies are not discussed.

The adsorption of thiamin, riboflavin, and niacin, components of the B vitamin present in Aspilia Africana extract, was investigated [50]. Those molecules may adsorb through a "soft epitaxy" mechanism on the metal surface, with accommodation of the molecular backbone in characteristic epitaxial grooves on the metal surface (Fig. 5.3).

Methionine derivatives adsorption on Fe(110) was studied by force field in the presence of water solvent [46]. Methionine is found to adsorb strongly on Fe(110) with $E_{ads} = -290$ kJ mol^{-1}. Unfortunately, the final structures were not discussed in terms of bonds formed between Fe and methionine, and the moiety(ies) responsible for the adsorption were not identified (Fig. 5.4).

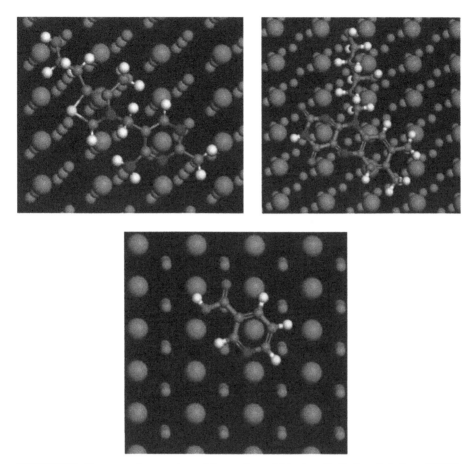

FIGURE 5.3   Representative snapshots of the top views of thiamine, riboflavin and niacin on Al(110) surface, emphasizing the soft epitaxial adsorption mechanism. With permission from Mejeha et al. [50].

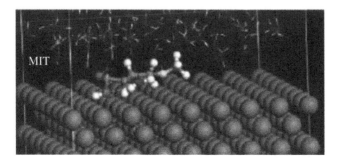

FIGURE 5.4   Methionine on Fe(110) after MD. With permission from Khaled [46].

To summarize, classical force fields allow identifying at low cost the organic inhibitor/surface interaction, and are useful for calculating the weak interactions between the molecule and the surface. It is interesting to explore how the organic molecule comes close to the surface in the presence of solvent. However, the studies presented earlier all suffer from the lack of benchmarking with more accurate DFT or MP2 methods. In particular, in the mentioned papers, force fields are determined per element in a definite compound (e.g., bulk solid or liquid water), but are not checked at the interfaces. Therefore, all the results should be carefully verified, especially the quantitative results of adsorption energies that might be strongly overestimated. FFs also suffer from a serious drawback, as no bonds between the molecule and the surface can be created. Note that more complete works concerning solid–liquid interfaces and adsorption of molecules in the presence of solvent have been performed by the scientific community of geochemistry, or concerning biomedical devices, sensors, using classical MD. In these works MD simulations are used to describe solvent organization near the surface, the molecule binding mode to the surface and the role of coadsorbed ions. Up to now, to our knowledge, such studies have not yet been applied to the metal/inhibitor interaction. Some illustrative examples can be found in Ref. [56].

In the next paragraph, we describe studies of inhibitor–surface interactions performed with quantum chemistry methods.

5.2.2 Organic Inhibitor/Surface Interaction Studied by Quantum Methods

As stated in the Introduction, a more sophisticated approach consists in using quantum mechanical methods to characterize both the molecule, the surface and their interaction. Theoretical *ab initio* tools have been extensively used to study inorganic surfaces in the recent years. Structural features of inorganic (metal or oxide) surfaces are well described by *ab initio* quantum mechanical (QM) methods, most often based on density functional theory (DFT). At variance with classical methods, quantum mechanics-based methods deal with bond breaking and making. Covalent, ionic, electrostatic, dipole–dipole interactions, and hydrogen bonds are well represented by those methods, especially by hybrid functionals. Recently, dispersion forces have been included in DFT codes [57] (called here DFT-D), such as, for example, VASP5.2 [58], Quantum Expresso, CP2K, or SIESTA. Briefly, the D2 (respectively, D3) Grimme's approach considers the two-body term of the dispersive interaction at the sixth- (respectively, sixth- and eight-) order for any atom pair. More details about the methods can be found in Ref. [59] and references cited therein. Few works appeared in the literature that take dispersion forces into account in a DFT approach when studying organic molecules adsorbed on inorganic surfaces [57, 60–68]. In the periodic boundary condition (PBC) approach, it is possible to study the adsorption at several coverages, thus to estimate the balance between molecule–surface, molecule–molecule interactions in the formation of a monolayer.

There is an increasing number of *ab initio* PBC works devoted to the adsorption of inhibitor molecules on metal or oxidized metal surfaces. They are summarized in Table 5.3 for different metals, Al, Cu, Fe, Zn, and their oxides. This list is not exhaustive, but provides a good view of the emerging possibilities of *ab initio* modeling. The molecules considered as potential corrosion inhibitors are azoles, carboxylates, and amines. For each metal, the studies are presented in the order of increasing complexity of the interface: works presenting adsorption at low coverage are presented, then results on full layer formation. Finally, the solid/liquid interface and adsorption at this interface is presented.

TABLE 5.3   DFT-PBC studies describing adsorption of organic corrosion inhibitors on metal and oxidized surfaces: metal surface, molecule, methods, and coverages are reported

Surface	Molecules	Method	Coverage	Reference
$Cu_2O(0001)$	Amino-mercapto-thiadiazol and methyl-mercapto-thiadiazol	LDA	Low	[69]
Cu(111)Cu(100)Cu(110)	Benzotriazole	PBE-D	Low	[70]
Cu(111)	Imidazole, 1,2,3-triazole, tetrazole, pentazole	PBE	Low	[71]
Cu(111)	Benzotriazole	GGA	Low to full coverage	[72]
Cu(111)	Imidazole, triazole, tetrazole, pentazole	PW91	Low to full coverage	[73]
Cu(111)	Benzotriazole, BTAOH	B3LYP	Low	[74]
Cu(111)	3-Amino-1,2,4-triazole benzotriazole, 1-hydroxybenzotriazole	PBE-D	Low to full coverage	[75]
Al(111)	Imidazole, 1,2,3-triazole, tetrazole, pentazole	PBE	Low	[75]
γ-Al_2O_3	Methylamine	PW91	Low	[76]
γ-Al_2O_3	Glycine	PW91	Low	[77]
AlOOH	Glycine	PBE-D in the presence of solvent	Low	[78]
Fe(110)	Methionine Phenylalanine	PBE	Low	[79]
ZnO	Carboxylates of increasing size	PW91-D	Low to SAM	[80]
ZnO(0001)	Diaminoethane	PW91	Low	[81]
ZnO(10-10)	Urea	DFT	Low	[82, 83]
ZnO(0001)	Glycine	PW91	Low to full layer	[84]

5.2.2.1 Cu and Cu Oxide Surfaces

Works performed on Cu surfaces were devoted to Cu–benzotriazole interactions. Indeed, benzotriazole (BTAH or $C_6N_3H_5$) has been a well-known corrosion inhibitor for Cu since 1947, as evidenced by a British patent. BTAH adsorption on Cu_2O was investigated by Blajiev and Hubin with LDA [69]. Several initial geometries were considered. The adsorption of the molecule induces a breakup of the bulk oxide periodicity. It was deduced that interfacial phases between the metal and the oxide and between the oxide and the electrolyte may be created in the conditions in which BTAH is used. Strong chemical interactions exist between the oxide and adsorbed molecules.

The adsorption of benzotriazole was modeled on three different Cu surfaces, Cu(111) [70, 71], Cu(100) [70], and Cu(110) [70]. An important aspect of the interaction of polar organic molecules with metal surfaces is related to electrostatic dipole–dipole interactions, because molecules such as BTAH have a rather large dipole moment of 4 D. These interactions not only contribute to the overall molecule–surface bonding at low coverage but are also responsible for considerable long-range lateral interactions [85]. Therefore, the authors

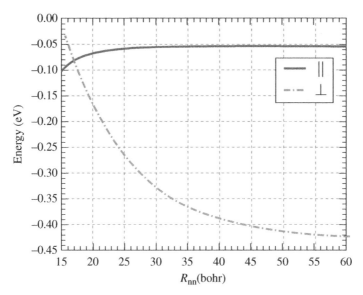

FIGURE 5.5 Electrostatic interaction energy, E_{dip}, of a layer of polarizable point-dipoles above the metal surface as a function of the nearest neighbor dipole–dipole distance for two limiting examples, that is, a layer of dipoles oriented parallel ($\parallel$) and perpendicular ($\perp$) to the surface. The plotted range of R_{nn} and the parameters of the polarizable point dipole model are chosen characteristically for BTAH ($m = 4.1$ D, $a = 91$ bohr^3, $d\perp$ im $= 7$ bohrs, and $d\parallel$ im $= 9$ bohrs). The two thin horizontal lines designate the zero coverage ($R_{nn} = N$) extrapolated values [70]. With permission from Peljhan and Kokalj [70].

have developed a method to extrapolate the obtained results at zero coverage. To better appreciate the extent of these interactions, Figure 5.5 displays the electrostatic energy, E_{dip}, as a function of the lateral nearest-neighbor distance for a crystalline layer of polarizable point-dipoles oriented either perpendicular or parallel to the surface with the parameters sensibly chosen to represent the BTAH molecule. It is seen that for perpendicular dipoles the lateral dipole–dipole repulsion is very important and long ranged, extending up to the nearest neighbor distance of about 60 bohrs (1 bohr = 0.52918 Å). On the other hand, for parallel dipoles, the lateral interactions are far less long ranged. They are important only for distances below about 25 bohrs, where the dipoles experience a slight attraction; note that at R_{nn} distances of about 15 bohrs, the parallel BTAH molecules start to feel lateral Pauli repulsion (due to the size of BTAH; the maximum interatomic distance in the BTAH molecule is 10.5 bohrs).

Chemisorption and physisorption modes were investigated. The terms "chemisorption" and "physisorption" are used to distinguish the type of molecule–surface interaction. While the chemisorption corresponds to the formation of N–Cu chemical bonds (the N–Cu bond lengths are in the range between 2.0 and 2.14 Å), which are optimized when the molecule plane is perpendicular to the surface (see Fig. 5.6), physisorption is driven by van der Waals dispersion forces, and the interaction between the molecule and the surface is optimized in a parallel orientation at 2.6 Å from the surface (Fig. 5.7). No orbital overlap between the π-electrons and Cu orbitals occur.

(a) (b) (c)

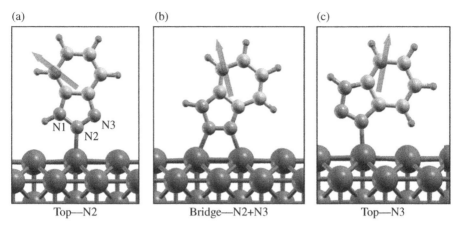

Top—N2   Bridge—N2+N3   Top—N3

FIGURE 5.6 Optimized structures of BTAH adsorbed on the Cu(110) surface along the [1–10] direction with the gas-phase dipole-moment vectors shown as arrows pointing from the barycenter of the molecules. The numbering of N atoms as used in this work is indicated in (a): Top-N2; (b): Bridge-N2+N3; (c):Top-N3. Similar structures are formed also on Cu(111) and Cu(100). With permission from Peljhan and Kokalj [70].

(a) (b)

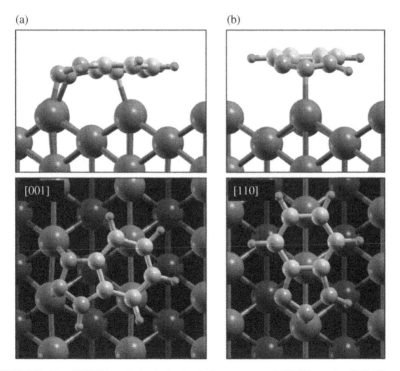

FIGURE 5.7 The PBE-D0 optimized physisorbed structures of BTAH on the Cu(110) surface oriented along the [001] and [1–10] directions. Surface Cu are in light grey, Cu of the underlying layers in dark grey. With permission from Peljhan and Kokalj [70].

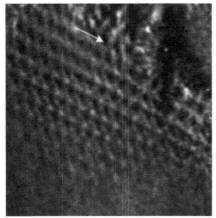
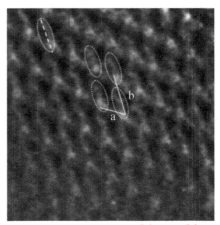

10 nm × 10 nm   5.8 nm × 5.8 nm

FIGURE 5.8   (a) *In situ* STM image after chemisorption of 5-MPhTT molecules on Cu ($V_{BIAS} = -50$ mV, $I_T = 1$ nA). The molecular structure of the adsorbed layer at the terrace edges is visible. The closely packed rows run along the steps as shows the arrow. (b) High-resolution STM image of 5-MPhTT adlayer structure on Cu(111) surface ($V_{BIAS} = -50$ mV, $I_T = 1$ nA). Some individual 5-MPhTT molecules are indicated by dotted lines. The intermolecular distances are $a \approx 0.7$ nm and $b \approx 1.0$ nm. With permission from Szocs et al. [86].

Molecules chemisorb on the surfaces with energies ranging between −0.5 and −0.9 eV (Fig. 5.7). On the Cu(111) surface, the dipolar interactions are dominant, but their relative importance goes down when passing to more open surfaces because of increasing chemical contribution. The adsorption is stronger on low coordinated sites. Physisorbed molecules (Fig. 5.8) interact rather strongly with the surface, by about −0.7 eV on the Cu(111) and Cu(100), and stabilize at 2.6 Å from the surface. On the open Cu(110) surface, the adsorption is even higher, −1.3 eV, with a smaller distance to the surface, 2.3 Å, as this surface is reactive enough to disturb the molecular π-system and to form chemisorption bonds with the molecule.

While the electrostatic dipole-to-image-dipole interaction is dominant on Cu(111), its relative importance reduces as passing to more open surfaces, because of increasing chemical contribution. The considerably enhanced chemisorption strength on low coordinated sites is correlated with their up-shifted energy of the *d*-band center causing a stronger hybridization between the metal *d* states and molecular orbitals.

In another work by the same team [34], the adsorption of different azoles (imidazole, triazole, tetrazole, and pentazole) was investigated on Cu(111). With increasing the number of nitrogen atoms in the azole ring, the molecules become more electronegative and chemically harder, the latter resulting in diminished molecule–surface bond strength. Thus the molecular reactivity indicators emerging from the theoretical formalization of the HSAB principle can be utilized to explain the adsorption energy trends.

More complete works consider the self-assembly of the molecules on the copper surface. Note that this is a difficult task as experimental structural data are scarce. Cho et al. studied BTAH adsorption on the clean Cu(110) and oxygen reconstructed Cu(110)—2 × 1 surfaces [87]. The atomically resolved scanning tunneling microscope (STM) images showed that BTA adsorbed on the clean Cu(110)—1 × 1 surface forms a c(4 × 2) structure, in agreement with the c(4 × 2) LEED pattern. On the other hand, STM images of BTA adsorbed on the oxygen induced Cu(110)—2 × 1 indicate a fully disordered structure. Kalman et al. [86] studied the adsorption of 5-mercapto-1-phenyl-tetrazole (5-PhTT) on Cu(111) and evidenced the full layer (formed at −70 mV vs SCE) as well as the individual molecules adsorbed at the surface in this layer (Fig. 5.8).

STM images of benzotriazole (BTAH) on Cu(100) electrodes in sulfuric acid revealed an ordered monolayer of BTAH at potentials less than or equal to −0.6 V versus sulfate electrode, which was attributed to a chemisorbed adlayer of BTAH molecules [7]. Similar results were obtained in HCl solution in the double-layer range up to potentials of −0.6 V versus SCE [7, 88]. In HCl solution, at potentials around −0.6 V, this structure is replaced by a c(2 × 2) Cl⁻ adlayer, which has the same atomic and long-range structure as found for BTAH-free HCl solution. Upon further potential increase to potentials greater than −0.35 V STM and electrochemical measurements indicate the onset of Cu dissolution, while the surface is still covered by the c(2 × 2) Cl⁻ adlayer. At slightly higher potentials (>−0.3 V) STM, IR, and electrochemical data point to the formation of a thick, inhibiting Cu(I)BTA film on the Cu surface. These results suggest that the surface may be protected by chemisorbed BTA or by a Cu(I)BTA film. These experimental observations are important for elaborating the theoretical model of corrosion inhibition.

Jiang and Adams [72] calculated the adsorption energy of a BTAH molecule and a BTA⁻ charged species on a Cu(111) surface by using first principle DFT calculations (VASP). It was found that BTAH can be physisorbed (<0.1 eV) or weakly chemisorbed (−0.43 eV) onto Cu(111), and that the chemical bonds are formed through N sp2 lone pairs. The chemisorption of BTAH can be stabilized by reacting with OH⁻, forming an intermolecular hydrogen bond and two chemisorption bonds on the surface. Some additional strong N–H–N hydrogen bonds can be found between two adsorbates. A model of the first layer of BTAH/BTA⁻ on a Cu(111) surface was developed based on a hydrogen bond (N–H–N) network structure (Fig. 5.9). It was proposed that the film consisted in a hybrid hydrogen bond network embedded with some segments of Cu(I)-BTA polymeric layer with a 1 : 1 Cu-BTA stoichiometry and bidentate structure (see Fig. 5.9).

The same authors performed a similar study on BTAH adsorption on Cu$_2$O(111) using the GGA approach (PW91) [73]. It was found that BTAH can be strongly chemisorbed onto the surface and that the chemisorption is due to the combined effect of a chemical bond between surface copper cation acting as a Lewis acid site, and nitrogen sp2 lone pairs and an additional hydrogen bond with a surface oxygen anion through C–H or N–H protons. The most stable state of adsorption is shown in Figure 5.10 at low and full coverages. It is observed that the energy of adsorption increases with coverage, due to the lateral interactions between adsorbed molecules.

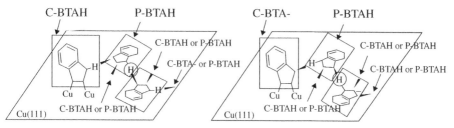

FIGURE 5.9 Scheme representative of the hydrogen bond network formed between BTAH and BTA molecules forming a layer on Cu(111). With permission from Jiang and Adams [72].

(a)   (b)

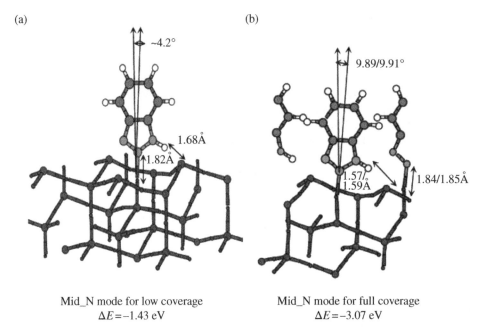

Mid_N mode for low coverage   Mid_N mode for full coverage
$\Delta E = -1.43$ eV $\Delta E = -3.07$ eV

FIGURE 5.10 The minimum energy configurations and energies for BTAH adsorptions on Cu_2O (111), at low (a) and full (b) coverages. With permission from Jiang et al. [73].

Milosev et al. performed a series of experimental and theoretical studies on the azoles inhibition properties for Cu [27, 34, 74, 75, 89–92]. The inhibition of copper corrosion in 3% NaCl solution was studied by using a well-known inhibitor, benzotriazole (BTAH) and its derivative 1-hydroxybenzotriazole (BTAOH) [74]. Corrosion parameters and inhibition effectiveness were determined experimentally. It was shown that benzotriazole is a more effective inhibitor of the corrosion of copper in chloride media than is 1-hydroxybenzotriazole.

Whereas in the presence of BTAH, a protective Cu-BTA layer is formed on the Cu surface, in the presence of BTAOH a thick, poorly protective layer is formed, which readily dissolves in chloride solution. BTAH and BTAOH have very similar electronic properties, suggesting they should display similar chemisorption behavior. However, markedly different inhibition effects are observed for the two molecules, and this cannot be explained in terms of molecular electronic properties. To reconcile these apparently contradictory features, the authors considered physisorption, H-bonding, and some specific steric factors of BTAH and BTAOH. In considering the molecular crystals of BTAH and BTAOH, it appeared that due to its planar structure BTAH may form planar H-bonded polymers of favorable geometry for the physisorption, so that a thin and compact layer can be formed (Fig. 5.11a). Summing together physisorption (molecule–surface interaction) and H-bond molecule–molecule interaction) energies, the SAM formation is 1.2 eV exothermic. On the other hand, such geometrical arrangement is not possible for BTAOH due to its nonplanar structure (Fig. 5.11b).

The structures presented in Figure 5.11 are not the result of quantum calculations, but clearly show the importance of taking into consideration self-assembling properties of the molecule, as the prerequisite condition for inhibition is the formation of a dense organic layer.

(a)   (b)

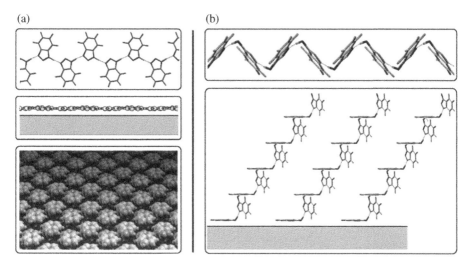

FIGURE 5.11   H-bonded polymers of BTAH (a) and BTAOH (b) taken from molecular crystals. No optimization is performed. The structure of the former is planar (a, top-panel) and therefore favorable for physisorption on any surface (a, middle panel), while the latter forms a zig-zag structure (b, top-panel). Tentative structure of planar physisorbed BTAH polymers packed so as to form compact and thin film (a, bottom panel). An analog BTAOH film would be thicker and less compact (side view shown in b, bottom panel). With permission from Finsgar et al. [74].

Three corrosion inhibitors for copper, 3-amino-1,2,4-triazole (ATA), benzotriazole (BTAH), and 1-hydroxybenzotriazole (BTAOH), were investigated by corrosion experiments and atomistic computer simulations [75]. The trend of copper corrosion inhibition effectiveness of the three inhibitors in near-neutral chloride solution was determined experimentally as BTAH > ATA > BTAOH. An exhaustive analysis of the possible interactions between the molecules (in their neutral or deprotonated form) and the surface was done with PBE-D [75]. Physisorption, chemisorption, self-assembly as well as organometallic polymer formation at the Cu(111) surface were considered. The results are reported in Table 5.4.

From these results, it appears that chemisorption of the radical adduct is the most stable state at low coverage. By contrast, the formation of a full layer occurs in the form of a physisorbed layer or of an organometallic layer.

The superior inhibiting action of BTAH and ATA is a result of their ability to form strong N–Cu chemical bonds in deprotonated form. While these bonds are not as strong as the Cl–Cu bonds, the presence of solvent favors the adsorption of inhibitor molecules onto the surface due to stronger solvation of the Cl$^-$ anions (see paragraph 5.2.2.5 for taking into account solvation free energies). Moreover, benzotriazole displays the largest affinity among the three inhibitors to form intermolecular aggregates, such as the [BTA– Cu]$_n$ polymeric complex. This is another factor contributing to the stability of the protective inhibitor film on the surface, thus making benzotriazole an outstanding corrosion inhibitor for copper. These findings cannot be anticipated on the basis of inhibitors' molecular electronic properties alone, thus emphasizing the importance of a rigorous modeling of the interactions between the components of the corrosion system in corrosion inhibition studies.

TABLE 5.4   Energies of adsorption of BTA, ATA, and BTAOH with Cu(111) as calculated with DFT-D^a

Molecule	Coverage	Adsorption mode	Eads (eV)
ATAH	1/16	Chemisorption	−0.60
BTAH			−0.40
BTAOH			−0.53
ATAH	1/16	Physisorption	−0.56
BTAH			−0.72
BTAOH			−0.97
ATA°	1/16	Chemisorption	−2.22
BTA°			−2.78
BTAO°			−1.65
ATA	Full layer	Physisorption + SAM	−1.12
BTAH			−1.23
Cu-ATA	Full layer	Organometallic polymer	−1.83
Cu-BTA			−2.98
Cu-BTAO			−1.55

^{a}Data from Kokalj et al. [75].

5.2.2.2 Al and Al Oxides   Benzotriazole, imidazole, tetrazole, and pentazole were adsorbed on Al(111) in the same spirit as the study on Cu(111) [71]. The bonding of adsorbates on Al(111) is not very different from that on Cu(111), with the trend imidazole > triazole > tetrazole > pentazole.

Using first-principle DFT calculations, molecular adsorption of methylamine (CH_3NH_2) at the γ-Al_2O_3(0001) surface was studied [76]. The adsorption structure and bonding nature is documented by the adsorption-induced changes in the electron density and in the projected density of states (Fig. 5.12). This figure evidences that methylamine binds to exposed surface cations via the N lone pair orbital, forming a polar covalent bond. Indeed, the energy of the N 2p electrons separated from the surface (in dotted line, in Fig. 5.12) is shifted to a lower energy (stabilized) and overlaps the Al 3p states when the molecule is adsorbed on the surface.

Aminoacids exhibit both amine and carboxylate moieties, which are both interesting functions for corrosion inhibition, and are therefore interesting candidates for corrosion inhibition [93]. Arrouvel et al. studied the adsorption of glycine on anhydrous and hydroxylated γ-Al_2O_3 [77]. It was shown that adsorption through the formation of an Al–OCO adduct of polar covalent nature is more stable than the Al–NH_2 adduct. In the outer sphere mode, a hydrogen bond network is formed between glycine and surface hydroxyls. Inner sphere adsorption implies the formation of cooperative hydrogen bonds and polar covalent bonds and is slightly exothermic.

5.2.2.3 Iron, Steel, and Fe Oxides   In a study of methionine and phenylalanine adsorption on Fe(110), Oguzie et al. [79] described a combined electrochemical and first-principle density functional study to probe the adsorption behavior and the corrosion inhibiting power of methionine (Met) and phenylalanine (Phe) on polycrystalline (grain size 50 µm) and nanocrystalline (grain size 39 nm) iron in acidic solution. This allows investigating the role of the surface defect density on the molecular adsorption (with the help of theoretical calculations) and inhibition efficiency (measured experimentally). Met functioned as a better inhibitor for both Fe

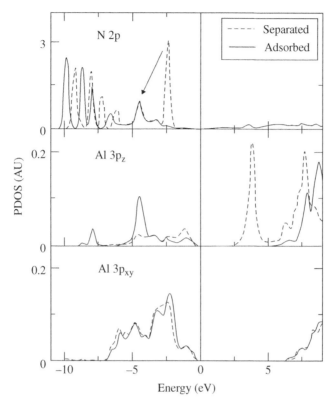

FIGURE 5.12   Projected density of states (PDOS) of N 2p and Al 3p electrons of the methylamine molecule and Al_2O_3 surface, separated (- - -) and adsorbed (—). With permission from Borck et al. [76].

microstructures, and was more favorably adsorbed on the nanocrystalline surface. The comparable values of the computed adsorption energies (−94.2 and −86.6 kcal mol^{-1} for Phe and Met, respectively) as well as the stable adsorption orientations of both molecules on Fe suggest a controlling influence of a soft epitaxial adsorption mechanism in which C, N, O, and S atoms of the molecules align with epitaxial grooves on the Fe lattice. Interestingly, for Met the thiol group imparts an added ability for covalent interaction with Fe. This explains why Met has a better inhibitor efficiency on nanocrystalline Fe, as the high defect population of nanocrystalline surfaces provides an abundance of active sites for bonding interactions with adsorbates having suitable electronic structures. At the opposite end, the diminished inhibition efficiency of Phe observed on the nanocrystalline Fe surface was related to disruption of the epitaxial patterns on the lattice as the surface becomes increasingly defective, leading to weaker adsorption. The general trends of inhibitor effect due to physisorption on flat surfaces and chemisorption at low coordinated sites corroborate the conclusions obtained for the Cu surfaces.

5.2.2.4   ZnO   On Zn-terminated and O-terminated ZnO (0001) surfaces (noted Zn–ZnO and O–ZnO), diaminoethane (DAE) adsorption is weak and involves mostly weak forces [81]. On the Zn–ZnO surface, a single Zn–N bond is formed. On the O-terminated surface, only weak bonds are formed between the amine and the surface. These results were obtained with a pure GGA approach. A proper description of the van der Waals forces could be useful to better describe the interface.

The adsorption of carboxylic acids (CAs) on the Zn–ZnO(0001) surface was studied by DFT calculations including dispersion forces (DFT + D) [80]. Carboxylic acids of formula $CH_3(CH_2)n-2COOH$ (where $n = 1–10$ is the total number of C atoms in the CA) were considered. Comparing different possible adsorption modes, it was concluded that the most likely mechanism for CA adsorption on the Zn–ZnO(0001) surface is dissociative bridging adsorption with the carboxylate group attached to 2 Zn atoms and the proton transferred to the neighboring Zn atom, forming a Zn–H bond.

The effect of the chain length and chain orientation on the SAM formation energy was investigated, as illustrated in Figure 5.13 and Table 5.5.

Adsorption at low coverage was performed in order to estimate the energy of formation of the polar covalent Zn–OCO–Zn bond with the surface. It decreases when going from formic acid (–1.62 eV) to propionic acid (–2.11 eV) and then reaches a plateau, as shown in Table 5.5.

Then the SAM formation was studied. To this end, the orientation of the molecule with respect to the surface and the size of the aliphatic chain were incrementally changed. A transition from perpendicular to tilted structures was calculated for $n = 7$, after which the tilt angle to the surface is 35° (see Fig. 5.13). This angle optimizes the lateral–lateral interactions. The adsorption energy decreases with increasing chain length (to reach the value of –2.77 eV for $n = 9$). This was attributed to the stabilization by van der Waals forces through lateral interactions.

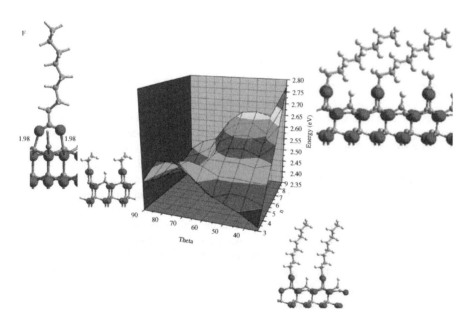

FIGURE 5.13 Left: Optimized geometry of nonanoic acid adsorbed on ZnO in a dissociative bridging configuration. Center: Absolute values of the Energies of SAM formation of carboxylic acids on ZnO as a function of the molecule size (n = number of C atoms) and of the tilt angle (theta) with respect to the surface. With permission from Islam et al. [80].

TABLE 5.5 Energies of adsorption of carboxylic acids of increasing sizes on the ZnO surface (low coverage, 2.6 CA/nm²), from Islam et al. [80]

n	Formula and name	Standard deprotonation enthalpy (kJ mol⁻¹) and pK_A (experimental value) [94]	Calculated vertical deprotonation energy (eV)	Adsorption energy (eV)
1.	HCOOH, formic (methanoic) acid	1444 3.74	3.02	−1.62
2.	CH_3–COOH, acetic (ethanoic) acid	1458 4.76	3.16	−1.61
3.	CH_3–CH_2–COOH, propionic (propanoic) acid	1453 4.86		−2.11
4.	CH_3–C_2H_4–COOH, butyric (butanoic) acid	1448 4.83		−2.07
5.	CH_3–C_3H_6–COOH, valeric (pentanoic) acid	4.82		−2.08
6.	CH_3–C_4H_8–COOH, caproic (hexanoic) acid	4.88		−2.05
7.	CH_3–C_5H_{10}–COOH, enanthic (heptanoic) acid		3.39	−2.04
8.	CH_3–C_6H_{12}–COOH, caprylic (octanoic) acid	4.89		−2.03
9.	CH_3–C_7H_{14}–COOH, pelargonic (nonanoic) acid	4.96	3.52	−2.08
10.	CH_3–C_8H_{16}–COOH, capric (decanoic) acid		3.58	−2.14

The mode of adsorption is dissociative bridging as shown in Figure 5.13.

Glycine adsorption on the same surface, Zn–ZnO(0001), was also investigated from low to high coverages [84]. Whatever the coverage, the dissociation of glycine to form a glycinate ion is favored. At low coverage, the most favorable conformation is obtained when glycine adsorbs parallel to the surface and maximizes the bonds with the Zn–ZnO surface. At the monolayer coverage, glycine may organize at the surface into an epitaxial glycinate monolayer in which each glycinate forms two Zn–O bonds and one Zn–N bond, thus adopting the surface structure of Zn–ZnO. An organometallic bilayer can also be formed at the Zn–ZnO surface. The organic layer near the surface adopts the honeycomb surface structure of Zn–ZnO. In the outermost layer, a Zn–glycinate complex is formed which adopts a gas phase-like conformation, in which the Zn ions are chelated by glycine through the carboxylic and amine function.

First principles calculations based on DFT were used to investigate the adsorption of urea onto a nonpolar ZnO(10–10) [83] surface. The results indicated that molecular urea adsorption was favored, and that stable adsorption products were formed through the reaction between nitrogen atom or oxygen atom from urea and zinc atom on the surface. The adsorption energy was −1.48 and −1.41 eV, respectively.

5.2.2.5 Solvent Effects: Free Energy Calculations and the "Real" Solid/Liquid Interface
An important question arising from DFT calculations of the oxide/vacuum interface performed at 0 K, is the extrapolation of the results to a more realistic solid–liquid

interface at ambient temperature. Indeed, different mechanisms may account for adsorption in UHV (at low temperature) and in solution (at room temperature). The influence of the solvent can be considered *a posteriori*. The adsorption free energy of a molecule (mol) at the water/solid interface can be calculated by employing a thermodynamic cycle where, to the electronic energy of adsorption calculated at the solid vacuum interface at 0 K, including ZPE corrections, free energy contributions from temperature (F_{vib}, F_{rot}, and F_{trans}) and solvent (G_{solv}) are added (more details can be found in Refs. [95] and [59]). Such a thermodynamic approach allows avoiding the costly calculations at the solid–liquid interface.

The solid–liquid interface has been explicitly considered only in a small number of works, because of the extensive cost of *ab initio* MD calculations and of the very small time periods simulated during a calculation (some pico-seconds). It is only recently that solid/liquid interfaces have begun to be investigated by means of DFT-based MD methods, with both the solid and the liquid represented at the same explicit level of theory [96–107]. As already explained, the use of DFT allows studying bond-breaking and -making. Indeed, the surface reactivity of oxidized surfaces toward organic species is strongly related to the nature of the surface species, especially the ligand (H)OH exchange processes, also called inner-sphere adsorption. This concept as introduced by Kummert and Stumm [108–110] who proposed a surface coordination model to explain the specific interaction of organic anions with hydrous oxides and indicated that organic anions might replace the surface hydroxyl groups of hydrous oxides by ligand exchange. The ligand exchange reaction can be expressed as follows:

$$M\text{-}OH + L^- \rightarrow M\text{-}L + OH^-$$

Based on this model, the density of hydroxyl groups on the surfaces of different oxides could strongly affect the adsorption capacity. There is another adsorption mode to consider, the outer-sphere adsorption, in which no ligand exchange reaction takes place, and the organic species is instead held close to the surface through hydrogen bonding and/or electrostatic interactions. One key question for organic molecular adsorption on surfaces is therefore the occurrence of inner *versus* outer sphere adsorption behaviors, which are usually considered as site-specific (inner sphere) versus non site-specific (outer sphere) adsorption.

To explore this question, the adsorption of glycine on AlOOH was recently investigated at the interface with water [78, 97]. This study confirmed the occurrence of an inner sphere Al–O–C–O bond that was predicted from calculations at the interface with vacuum, and from numerous experimental studies of carboxylic acid adsorption on alumina polymorphs (see Ref. [78] and references therein). It was found that inner sphere adsorption (shown in Fig. 5.14, right), with −161.6 and −113.6 kJ mol^{-1} for the anionic and zwitterionic species, respectively is significantly more stable as compared to outer sphere adsorption (−20.5 kJ mol^{-1} (Fig. 5.14, left)). Beyond the result itself, it is important to note that complex events can now be handled with *ab initio* methods because they are able to model all types of forces in reasonable agreement with experiment.

Such an approach was used in a recent AIMD study of the interaction of a gallic acid layer of density 3.6 molecule nm^{-2} on AlOOH, a model of Al immersed in aqueous solution, with liquid water (Ribeiro, T.; Motta, A.; Marcus, P.; Gaigeot, MP.; Lopez, X.; Costa, D., Formation of the OOH center dot radical at steps of the boehmite surface and its inhibition by gallic acid: A theoretical study including DFT-based dynamics, J. Inorg. Biochem., (2013) 128, 164–173.). During the time of the dynamics (10 ps), the organic layer was stable and no water penetration was observed. It was also shown that the first water layer above the organic layer is strongly oriented

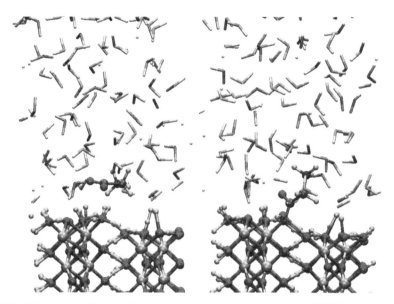

FIGURE 5.14 Supercells used for the study of glycine adsorption on AlOOH boehmite at the interface with water in the outer-sphere adsorption mode (left), and inner sphere adsorption mode (right). With permission from Motta et al. [78].

with O-down, being H-bond receiver from the gallic acid molecules. This ordering suggests a strong water–organic film interaction that may help in blocking the access of corrosive species.

In addition to the solvent contributions, the electrochemical potential can be modeled. Application of an external electric field within a metal/vacuum interface model has been used to investigate the impact of potential alteration on the adsorption process [111, 112]. Although this approach can model the effects of the electrical double layer, it does not consider the adsorbate–solvent, solvent–solvent, and solvent–metal interactions at the electrode–electrolyte interface. In another approach, Nørskov and co-workers model the electrochemical environment by changing the number of electrons and protons in a water bilayer on a Pt(111) surface [113–115]. Jinnouchi and Anderson used the modified Poisson–Boltzmann theory and DFT to simulate the solute–solvent interaction to integrate a continuum approach to solvation and double layer affects within a DFT system [116–120]. These methods differ in the approximations made to represent the electrochemical interface, as the time and length scales needed for a fully quantum mechanical approach are unreachable.

In the simplest approach, analogous to the method used by Nørskov et al. [115] and Anderson and Kang [120], anion adsorption is taken to occur with the transfer of an electron, and the energy of the electron transferred varies linearly with potential, whereas the adsorbate–metal system energy is potential independent. The impact of an electric field during the adsorption process is tested using an applied external electric field model, both with and without inclusion of a water bilayer at the adsorbate–metal interface. This method was used to study the adsorption of azoles on Cu [75]. Intermolecular interactions with interfacial solvent molecules within a charged double-layer model are simulated with the double reference method developed by Neurock and co-workers [121–131]. The results of these DFT methods are used to simulate the linear sweep voltammogram and to predict the surface coverage of adsorbate as a function of surface potential. The simulated voltammetric data are compared with experimental data from the literature. Other works report the adsorption of organic species on metal surfaces under electrochemical potential [132]; but to our knowledge, this sophisticated method has not yet been used for the formation of an inhibition layer.

5.3 CONCLUSIONS AND FUTURE DIRECTIONS

Corrosion inhibition is undoubtedly an area where molecular modeling can significantly contribute to better understanding, and also be used as a tool for the selection of adequate inhibitors. Quantum chemistry has allowed the characterization of possible inhibitor molecules through their intrinsic properties, HOMO, LUMO, and electronegativity signatures. Further calculations considering the interaction between molecule and surface have shown that only the isolated molecule–surface interaction is correctly described by the study of the respective positions of the HOMO–LUMO gap versus the Fermi level of the surface. The examination of literature shows that a good understanding of the systems can be achieved when several surfaces and several molecules are compared. They have shown that on metal surfaces as Cu, the dipole-to-image-dipole interaction is determining for the adsorption on dense surfaces and less important for open surfaces where chemical interaction is stronger.

The situation is more complex when considering the formation of a full inhibitor layer on a metal or oxide surface. Indeed, other parameters such as self-assembly properties and possible organometallic layer formation become predominant. Here, the molecular properties derived from hard–soft/acid–base theory are not always sufficient to describe the layer formation.

The onset of theoretical studies also suggests that the inhibition properties are due to soft epitaxy adsorption and organic layer formation on the dense metal surfaces, whereas on stepped or nanocrystalline surfaces, such a homogeneous layer cannot be formed and molecules able to form a covalent bond with the surface will have better inhibition power.

An important step has been performed recently in modelling explicitly the surface–water interface through *ab initio* methods, and is beginning to be applied to the inhibition properties of organic layers. Such an approach could be strongly aided by benchmark experimental atomic structural studies providing information on the local structure and organization of the organic molecules in the SAMs.

Future directions include modeling explicitly the solvent and the electrochemical potential, the metal and alloy surface layers, including the formation of a thin oxide film covering the metal surface, understanding the role of the solvent in the adsorption process, and the study of organic molecules of increasing size and complexity. The role of surface defects, of water organization near the surface, and of the presence of inorganic ions in the solution also has to be considered, as all these parameters have a major effect on the organic/bio-inorganic interface.

NOMENCLATURE

ATA	3-amino-1,2,4-triazole
BTAH	Benzotriazole ($C_6H_5N_3$)
BTAOH	1-hydroxybenzotriazole
COMPASS	condensed phase optimized molecular potentials for atomistic simulations studies

COSMO	"COnductor-like Screening MOdel," a calculation method for determining the electrostatic interaction of a molecule with a solvent
CP2K	a program to perform atomistic and molecular simulations of solid state, liquid, molecular, and biological systems www.cp2k.org
DFT	density functional theory
DFT-D	DFT including dispersion forces using the Grimme scheme
DFT-VW	van der Waals corrected DFT
FF	force field
HOMO	highest occupied molecular orbital
LUMO	lowest unoccupied molecular orbital
MC	Monte Carlo
MD	molecular dynamics
OATP	o-aminothiophenol
PBC	periodic boundary conditions
PCM	polarized continuum model a method in computational chemistry to model solvation effects
P-DOS	projected density of states
QSAR	quantitative structure–activity relationship
Quantum Expresso	an integrated suite of open-source computer codes for electronic-structure calculations and materials modeling at the nanoscale http://www.quantum-espresso.org/
SAM	self-assembled monolayer
SIESTA	(Spanish initiative for electronic simulations with thousands of atoms) is both a method and its computer program implementation, to perform electronic calculations http://www.uam.es/siesta
VASP	Vienna *ab initio* Simulation Package

REFERENCES

1. Trabanelli, G., Corrosion Inhibitors, in *Corrosion Mechanisms*. F. Mansfeld, Editor, Marcel Dekker, New York, 1987.

2. Schwenk, W., in: Leidheiser, H., Jr. Editor, *Corrosion Control by Organic Coatings*, NACE, Houston, TX, 1981, p. 103.

3. Rohwerder, M., G. Grundmeier, Corrosion Prevention by Adsorbed Organic Monolayers and Ultrathin Plasma Polymer Films, in *Corrosion Mechanisms in Theory and Practice*, P. Marcus, Editor, CCR Press, Boca Raton, FL, 2012.

4. Gece, G., Drugs: A review of promising novel corrosion inhibitors. *Corrosion Science*, 2011. 53 (12): p. 3873–3898.

5. Raja, P.B. and M.G. Sethuraman, Natural products as corrosion inhibitor for metals in corrosive media—A review. *Materials Letters*, 2008. 62(1): p. 113–116.

6. Gece, G., The use of quantum chemical methods in corrosion inhibitor studies. *Corrosion Science*, 2008. 50(11): p. 2981–2992.

7. Ashassi-Sorkhabi, H., B. Shaabani, and D. Seifzadeh, Corrosion inhibition of mild steel by some Schiff base compounds in hydrochloric acid. *Applied Surface Science*, 2005. 239(2): p. 154–164.

8. Ashassi-Sorkhabi, H., B. Shaabani, and D. Seifzadeh, Effect of some pyrimidinic Schiff bases on the corrosion of mild steel in hydrochloric acid solution. *Electrochimica Acta*, 2005. 50 (16–17): p. 3446–3452.

9. Bentiss, F., M. Lebrini, M. Lagrenee, M. Traisnel, A. Elfarouk, and H. Vezin, The influence of some new 2,5-disubstituted 1,3,4-thiadiazoles on the corrosion behaviour of mild steel in 1M HCl solution: AC impedance study and theoretical approach. *Electrochimica Acta*, 2007. 52 (24): p. 6865–6872.

10. Bentiss, F., M. Traisnel, H. Vezin, and M. Lagrenee, Linear resistance model of the inhibition mechanism of steel in HCl by triazole and oxadiazole derivatives: Structure–activity correlations. *Corrosion Science*, 2003. 45(2): p. 371–380.

11. Khaled, K.F., K. Babic-Samardzija, and N. Hackerman, Theoretical study of the structural effects of polymethylene amines on corrosion inhibition of iron in acid solutions. *Electrochimica Acta*, 2005. 50(12): p. 2515–2520.

12. Lebrini, M., M. Traisnel, M. Lagrenee, B. Mernari, and F. Bentiss, Inhibitive properties, adsorption and a theoretical study of 3,5-bis(n-pyridyl)-4-amino-1,2,4-triazoles as corrosion inhibitors for mild steel in perchloric acid. *Corrosion Science*, 2008. 50(2): p. 473–479.

13. Sanyal, B., Organic compounds as corrosion inhibitors in different environments—A review. *Progress in Organic Coatings*, 1981. 9(2): p. 165–236.

14. Yang, W., R.G. Parr, and R. Pucci, Electron density, Kohn–Sham frontier orbitals, and Fukui functions. *The Journal of Chemical Physics*, 1984. 81(6): p. 2862.

15. Parr, R.G. and W. Yang, Density functional approach to the frontier-electron theory of chemical reactivity. *Journal of the American Chemical Society*, 1984. 106(14): p. 4049–4050.

16. Khaled, K.F., The inhibition of benzimidazole derivatives on corrosion of iron in 1 M HCl solutions. *Electrochimica Acta*, 2003. 48(17): p. 2493–2503.

17. Khaled, K.F., Experimental and theoretical study for corrosion inhibition of mild steel in hydrochloric acid solution by some new hydrazine carbodithioic acid derivatives. *Applied Surface Science*, 2006. 252(12): p. 4120–4128.

18. El Ashry, E.S.H., A. El Nemr, S.A. Essawy, and S. Ragab, Corrosion inhibitors part V: QSAR of benzimidazole and 2-substituted derivatives as corrosion inhibitors by using the quantum chemical parameters. *Progress in Organic Coatings*, 2008. 61(1): p. 11–20.

19. Bentiss, F., B. Mernari, M. Traisnel, H. Vezin, and M. Lagrenee, On the relationship between corrosion inhibiting effect and molecular structure of 2,5-bis(n-pyndyl)-1,3,4-thiadiazole derivatives in acidic media Ac impedance and DFT studies. *Corrosion Science*, 2011. 53(1): p. 487–495.

20. Lukovits, I., A. Shaban, and E. Kalman, Thiosemicarbazides and thiosemicarbazones: Non-linear quantitative structure-efficiency model of corrosion inhibition. *Electrochimica Acta*, 2005. 50(20): p. 4128–4133.

21. Khalil, N., Quantum chemical approach of corrosion inhibition. *Electrochimica Acta*, 2003. 48 (18): p. 2635–2640.

22. Behpour, M., S.M. Ghoreishi, N. Soltani, M. Salavati-Niasari, M. Hamadanian, and A. Gandomi, Electrochemical and theoretical investigation on the corrosion inhibition of mild steel by thiosalicylaldehyde derivatives in hydrochloric acid solution. *Corrosion Science*, 2008. 50(8): p. 2172–2181.

23. Gece, G. and S. Bilgic, A theoretical study on the inhibition efficiencies of some amino acids as corrosion inhibitors of nickel. *Corrosion Science*, 2010. 52(10): p. 3435–3443.

24. El Ashry, E.S.H. and S.A. Senior, QSAR of lauric hydrazide and its salts as corrosion inhibitors by using the quantum chemical and topological descriptors. *Corrosion Science*, 2011. 53(3): p. 1025–1034.

25. Obi-Egbedi, N.O., I.B. Obot, and M.I. El-Khaiary, Quantum chemical investigation and statistical analysis of the relationship between corrosion inhibition efficiency and molecular structure of

xanthene and its derivatives on mild steel in sulphuric acid. *Journal of Molecular Structure*, 2011. 1002(1–3): p. 86–96.

26. Arslan, T., F. Kandemirli, E.E. Ebenso, I. Love, and H. Alemu, Quantum chemical studies on the corrosion inhibition of some sulphonamides on mild steel in acidic medium. *Corrosion Science*, 2009. 51(1): p. 35–47.

27. Lesar, A. and I. Milosev, Density functional study of the corrosion inhibition properties of 1,2,4-triazole and its amino derivatives. *Chemical Physics Letters*, 2009. 483(4–6): p. 198–203.

28. Arshadi, M.R., M. Lashgari, and G.A. Parsafar, Cluster approach to corrosion inhibition problems: Interaction studies. *Materials Chemistry and Physics*, 2004. 86(2–3): p. 311–314.

29. Jamalizadeh, E., S.M.A. Hosseini, and A.H. Jafari, Quantum chemical studies on corrosion inhibition of some lactones on mild steel in acid media. *Corrosion Science*, 2009. 51(6): p. 1428–1435.

30. Mousavi, M., M. Mohammadalizadeh, and A. Khosravan, Theoretical investigation of corrosion inhibition effect of imidazole and its derivatives on mild steel using cluster model. *Corrosion Science*, 2011. 53(10): p. 3086–3091.

31. Camacho, R.L., E. Montiel, N. Jayanthi, T. Pandiyan, and J. Cruz, DFT studies of alpha-diimines adsorption over Fe-n surface (n = 1, 4, 9 and 14) as a model for metal surface coating. *Chemical Physics Letters*, 2010. 485(1–3): p. 142–151.

32. Lashgari, M., Theoretical challenges in understanding the inhibition mechanism of aluminum corrosion in basic media in the presence of some p-phenol derivatives. *Electrochimica Acta*, 2011. 56(9): p. 3322–3327.

33. Lashkari, M. and M.R. Arshadi, DFT studies of pyridine corrosion inhibitors in electrical double layer: Solvent, substrate, and electric field effects. *Chemical Physics*, 2004. 299(1): p. 131–137.

34. Kokalj, A., N. Kovacevic, S. Peljhan, M. Finsgar, A. Lesar, and I. Milosev, Triazole, benzotriazole, and naphthotriazole as copper corrosion inhibitors: I. Molecular electronic and adsorption properties. *Chemphyschem*, 2011. 12(18): p. 3547–3555.

35. Miertuš, S., E. Scrocco, and J. Tomasi, Electrostatic interaction of a solute with a continuum. A direct utilization of *ab initio* molecular potentials for the prevision of solvent effects. *Chemical Physics*, 1981. 55(1): p. 117–129.

36. Raymand, D., A.C.T. van Duin, W.A. Goddard, K. Hermansson, and D. Spangberg, Hydroxylation structure and proton transfer reactivity at the zinc oxide–water interface. *Journal of Physical Chemistry C*, 2011. 115(17): p. 8573–8579.

37. Jeon, B., S. Sankaranarayanan, A.C.T. van Duin, and S. Ramanathan, Atomistic insights into aqueous corrosion of copper. *Journal of Chemical Physics*, 2011. 134(23).

38. Kong, D.S., S.L. Yuan, Y.X. Sun, and Z.Y. Yu, Self-assembled monolayer of o-aminothiophenol on Fe(110) surface: A combined study by electrochemistry, in situ STM, and molecular simulations. *Surface Science*, 2004. 573(2): p. 272–283.

39. Hagler, A.T., E. Huler, and S. Lifson, Energy functions for peptides and proteins. I. Derivation of a consistent force field including the hydrogen bond from amide crystals. *Journal of the American Chemical Society*, 1974. 96(17): p. 5319–5327.

40. Hagler, A.T. and S. Lifson, Energy functions for peptides and proteins. II. The amide hydrogen bond and calculation of amide crystal properties. *Journal of the American Chemical Society*, 1974. 96(17): p. 5327–5335.

41. Kitson, D.H. and A.T. Hagler, Theoretical studies of the structure and molecular dynamics of a peptide crystal. *Biochemistry*, 1988. 27(14): p. 5246–5257.

42. Kitson, D.H. and A.T. Hagler, Catalysis of a rotational transition in a peptide by crystal forces. *Biochemistry*, 1988. 27(19): p. 7176–7180.

43. Dauber-Osguthorpe, P., V.A. Roberts, D.J. Osguthorpe, J. Wolff, M. Genest, and A.T. Hagler, Structure and energetics of ligand binding to proteins: *Escherichia coli* dihydrofolate reductase-trimethoprim, a drug–receptor system. *Proteins*, 1988. 4(1): p. 31–47.

44. Zhang, Z., S.H. Chen, Y.H. Li, S.H. Li, and L. Wang, A study of the inhibition of iron corrosion by imidazole and its derivatives self-assembled films. *Corrosion Science*, 2009. 51(2): p. 291–300.

45. Feng, L.J., H.Y. Yang, and F.H. Wang, Experimental and theoretical studies for corrosion inhibition of carbon steel by imidazoline derivative in 5% NaCl saturated Ca(OH)(2) solution. *Electrochimica Acta*, 2011. 58: p. 427–436.

46. Khaled, K.F., Monte Carlo simulations of corrosion inhibition of mild steel in 0.5M sulphuric acid by some green corrosion inhibitors. *Journal of Solid State Electrochemistry*, 2009. 13(11): p. 1743–1756.

47. Sun, H., P. Ren, and J.R. Fried, The COMPASS force field: Parameterization and validation for phosphazenes. *Computational and Theoretical Polymer Science*, 1998. 8(1–2): p. 229–246.

48. Khaled, K.F., Molecular simulation, quantum chemical calculations and electrochemical studies for inhibition of mild steel by triazoles. *Electrochimica Acta*, 2008. 53(9): p. 3484–3492.

49. Oguzie, E.E., C.E. Ogukwe, J.N. Ogbulie, F.C. Nwanebu, C.B. Adindu, I.O. Udeze, K.L. Oguzie, and F.C. Eze, Broad spectrum corrosion inhibition: Corrosion and microbial (SRB) growth inhibiting effects of *Piper guineense* extract. *Journal of Materials Science*, 2012. 47(8): p. 3592–3601.

50. Mejeha, I.M., M.C. Nwandu, K.B. Okeoma, L.A. Nnanna, M.A. Chidiebere, F.C. Eze, and E.E. Oguzie, Experimental and theoretical assessment of the inhibiting action of Aspilia africana extract on corrosion aluminium alloy AA3003 in hydrochloric acid. *Journal of Materials Science*, 2012. 47(6): p. 2559–2572.

51. Rappe, A.K., C.J. Casewit, K.S. Colwell, W.A. Goddard, and W.M. Skiff, UFF, a full periodic table force field for molecular mechanics and molecular dynamics simulations. *Journal of the American Chemical Society*, 1992. 114(25): p. 10024–10035.

52. Casewit, C.J., K.S. Colwell, and A.K. Rappe, Application of a universal force field to organic molecules. *Journal of the American Chemical Society*, 1992. 114(25): p. 10035–10046.

53. Casewit, C.J., K.S. Colwell, and A.K. Rappe, Application of a universal force field to main group compounds. *Journal of the American Chemical Society*, 1992. 114(25): p. 10046–10053.

54. Akalezi, C.O., C.K. Enenebaku, C.E. Ogukwe, and, E.E. Oguzie, Corrosion inhibition of aluminium pigments in aqueous alkaline medium using plant extracts. *Environment and Pollution*, 2012. 1(2): p. 45–60.

55. Lifson, S., A.T. Hagler, and P. Dauber, Consistent force field studies of intermolecular forces in hydrogen-bonded crystals. 1. Carboxylic acids, amides, and the C:O.cntdot..cntdot.. cntdot.H- hydrogen bonds. *Journal of the American Chemical Society*, 1979. 101(18): p. 5111–5121.

56. Rimola, A., D. Costa, M. Sodupe, J.-F. Lambert, and P. Ugliengo, Silica surface features and their role in the adsorption of biomolecules: Computational modeling and experiments. Chemical Reviews, 2013. 113, 4216–4313.

57. Grimme, S., Semiempirical GGA-type density functional constructed with a long-range dispersion correction. *Journal of Computational Chemistry*, 2006. 27(15): p. 1787–1799.

58. Kresse, G. and J. Hafner, Ab-initio molecular-dynamics simulation of the liquid-metal amorphous-semiconductor transition in germanium. *Physical Review B*, 1994. 49(20): p. 14251–14269.

59. Garrain, P.A., D. Costa, and P. Marcus, Biomaterial–biomolecule interaction: DFT-D study of glycine adsorption on Cr_2O_3. *Journal of Physical Chemistry C*, 2011. 115(3): p. 719–727.

60. Chakarova-Kack, S.D., O. Borck, E. Schroder, and B.I. Lundqvist, Adsorption of phenol on graphite(0001) and alpha-Al_2O_3(0001): Nature of van der Waals bonds from first-principles calculations. *Physical Review B*, 2006. 74(15). 155402.

61. Chen, W., C. Tegenkamp, H. Pfnur, and T. Bredow, The interplay of van der Waals and weak chemical forces in the adsorption of salicylic acid on NaCl(001). *Physical Chemistry Chemical Physics*, 2009. 11(41): p. 9337–9340.

62. Chen, W., C. Tegenkamp, H. Pfnur, and T. Bredow, Insight from first-principles calculations into the interactions between hydroxybenzoic acids and alkali chloride surfaces. *Journal of Physical Chemistry C*, 2010. 114(1): p. 460–467.

63. Reckien, W., B. Kirchner, F. Janetzko, and T. Bredow, Theoretical investigation of formamide adsorption on Ag(111) surfaces. *Journal of Physical Chemistry C*, 2009. 113(24): p. 10541–10547.

64. Ruiz, V.G., W. Liu, E. Zojer, M. Scheffler, and A. Tkatchenko, Density–functional theory with screened van der Waals interactions for the modeling of hybrid inorganic–organic systems. *Physical Review Letters*, 2012. 108(14). 146103.

65. Tkatchenko, A., R.A. DiStasio, R. Car, and M. Scheffler, Accurate and efficient method for many-body van der Waals interactions. *Physical Review Letters*, 2012. 108(23). 236402.

66. Tkatchenko, A., L. Romaner, O.T. Hofmann, E. Zojer, C. Ambrosch-Draxl, and M. Scheffler, Van der Waals interactions between organic adsorbates and at organic/inorganic interfaces. *MRS Bulletin*, 2010. 35(6): p. 435–442.

67. Antony, J. and S. Grimme, Density functional theory including dispersion corrections for intermolecular interactions in a large benchmark set of biologically relevant molecules. *Physical Chemistry Chemical Physics*, 2006. 8(45): p. 5287–5293.

68. Di Valentin, C. and D. Costa, Anatase TiO_2 surface functionalization by alkylphosphonic acid: A DFT+D study. *Journal of Physical Chemistry C*, 2012. 116(4): p. 2819–2828.

69. Blajiev, O. and A. Hubin, Inhibition of copper corrosion in chloride solutions by amino-mercapto-thiadiazol and methyl-mercapto-thiadiazol: An impedance spectroscopy and a quantum-chemical investigation. *Electrochimica Acta*, 2004. 49(17–18): p. 2761–2770.

70. Peljhan, S. and A. Kokalj, DFT study of gas-phase adsorption of benzotriazole on Cu(111), Cu(100), Cu(110), and low coordinated defects thereon. *Physical Chemistry Chemical Physics*, 2011. 13(45): p. 20408–20417.

71. Kovacevic, N. and A. Kokalj, DFT study of interaction of azoles with Cu(111) and Al(111) surfaces: Role of azole nitrogen atoms and dipole–dipole interactions. *Journal of Physical Chemistry C*, 2011. 115(49): p. 24189–24197.

72. Jiang, Y. and J.B. Adams, First principle calculations of benzotriazole adsorption onto clean Cu(111). *Surface Science*, 2003. 529(3): p. 428–442.

73. Jiang, Y., J.B. Adams, and D.H. Sun, Benzotriazole adsorption on $Cu_2O(111)$ surfaces: A first-principles study. *Journal of Physical Chemistry B*, 2004. 108(34): p. 12851–12857.

74. Finsgar, M., A. Lesar, A. Kokalj, and I. Milosev, A comparative electrochemical and quantum chemical calculation study of BTAH and BTAOH as copper corrosion inhibitors in near neutral chloride solution. *Electrochimica Acta*, 2008. 53(28): p. 8287–8297.

75. Kokalj, A., S. Peljhan, M. Finsgar, and I. Milosev, What determines the inhibition effectiveness of ATA, BTAH, and BTAOH corrosion inhibitors on copper? *Journal of the American Chemical Society*, 2010. 132(46): p. 16657–16668.

76. Borck, O., P. Hyldgaard, and E. Schroder, Adsorption of methylamine on alpha-Al_2)O_3)(0001) and alpha-Cr_2)O_3)(0001): Density functional theory. *Physical Review B*, 2007. 75 (3). 035403.

77. Arrouvel, C., B. Diawara, D. Costa, and P. Marcus, DFT periodic study of the adsorption of glycine on the anhydrous and hydroxylated (0001) surfaces of alpha-alumina. *Journal of Physical Chemistry C*, 2007. 111(49): p. 18164–18173.

78. Motta A., M.-P. Gaigeot, Costa D., AIMD evidence of inner sphere adsorption of glycine on a stepped (101) boehmite AlOOH surface. *Journal of Physical Chemistry C*, 2012. 116: p. 23418–23427.

79. Oguzie, E.E., Y. Li, S.G. Wang, and F.H. Wang, Understanding corrosion inhibition mechanisms-experimental and theoretical approach. *RSC Advances*, 2011. 1(5): p. 866–873.

80. Islam, M.M., B. Diawara, P. Marcus, and D. Costa, Synergy between iono-covalent bonds and van der Waals interactions in SAMs formation: A first-principles study of adsorption of carboxylic acids on the Zn-ZnO(0001) surface. *Catalysis Today*, 2011. 177(1): p. 39–49.

81. Irrera, S., D. Costa, K. Ogle, and P. Marcus, Molecular modelling by DFT of 1,2-diaminoethane adsorbed on the Zn-terminated and O-terminated, anhydrous and hydroxylated ZnO (0001) surface. *Superlattices and Microstructures*, 2009. 46(1–2): p. 19–24.

82. Gao, Y.Y., N. Zhao, W. Wei, and Y.H. Sun, *Ab initio* DFT study of urea adsorption and decomposition on the ZnO (10(1)over-bar0) surface. *Computational and Theoretical Chemistry*, 2012. 992: p. 1–8.

83. Tang, W.D., Y.Y. Gao, W. Wei, and Y.H. Sun, Adsorption of urea onto a ZnO(10(1)over-bar0) surface. *Acta Physico-Chimica Sinica*, 2010. 26(5): p. 1373–1377.

84. Irrera, S., D. Costa, and P. Marcus, DFT periodic study of adsorption of glycine on the (0001) surface of zinc terminated ZnO. *Journal of Molecular Structure–Theochem*, 2009. 903(1–3): p. 49–58.

85. Kokalj, A., Electrostatic model for treating long-range lateral interactions between polar molecules adsorbed on metal surfaces. *Physical Review B*, 2011. 84(4). 045418.

86. Szocs, E., I. Bako, T. Kosztolanyi, I. Bertoti, and E. Kalman, EC-STM study of 5-mercapto-1-phenyl-tetrazole adsorption on Cu(111). *Electrochimica Acta*, 2004. 49(9–10): p. 1371–1378.

87. Cho, K., J. Kishimoto, T. Hashizume, H.W. Pickering, and T. Sakurai, Adsorption and film growth of BTA on clean and oxygen adsorbed Cu(110) surfaces. *Applied Surface Science*, 1995. 87–88 (1–4): p. 380–385.

88. Vogt, M.R., R.J. Nichols, O.M. Magnussen, and R.J. Behm, Benzotriazole adsorption and inhibition of Cu(100) corrosion in HCl: A combined in situ STM and in situ FTIR spectroscopy study. *Journal of Physical Chemistry B*, 1998. 102(30): p. 5859–5865.

89. Milosev, I. and T. Kosec, Electrochemical and spectroscopic study of benzotriazole films formed on copper, copper–zinc alloys and zinc in chloride solution. *Chemical and Biochemical Engineering Quarterly*, 2009. 23(1): p. 53–60.

90. Kosec, T., D.K. Merl, and I. Milosev, Impedance and XPS study of benzotriazole films formed on copper, copper–zinc alloys and zinc in chloride solution. *Corrosion Science*, 2008. 50(7): p. 1987–1997.

91. Finsgar, M., I. Milosev, and B. Pihlar, Inhibition of copper corrosion studied by electrochemical and EQCN techniques. *Acta Chimica Slovenica*, 2007. 54(3): p. 591–597.

92. Finsgar, M. and I. Milosev, Corrosion study of copper in the presence of benzotriazole and its hydroxy derivative. *Materials and Corrosion*, 2011. 62(10): p. 956–966.

93. Helal, N.H. and W.A. Badawy, Environmentally safe corrosion inhibition of Mg–Al–Zn alloy in chloride free neutral solutions by amino acids. *Electrochimica Acta*, 2011. 56(19): p. 6581–6587.

94. Chong, S.V. and H. Idriss, The reactions of carboxylic acids on $UO_2(111)$ single crystal surfaces. Effect of gas-phase acidity and surface defects. *Surface Science*, 2002. 504(1–3): p. 145–158.

95. Bouzoubaa, A., D. Costa, B. Diawara, N. Audiffren, and P. Marcus, Insight of DFT and atomistic thermodynamics on the adsorption and insertion of halides onto the hydroxylated NiO(111) surface. *Corrosion Science*, 2010. 52(8): p. 2643–2652.

96. Gaigeot, M.P., M. Sprik, and M. Sulpizi, Oxide/water interfaces: How the surface chemistry modifies interfacial water properties. *Journal of Physics—Condensed Matter*, 2012. 24(12). 124106.

97. Motta, A., M.P. Gaigeot, and D. Costa, *Ab initio* molecular dynamics study of the AlOOH boehmite/water interface: Role of steps in interfacial Grotthus proton transfers. *Journal of Physical Chemistry C*, 2012. 116(23): p. 12514–12524.

98. Chernyshova, I.V., S. Ponnurangam, and P. Somasundaran, Adsorption of fatty acids on iron (Hydr)oxides from aqueous solutions. *Langmuir*, 2011. 27(16): p. 10007–10018.

99. Mason, S.E., T.P. Trainor, and A.M. Chaka, Hybridization-reactivity relationship in Pb(II) adsorption on alpha-Al2O3-water interfaces: A DFT Study. *Journal of Physical Chemistry C*, 2011. 115 (10): p. 4008–4021.

100. Panagiotou, G.D., T. Petsi, K. Bourikas, C.S. Garoufalis, A. Tsevis, N. Spanos, C. Kordulis, and A. Lycourghiotis, Mapping the surface (hydr)oxo-groups of titanium oxide and its interface with an aqueous solution: The state of the art and a new approach. *Advances in Colloid and Interface Science*, 2008. 142(1–2): p. 20–42.

101. Zhang, Z., P. Fenter, S.D. Kelly, J.G. Catalano, A.V. Bandura, J.D. Kubicki, J.O. Sofo, D.J. Wesolowski, M.L. Machesky, N.C. Sturchio, and M.J. Bedzyk, Structure of hydrated Zn2+ at the rutile TiO2(110)-aqueous solution interface: Comparison of X-ray standing wave, X-ray absorption spectroscopy, and density functional theory results. *Geochimica Et Cosmochimica Acta*, 2006. 70(16): p. 4039–4056.

102. Leung, K., I.M.B. Nielsen, and L.J. Criscenti, Elucidating the bimodal acid–base behavior of the water–silica interface from first principles. *Journal of the American Chemical Society*, 2009. 131(51): p. 18358–18365.

103. Jonsson, E.O., K.S. Thygesen, J. Ulstrup, and K.W. Jacobsen, *Ab initio* calculations of the electronic properties of polypyridine transition metal complexes and their adsorption on metal surfaces in the presence of solvent and counterions. *Journal of Physical Chemistry B*, 2011. 115(30): p. 9410–9416.

104. Adeagbo, W.A., N.L. Doltsinis, K. Klevakina, and J. Renner, Transport processes at alpha–quartz–water interfaces: Insights from first-principles molecular dynamics simulations. *Chemphyschem*, 2008. 9(7): p. 994–1002.

105. Sulpizi, M., M.P. Gaigeot, and M. Sprik, The silica–water interface: How the silanols determine the surface acidity and modulate the water properties. *Journal of Chemical Theory and Computation*, 2012. 8(3): p. 1037–1047.

106. Rignanese, G.M., J.C. Charlier, and X. Gonze, First-principles molecular-dynamics investigation of the hydration mechanisms of the (0001) alpha-quartz surface. *Physical Chemistry Chemical Physics*, 2004. 6(8): p. 1920–1925.

107. Kummert, R. and W. Stumm, The surface complexation of organic acids on hydrous γ-Al$_2$O$_3$. *Journal of Colloid and Interface Science*, 1980. 75(2): p. 373–385.

108. Stumm, W., R. Kummert, and L. Sigg, A ligand exchange model for the adsorption of inorganic and organic ligands at hydrous oxide interfaces *Croatica Chemica Acta*, 1980. 53(2): p. 291–312.

109. Stumm, W., The Inner-Sphere Surface Complex—A Key to Understanding Surface Reactivity, in *Aquatic Chemistry: Interfacial and Interspecies Processes*, C.P. Huang, C.R. Omelia, and J.J. Morgan, Editors. 1995, American Chemical Society: Washington, DC. p. 1–32.

110. Panchenko, A., M.T.M. Koper, T.E. Shubina, S.J. Mitchell, and E. Roduner, *Ab initio* calculations of intermediates of oxygen reduction on low-index platinum surfaces. *Journal of The Electrochemical Society*, 2004. 151(12): p. A2016–A2027.

111. Hyman, M.P. and J.W. Medlin, Theoretical study of the adsorption and dissociation of oxygen on Pt(111) in the presence of homogeneous electric fields. *Journal of Physical Chemistry B*, 2005. 109(13): p. 6304–6310.

112. Skulason, E., G.S. Karlberg, J. Rossmeisl, T. Bligaard, J. Greeley, H. Jonsson, and J.K. Nørskov, Density functional theory calculations for the hydrogen evolution reaction in an electrochemical double layer on the Pt(111) electrode. *Physical Chemistry Chemical Physics*, 2007. 9(25): p. 3241–3250.

113. Rossmeisl, J., E. Skulason, M.E. Bjorketun, V. Tripkovic, and J.K. Nørskov, Modeling the electrified solid–liquid interface. *Chemical Physics Letters*, 2008. 466(1–3): p. 68–71.

114. Nørskov, J.K., J. Rossmeisl, A. Logadottir, L. Lindqvist, J.R. Kitchin, T. Bligaard, and H. Jonsson, Origin of the overpotential for oxygen reduction at a fuel-cell cathode. *Journal of Physical Chemistry B*, 2004. 108(46): p. 17886–17892.

115. Anderson, A.B., J. Uddin, and R. Jinnouchi, Solvation and zero-point-energy effects on OH(ads) reduction on Pt(111) electrodes. *Journal of Physical Chemistry C*, 2010. 114(35): p. 14946–14952.

116. Jinnouchi, R. and A.B. Anderson, Aqueous and surface redox potentials from self-consistently determined Gibbs energies. *Journal of Physical Chemistry C*, 2008. 112(24): p. 8747–8750.

117. Jinnouchi, R. and A.B. Anderson, Electronic structure calculations of liquid–solid interfaces: Combination of density functional theory and modified Poisson–Boltzmann theory. *Physical Review B*, 2008. 77(24).

118. Tian, F., R. Jinnouchi, and A.B. Anderson, How potentials of zero charge and potentials for water oxidation to OH(ads) on Pt(111) electrodes vary with coverage. *Journal of Physical Chemistry C*, 2009. 113(40): p. 17484–17492.

119. Anderson, A.B. and D.B. Kang, Quantum chemical approach to redox reactions including potential dependence: Application to a model for hydrogen evolution from diamond. *Journal of Physical Chemistry A*, 1998. 102(29): p. 5993–5996.

120. Yeh, K.-Y., S.A. Wasileski, and M.J. Janik, Electronic structure models of oxygen adsorption at the solvated, electrified Pt(111) interface. *Physical Chemistry Chemical Physics*, 2009. 11(43): p. 10108–10117.

121. Filhol, J.S. and M. Neurock, Elucidation of the electrochemical activation of water over Pd by first principles. *Angewandte Chemie-International Edition*, 2006. 45(3): p. 402–406.

122. Janik, M.J., C.D. Taylor, and M. Neurock, First-principles analysis of the initial electroreduction steps of oxygen over Pt(111). *Journal of The Electrochemical Society*, 2009. 156(1): p. B126–B135.

123. Rossmeisl, J., J.K. Norskov, C.D. Taylor, M.J. Janik, and M. Neurock, Calculated phase diagrams for the electrochemical oxidation and reduction of water over Pt(111). *Journal of Physical Chemistry B*, 2006. 110(43): p. 21833–21839.

124. Taylor, C.D., R.G. Kelly, and M. Neurock, Theoretical analysis of the nature of hydrogen at the electrochemical interface between water and a Ni(111) single-crystal electrode. *Journal of The Electrochemical Society*, 2007. 154(3): p. F55–F64.

125. Taylor, C.D., The transition from metal–metal bonding to metal–solvent interactions during a dissolution event as assessed from electronic structure. *Chemical Physics Letters*, 2009. 469 (1–3): p. 99–103.

126. Taylor, C.D., M.J. Janik, M. Neurock, and R.G. Kelly, *Ab initio* simulations of the electrochemical activation of water. *Molecular Simulation*, 2007. 33(4–5): p. 429–436.

127. Taylor, C.D., R.G. Kelly, and M. Neurock, A first-principles analysis of the chemisorption of hydroxide on copper under electrochemical conditions: A probe of the electronic interactions that control chemisorption at the electrochemical interface. *Journal of Electroanalytical Chemistry*, 2007. 607(1–2): p. 167–174.

128. Taylor, C.D., R.G. Kelly, and M. Neurock, First-principles prediction of equilibrium Potentials for water activation by a series of metals. *Journal of The Electrochemical Society*, 2007. 154(12): p. F217–F221.

129. Taylor, C.D., M. Neurock, and J.R. Scully, First-principles investigation of the fundamental corrosion properties of a model Cu(38) nanoparticle and the (111), (113) surfaces. *Journal of The Electrochemical Society*, 2008. 155(8): p. C407–C414.

130. Taylor, C.D., M. Neurock, and J.R. Scully, A first-principles model for hydrogen uptake promoted by sulfur on Ni(111). *Journal of The Electrochemical Society*, 2011. 158(3): p. F36–F44.

131. Savizi, I.S.P. and M.J. Janik, Acetate and phosphate anion adsorption linear sweep voltammograms simulated using density functional theory. *Electrochimica Acta*, 2011. 56(11): p. 3996–4006.

6

THERMODYNAMICS OF PASSIVE FILM FORMATION FROM FIRST PRINCIPLES

MICHAEL F. FRANCIS[1,2] AND EDWARD F. HOLBY[3]

[1]*École Polytechnique Fédérale de Lausanne (EPFL), STI IGM LAMMM, Lausanne, Switzerland*
[2]*Brown University, School of Engineering, Providence, Rhode Island, USA*
[3]*Materials Science and Technology Division, MST-6, Los Alamos National Laboratory, Los Alamos, NM, USA*

6.1 INTRODUCTION

The oxide film is often the first line of defense a material has against the corrosive influences of its environment. A passive oxide film forms upon oxygen adsorption from the environment. These films can be beneficial, serving to slow or block metal dissolution and further corrosion. This effect is called oxide passivation. On the other hand, metals may continue to oxidize/corrode, even after an oxide film has formed or this film may break down leading to active corrosion. Because of these drastically different outcomes, an understanding of how these films are formed under different environmental conditions is needed to understand how to mitigate corrosion. The focus of this chapter is the use of quantum mechanics simulations to understand the thermodynamics of passive film formation. In particular, we look at calculations of first-principles phase diagrams as a function of environmental conditions and how these studies fit with experimental data. We focus on Pd, Mg, and Pt metal systems as they are well studied, represent different types of oxide film formation, and fall within the authors' areas of expertise.

The widely accepted processes through which oxides form as stated in the literature are [1]:

1. Dissociative adsorption of oxygen on metal surfaces
2. Formation of chemisorbed oxygen surface phases

Molecular Modeling of Corrosion Processes: Scientific Development and Engineering Applications, First Edition.
Edited by Christopher D. Taylor and Philippe Marcus.
© 2015 John Wiley & Sons, Inc. Published 2015 by John Wiley & Sons, Inc.

3. Formation of a thin oxide film typically from one to several monolayers thick and possibly distinct from bulk oxides

4. Growth of the oxide film toward bulk.

Oxide film formation is a function of both system thermodynamics (governed by environmental conditions) and system kinetics. While kinetics can play a key role, especially in step 4, we focus on the thermodynamic processes, which gives a metric of relative stability, indicating if and when certain phases are stable under varied environmental conditions. Such a thermodynamic focus can provide a great deal of information regarding the likely oxide formation pathways (including oxide film and intermediary structures). A thermodynamic study will not be able to determine the presence of metastable phases (kinetically stabilized phases) so the combination of the thermodynamic understanding with kinetics may be needed in order to fully understand a given oxide system.

The field of first-principles calculations has matured over the past two decades to the point where these simulations are invaluable in guiding and interpreting experiments. While we will here focus on the thermodynamic quantities obtained using firstprinciples in order to study phase stability, a whole host of atomic and electronic structural properties can now be predicted with acceptable degrees of accuracy. Our goal in this chapter is to familiarize the reader with the basics of first-principles calculations (in particular density functional theory, abbreviated as DFT, and thermodynamic models that incorporate it) and show how these techniques have been applied to oxide systems in the literature. This chapter is outlined as follows:

- Background on oxide/experimental comparison (structural prediction for motivation)
- Outline of quantum mechanics methodology
- Outline of thermodynamics methodology
  - ○ Mg surface phase diagram examples
  - ○ Pt surface phase diagram examples
- The future of first-principles oxide modeling

6.2   BACKGROUND ON OXIDE FORMATION

Oxide formation is a tremendously large field of study that is typically broken down by the particular metal of interest. There are several general theories for how oxides are formed including the Tamman–Pilling–Bedworth parabolic growth law [2–4], Wagner theory [5, 6], Cabrera–Mott theory [7], and Fehlner–Mott theory [8]. These models are predicated on the existence of an oxide/metal interface already preexisting and so do not cover the initial stages of oxidation during which this interface is created. There seem to be a wide variety of metal dependent behaviors in this approximately 1 monolayer (ML, 1 ML corresponding in our nomenclature as one oxygen per surface metal atom) regime [1, 4, 9]. Despite decades of research, there still remains much that is unknown about oxide formation, particularly in this approximately 1 ML coverage region. First-principles modeling provides an immensely useful way of studying these systems at the atomic scale. To illustrate this point, we next discuss how firstprinciples have been used to study surface phases of the O^*/Pd system.

6.3 COMPARISON WITH EXPERIMENT

Comparison of first-principles structural predictions with experiment serves to benchmark both methods as well as to link atomic and electronic structural properties with experimental observables. Several experimental techniques that have been developed in order to study surfaces which have been applied to oxide growth. We here consider examples of direct comparison of experiments and first-principles structure calculations of Pd oxide surfaces in order to motivate the reader and suggest other ways in which these simulations may be useful to those doing experiments beyond the phase diagram generation approaches discussed later.

Surface X-ray diffraction (SXRD) is a powerful technique for looking at the atomic structure of oxide surfaces but is often complicated by a lack of fitting models for the diffraction patterns. Bulk oxide models are an obvious choice for fitting, but surface oxides do not often fit the bulk structures (in addition to this example, see the following example for $O^*/Pt(111)$ 1 ML structures that vary from bulk PtO_x orderings as well). In many cases, the surface

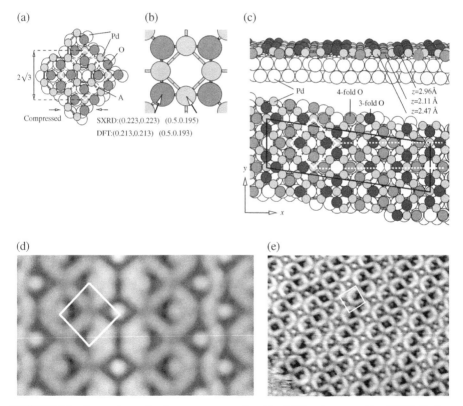

FIGURE 6.1 Comparison of surface structure predicted by DFT and measured by SXRD. Surface structure model from DFT used in fitting SXRD data (a, b, and c) with small light circles representing Pd and larger circles representing O (with gray scale showing coordination environment). (d) STM prediction from DFT model in (a, b, and c) based on the Tersoff–Hamann approximation [11], and experimentally observed STM image (e) from reference [10]. This work is an example of the early success of DFT in the field of oxide formation and how DFT can be used to complement and augment experimental studies. Reproduced with permission from Lundgren et al. [10]. © American Physical Society.

oxide phase is complicated involving surface reconstruction and large unit cells (long-range patterning on the surface). DFT can help provide the models needed in order to interpret SXRD data. One example of this comes from the work of Lundgren et al. [10]. In this work, the authors provide a DFT model for a two-dimensional (2D) oxide phase on the Pd(111) surface that is congruent with SXRD as well as scanning tunneling microscopy (STM) results (Fig. 6.1). Their findings show that 2D oxides that are completely independent of bulk oxides can be stable on metal surfaces. In the case of Pd(111), a Pd_5O_4 structure with a complex long-range ordering is found to be stable (shown in Fig. 6.1). There is a remarkable agreement between the experimentally observed SXRD and STM data and those predicted using the Pd_5O_4 structure model.

Another example is the direct comparison of DFT predicted STM Pd oxide images with STM experiments; the success of this investigation leads to further collaborative computational and experimental work leading to a detailed understanding of the structure of the Pd_xO_y surface across varied concentrations (as shown in Fig. 6.2). The striking correspondence of predicted structures and STM micrographs at an array of oxygen coverages demonstrates how DFT can complement experiments giving rise to an integrated theory/experiment research approach. It also suggests that DFT is a powerful tool for determination of phase diagrams, that is, what atomic structures of metal and oxygen will be stable as a

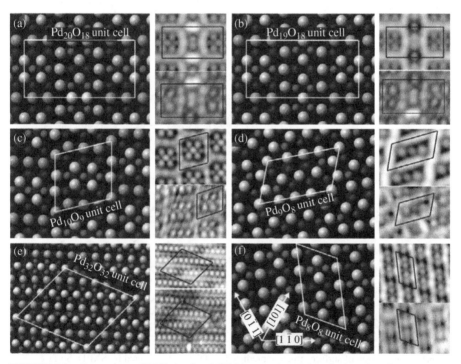

FIGURE 6.2   Comparison of DFT simulated STM images (top right) of selected surface structures (left) and comparison with measured STM micrographs (bottom right) taken from references [12, 13]. (a) $Pd_{20}O_{18}$, (b) $Pd_{19}O_{18}$, (c) $Pd_{10}O_9$, (d) Pd_9O_8, (e) $Pd_{32}O_{32}$, and (f) Pd_8O_8. Oxygen atoms are shown as smaller and darker red balls. The buckling of the surface layer in the structure models is indicated by the atom brightness. Reproduced with permission from Klikovits et al. [13]. © American Physical Society.

function of environment. With this motivation showing the power of DFT to both corroborate and aid in the interpretation of experimental data, we move on to the methodology employed when considering a variety of possible oxide surface phases.

6.4 METHODOLOGY FOR STUDYING OXIDE FILM FORMATION FROM FIRST PRINCIPLES

Before looking at further examples of how first-principles theoretical techniques are used in studying passive film formation, a general understanding of the methodologies behind these techniques and the thermodynamic models that they are fed into is required. There exist many levels of quantum mechanical approximation not limited to but including Hartree–Fock (HF) methods, configurational interaction (CI), coupled cluster (CC), Møller–Plesset perturbation theory (MP2, MP3, and MP4), and DFT. At each layer of calculation, there is a trade-off between computational cost and physical accuracy. Today, DFT calculations are the main quantum mechanical workhorse of the computational materials science community, and we will limit ourselves to describing the underpinnings of this method.

6.4.1 Quantum Mechanics Methodology

This section is meant as a primer for those not familiar with first-principles calculations and, more directly, DFT. It is not intended to be exhaustive but instead to provide a sense of how quantum mechanics is applied to oxide systems using computational methods. For more complete treatments of first-principles methods, please see references [14–16]. While sometimes perceived as a black box where atomic structures go in and system properties come out, a more detailed understanding of the advantages and limitations of DFT will help those working with computational methods and those working alongside computation professionals.

Quantum mechanics is necessary to identify locally stable ground states of a given molecular or solid-state system. These states are a function of atomic and electronic structure (organization of nuclei and electrons). The locally stable ground-state energy can be calculated and used with a thermodynamic model (next section) to determine that state's phase stability in a given environment. This is done through the solution of the time-independent Schrödinger equation:

$$\hat{H}\Psi = E\Psi \tag{6.1}$$

where $\hat{H}$ is the Hamiltonian operator of the system, Ψ is the quantum mechanical system wave function, and E is the internal energy of the system with wave function Ψ. The solution to the Schrödinger equation for realistic systems, however, can be incredibly difficult due to the many-body nature of these systems (many interacting nuclei and electrons). For these systems, a number of assumptions/approximations are often made in order to aid in the tractability of a solution.

A many-body quantum mechanical Hamiltonian operator consists of a sum of kinetic and potential energy terms with contributions from nuclei and electrons. The first approximation we will make that is typically made in the literature for this general Hamiltonian is the Born–Oppenheimer approximation in which electrons are considered to respond instantaneously to any motion of the nuclei and electrons are considered to be in their ground-state configuration [17]. This approximation is often justified by the large difference in mass between the nuclei and the electrons and the frequent dominance of ground-state effects in determining structures and properties. As a result of this approximation, if the kinetic energy of the nuclei is ignored, and one writes the various operators as $\hat{T}$ (the operators for electronic kinetic energy), $\hat{U}$ (electron–electron interaction), and $\hat{V}$ (electron–nuclei interaction), the Schrödinger equation can be written as

$$\hat{H}\Psi = \left[\hat{T} + \hat{U} + \hat{V}\right]\Psi = E\Psi \tag{6.2}$$

where Ψ now represents the stationary electronic quantum mechanical wave function describing the system. As a consequence of this approximation, the energies obtained will be valid for 0 K, but some thermodynamic model is needed to introduce thermal effects at real temperatures.

One key issue for solving this many-electron equation is the coupled nature of the $\hat{U}$ electronic potential operator. The energy of N-interacting (correlated) electrons in an external field cannot be broken into only the sum of N-noninteracting electrons in a simple and exact approach as of yet. Coulombic charge interaction, Pauli exclusion, and exchange interaction all contribute to the interactions of the coupled many-electron system. A solution of this many-body interaction problem necessitates approximations that are the crux of why there are so many differing approaches and approximations in solving Equation (6.2) in quantum chemistry as listed earlier. The validity of various approximations is often system specific. As such, interpretation of calculations requires an understanding of when and why various approximations can be used.

It was shown by Hohenberg and Kohn [18] that the ground-state wave function of an N-body electronic system is a unique function of a three-dimensional electron density, $\rho(\vec{r})$. This density is given by the electron probability (wave function squared) at a given spatial point, $\vec{r}$. As a consequence, the ground-state energy (and any quantum mechanical observable) can be written as a functional of this electron density instead of the wave function. A functional is similar to a mathematical function except that instead of a single numerical input value leading to a single numerical output value, a mathematical function is taken as input and a single numerical value is output. One example of a functional is an integral between two points. A function is input and a value, in this case the integral, is output. In the case of DFT, the system ground-state energy value is a functional of the electron density function. In this way, the problem of calculating the ground-state energy is reduced to finding the electron density that minimizes the energy functional. In doing so, we have moved from a 3N-dimensional problem to a three-dimensional one greatly simplifying the problem. Kohn and Sham [19] cast the functional problem such that, following Equation (6.2), the energy functional is written as

$$E\left[\rho(\vec{r})\right] = TE\left[\rho(\vec{r})\right] + UE\left[\rho(\vec{r})\right] + VE\left[\rho(\vec{r})\right] \tag{6.3}$$

Here, the wave function operators have been replaced with functionals of the electron density (as allowed by the Hohenberg–Kohn theorems). The third term (external potential term) is a simple integral over all space for the electron density of a given position interacting with the external potential at that same position. This is, again, the term due to electron–nuclei interaction. Thus, we are left with calculating the first and second terms of Equation (6.3) and, armed with such knowledge, can attempt to minimize the energy of Equation (6.3) by finding the ground-state electron density.

We here present a simplified view of how these functional terms are further divided. The overall approach used in DFT is to separate the terms that are known and/or easy to calculate and isolate those terms that are either unknown or not easy to calculate. The hope is that the known/easy-to-calculate terms dominate the system energy and that for the unknown/hard-to-calculate terms, a suitable approximation can be made that captures these energy contributions in a way that is computationally tractable. We first consider the kinetic energy term, $T[\rho(\vec{r})]$:

$$T[\rho(\vec{r})] = T_{NI}[\rho(\vec{r})] + T_C[\rho(\vec{r})] \tag{6.4}$$

Electron kinetic energy is broken down into two parts as shown in Equation (6.4). The first part is the kinetic energy of N-noninteracting single electrons, $T_{NI}[\rho(\vec{r})]$, which is nothing more than the sum of the kinetic energies of the individual electrons. This is a well-understood functional and not difficult to calculate based on the Laplacian of the electron density. The second term, $T_C[\rho(\vec{r})]$, accounts for the electron kinetic energy due to electron correlation. Two electrons with antiparallel spins interact and so do not move independent of one another, leading to correlated motion and a term in addition to the noninteracting kinetic energy. This second term is more complicated as no general solution is known and so will be put into the list of terms that need approximating:

$$U[\rho(\vec{r})] = U_H[\rho(\vec{r})] + U_X[\rho(\vec{r})] + U_C[\rho(\vec{r})] \tag{6.5}$$

The electron–electron potential energy is broken down into three parts. The first term, $U_H[\rho(\vec{r})]$, is called the Hartree potential and is comprised of the classical electrostatic interaction of electron pairs. This is one of the known terms that is not difficult to calculate (though it can be fairly time consuming). The next term, $U_X[\rho(\vec{r})]$, comes about due to the quantum mechanical effect when considering exchanging two electrons in a multiple-electron system. The inability of electrons with parallel spins to occupy the same local state leads to an effective repulsion potential between these electrons. This is called the exchange potential. This potential is put into our approximation list. The third potential term, $U_C[\rho(\vec{r})]$, comes about due to electron correlation. The sum of electrons interacting quantum mechanically with a mean-field charge distribution is not the same energetically as the many-body electron system being considered due to electronic correlation. This difference in potentials is the correlation potential and is put into the list of approximated terms.

Combining the aforementioned terms that are to be approximated due to electron exchange and correlation effects $(T_C[\rho(\vec{r})], U_C[\rho(\vec{r})],$ and $U_X[\rho(\vec{r})])$ into a single functional, $E_{XC}[\rho(\vec{r})]$, leads to the energy functional as written in Equation (6.6):

$$E[\rho(\vec{r})] = T_{NI}[\rho(\vec{r})] + U_H[\rho(\vec{r})] + \int V\rho(\vec{r})dr + E_{XC}[\rho(\vec{r})] \tag{6.6}$$

Much effort has been put into deriving accurate and computationally inexpensive exchange–correlation (XC) functional (discussed in the following), but it is still the major source of error in DFT implementations [20]. It is also an important part of why certain systems, especially those with strongly correlated electrons such as in actinides [21], are more difficult to study using DFT.

While the Hohenberg–Kohn theorems dictate that a given electron density describes the quantum mechanical ground state, Kohn and Sham introduced a pragmatic methodology for finding that density [22], thus solving Equation (6.6). Their approach includes the introducing of N-noninteracting single-electron wave functions called Kohn–Sham (KS) orbitals, $\phi(r)$. These orbitals have the property that their sum of the squares of the KS orbitals gives the electron density function:

$$\sum_i \left| \phi(\vec{r}) \right|^2 = \rho(\vec{r}) \tag{6.7}$$

Aside from this property of being able to produce the electron density, these functionals do not correspond to real electronic orbitals and so can be modeled with convenient function basis sets, which adds to the computational tractability of this approach. Equation (6.6) can be recast with the functionals grouped and used as operators on the KS orbitals:

$$V_{KS}\left[\rho(\vec{r})\right] = U_H\left[\rho(\vec{r})\right] + E_{XC}\left[\rho(\vec{r})\right] \tag{6.8}$$

$$\left[T_{NI}\left[\rho(\vec{r})\right] + \int V\left[\rho(\vec{r})\right]dr + V_{KS}\left[\rho(\vec{r})\right] \right]\phi_i(\vec{r}) = \varepsilon_i\phi_i(\vec{r}) \tag{6.9}$$

Here, the ε_i value is the orbital's energy eigenvalue. Equation (6.9) is remarkably similar to the original Schrödinger equation, Equation (6.2), but the wave functions have been replaced with the KS orbitals and the exchange and correlation terms have been isolated. Thus, we have replaced the N-body coupled electronic wave function with a collection of uncorrelated wave functions while at the same time defining precisely what the uncertain many-body terms in need of approximation are.

Because Equations (6.7) and (6.9) are coupled, the solution of the KS orbitals must be found in a self-consistent manner (implemented in the self-consistent field or SCF portions of DFT codes). An initial, sometimes random, guess of an electron density dictates the operators on the left side of Equation (6.9). Then, KS orbitals that satisfy Equation (6.9) are solved for. Using these KS orbitals, Equation (6.7) dictates a new electron density. If this density is different than the original density, this new density can be fed back into Equation (6.9) giving new operators, thus producing new KS orbitals that define a (possibly) new density. This loop is iterated until the density no longer varies to within some tolerance. The minimized energy eigenvalue is the locally stable system energy. Typically, this SCF is performed between steps that optimize nuclei positions and is used to locally optimize the atomic as well as electronic structure of the system.

6.4.2 Physical Accuracy of DFT and XC Functionals

The take-away from the aforementioned description of DFT is that certain aspects of the system energy are easy to calculate while others are more difficult (either being computationally expensive or including many-body solutions that remain unknown for all but the most simple of cases). These difficult terms are lumped together in the XC functional. XC functional is very important in most physical systems that one would wish to study. While it may not contribute significantly to the system energy, the terms included in this potential are often what lead to appropriate atomic cohesion that is important for both molecular and solid-state studies. A good review of XC functionals is given by Perdew and Schmidt [20], but we will here discuss XC functions typically used for oxide surface formation studies: the local density approximation (LDA), the generalized gradient approximation (GGA), Hubbard corrections (+U), and hybrid functionals. This review of XC functionals is not meant to be exhaustive but to give a rough working knowledge of these functionals (Fig. 6.3).

Before considering the functionals, it is important to understand what is meant by accuracy as pertaining to DFT. As our main goal is the ability to model physical systems to obtain system parameters without the need to measure them experimentally, then a comparison of how well DFT reproduces such parameters is needed (a measure called *physical accuracy*). Like any computational technique, DFT has its strengths and weaknesses. While physical accuracy can vary quite widely with different XC functionals, numerical accuracy (dependent on convergence criteria and basis sets utilized) can also have a great effect on predicted parameters. Such parameters include but are not limited to relative phase stability, system energy and geometry, vibrational frequencies, and cohesive and binding energies. If we assume that the numerical accuracy is sufficient, then some generalities regarding the physical accuracy of DFT can be drawn. DFT is particularly good at producing qualitative comparisons

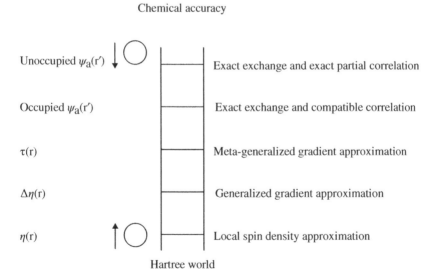

FIGURE 6.3   Perdew's "Jacob's Ladder" of DFT functional types (from reference [20]) with LDA and GGA making up the bottom rungs, stepping out of the "Hartree World" in which electrons are not correlated. The meta-GGA rung includes second derivatives of the electron density. Hybrid functionals fall into the exact exchange and compatible correlation rung. The exact exchange and exact partial correlations are not discussed here but include such methods as the random phase approximation (RPA) with suitable short-range corrections. Reproduced with permission from Perdew and Schmidt [20]. © American Institute of Physics.

between systems. An example of this is the relative stability of different phases. From a quantitative point of view, DFT particular excels (generally speaking) at structural properties such as lattice constant where even with LDA and GGA XC functionals are usually within a few percent of experimental data (but, as always, there are exceptions to this rule, particularly when dealing with strongly correlated electron systems). Vibrational DFT data versus experimental data have also been calculated for a wide range of systems and functionals [23] showing that DFT typically overestimates vibrational frequencies by 1–5% (which, in most cases, is sufficiently accurate for vibrational peak assignments). A more detailed discussion of physical accuracy is given in reference [14].

It is worth highlighting a few known issues with DFT that may affect those studying oxide formation. The first is the so-called CO/Pt(111) puzzle associated with CO adsorption on Pt(111) surfaces at low coverages, as discussed in reference [24]. In this system, DFT with an LDA or GGA functional predicts the wrong binding site for CO on the Pt(111) surface (three-fold site predicted versus experimentally observed top site). It has been determined that this failure stems from the inability of LDA/GGA to accurately predict HOMO/LUMO band gaps, a parameter that impacts CO binding on this surface [25, 26]. This is a general problem with lower functionals (LDA and GGA) in that they do not accurately predict excited states that are needed to predict band gaps and associated materials behavior (such as metallic, semiconductor, and insulator behavior). Hybrid functionals that include the exact exchange contribution, however, are more accurate in terms of band gap prediction and accurately predict the correct CO binding site from first principles [27]. The take-home message of this is that if you believe the band gap, or more generally any excited state behavior, is relevant for your system, using hybrid XC functionals, despite their increased computational expense, may be required.

Other examples of systems that general DFT often has issues with are those that have significant contributions from dispersion forces (van der Waals) and strongly correlated electrons. In the former case, several approximations for energy shifts and XC functional modification due to these forces have been suggested [28–31]. In the latter case, higher-level XC functionals (such as GGA+U and hybrid functionals) have been shown to more accurately predict system parameters such as stable structures and magnetic ordering [21, 32, 33].

One important note about different XC functionals is that more sophisticated approximations, which generally lead to more accurate results (but may in fact not), come at the cost of increased computational expense. As such, it is important to be realistic about the level of physical accuracy required for the parameter of interest and to pick an XC functional that fits your needs (and deadlines)! We will now discuss the four types of functionals typically used for studying oxide formation.

The simplest XC functional typically employed is LDA (or LSDA if spin polarization is also included). This XC functional uses only the local electronic density to calculate the energy contributions from exchange and correlation, often using a homogeneous electron gas as a starting point. This is due to the homogeneous electron gas having known exchange energies and, for limiting cases, known correlation energies as well. Advantages of this functional include relatively inexpensive computation (for DFT) and high accuracy for slowly varying densities (due to its derivation from the homogeneous electron gas). It does quite well even with systems of rapidly varying electron densities as well due to its derivation from a well-behaved/well-understood quantum mechanical system. It is usually a better choice for periodic systems (solid-state calculations) than for molecular systems, however, due to rapid electronic density variations in molecules.

The GGA builds on the LDA by not only incorporating the local density but also gradients in the local density. In GGA XC functionals, the gradient terms must be properly added so as not to overcorrect for the varying electronic density. Such an implementation has been made by Perdew, Burke, and Ernzerhof (GGA–PBE) in such a way that the formal properties of LDA are captured (accurate homogeneous electron gas) but density fluctuations are captured more accurately. This is done in a nonempirical fashion, and so, no a priori experimental data is needed for parameterization. GGA is, in practice, not much more computationally intensive than LDA and is often the choice for the large surface models utilized when studying oxide formation thermodynamics. Parameterizations for molecules, solids, or general systems are available in many software packages. While GGA is often used in the oxide formation literature, it is still far from exact, especially in the exchange terms. Electronic band gaps and spin states are often predicted incorrectly, and so, phenomena related to these properties often require a higher level of XC functional or a varied implementation of LDA/GGA.

One such varied implementation that has recently gained in popularity (especially for transition metals) is the use of LDA and GGA with the addition of a "Hubbard U correction" [34, 35], often denoted with a "+U." In this XC functional, the base LDA or GGA functional is used with the addition of a mean-field term for the Coulomb energy that aids in correcting improper self-interactions. This additional term can lead to more accurate spin states and band gaps and computationally isn't much more expensive than the underlying XC functional. One major drawback, however, is that there is no fundamental parameterization, so terms are often treated as adjustable parameters. This drastically hurts the predictive ability of the DFT model and means that some external experimental values must be included to help with parameterization. Fully self-consistent parameterizations of the Hubbard correction terms are a field of much current research and if shown to give accurate results with no a priori fitting needed would be incredibly useful for the study of metal oxides and oxide film formation.

The last group of XC functionals that we will discuss is called "hybrid" functionals. They get their name from partial inclusion of the exact HF exchange potential with the LDA or GGA functionals. These functionals include B3LYP, PBE0, and HSE03/HSE06. These represent some of the most accurate and sophisticated XC functionals that are practical in many atom systems and are most used when accurate band gap behavior is needed. This accuracy comes at a cost, however, with even the best implementations taking 10–1000X longer than the same structure run with LDA/GGA. The scaling to more electron systems is also an issue, and even with state-of-the-art computer clusters, larger systems are rather time consuming as hybrids do not scale as well as LDA/GGA methods due to the HF exchange calculation that must occur in parallel with the LDA/GGA XC portion. As computer power increases and algorithmic efficiency improves, however, it seems likely that many future studies of oxide formation, especially involving transition metals and actinides, will utilize these hybrid functionals. Recent advances in general-purpose graphics processing units (GPGPUs) have already been utilized to speed up hybrid implementations by up to 20X [36]. Such advances will enable the more ubiquitous use of these hybrid functionals on a more diverse set of larger systems, which will in turn lead to more accurate hybrid functionals in the future.

The exact fashion in which the KS implementation of DFT is solved is beyond the scope of this chapter and is dependent on the code used. Generally speaking, however, some basis set (a collection of functions meant to span the relevant KS orbitals) is utilized to describe the KS orbitals. These basis sets are one of the largest differences between modern quantum chemistry codes that use KS-DFT, and this choice can drastically affect accuracy and speed of computation for different molecular and solid-state systems.

Armed with the knowledge of DFT and XC functionals, we turn our attention to the practical matter of how DFT can be used to study oxide surfaces. We will discuss the general thermodynamic approaches used next. After that, we will present literature studies of O/Mg and O/Pt surfaces, showing some specific applications of these methods.

6.4.3 Thermodynamics and First-Principles Phase Diagrams

The importance of phase diagrams to materials scientists, physicists, and chemists is difficult to overemphasize as the physical, mechanical, and chemical properties of a material depend upon its underlying structure and composition. A phase diagram is a description of the equilibrium structure and composition of a material as a function of selected (often) exogenous variables such as pressure, temperature, voltage, and so on.

As a result of its importance, extensive research has been put into understanding the fundamental relationships between materials systems and their phase diagrams. Among the research approaches has been the derivation of phase diagrams from free energy diagrams [37–40]. A free energy diagram is nothing more than a representation of the Gibbs free energy as a function of the desired variable (pressure, temperature, voltage, etc.). These free energies are not always measured but sometimes calculated based on some underlying physical model. In general, it is always prudent to question the mathematical forms that free energy diagrams take; however, assuming that one has a reasonable approximation of free energy, the stable phase(s) will be those that ultimately minimize the free energy for a given composition. An example free energy diagram is shown in Figure 6.4; the unprimed free energies labeled G_0, G_2, G_4, G_6, and G_8 represent the free energies of pure compounds and the x-axis represents total concentration of some impurity. The emboldened line is the convex hull (comprised of tie lines between phases on the convex hull) and represents the free energy–concentration path the system will take. As some underlying variable is changed (e.g., concentration), the system will follow a path controlled by the double criterion of (1) minimizing the free energy for (2) a given equilibrium. The first criterion means minimizing the free energy, while the second criterion means that at every location in a mixture atoms of the same element must have the same chemical potential, regardless of the phase:

$$\mu = \frac{\partial g}{\partial x} \tag{6.10}$$

This makes clear that the two problems central to determining the phase diagram of a material are the determination of the convex hull [41–45] and the construction of the free energy diagram.

The same convex hull and free energy diagram methods that have been used in materials science and solid-state physics can and have been successfully used to understand surface phase stability. We do not here address the convex hull methods but rather the methods used to construct the free energy diagram sometimes referred to as first-principles thermodynamics (FPT), which has had successes not limited to but including the phase diagram of oxygen adsorbed to Ru, Pd, Pt, and Ag metal surfaces [46–51]. The goal of this section is to connect free energy expressions to physical parameters and to show how one can use DFT to parameterize these expressions and subsequently perform a phase stability analysis.

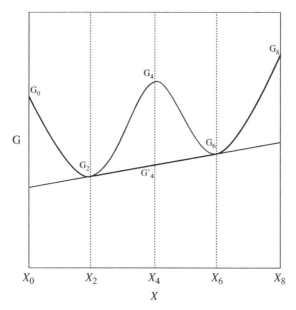

FIGURE 6.4   Example free energy diagram of free energy (G) versus composition (X). G_0 and G_8 represent endpoints. G_2 and G_6 stable phases. The emboldened line connecting G_0–G_2–G'_4–G_6–G_8 is the convex hull describing the stable phases as a function of average composition. Reproduced with permission from Lee et al. [41]. © Springer.

The formation of a surface oxide from bulk may begin with first the cleavage of the metal, followed by the adsorption of O_2 or another oxygen-bearing molecule, such as water, and subsequent restructuring (Fig. 6.5). To obtain an estimate of the surface formation energy, the free energy of an atom in the bulk phase, $F_{bulk}(T)$, may be compared to the free energy of a surface slab, $F_{slab}(N_{slab}, T)$.

Periodic DFT methods are well suited to the calculation of bulk and surface properties. The periodic simulation method requires that a slab of some finite thickness be simulated; the slab will have two surfaces of area, A_{slab}, exposed to the vacuum and a number of atoms N_{slab}. The consequences of this simulation method are that the surface formation energy can be written as

$$\gamma_{clean}(T) = \frac{F_{slab}(N_{slab}, T) - N_{slab}F_{bulk}(T)}{2A_{slab}} \tag{6.11}$$

The area term has a prefactor of 2, indicating that within the simulation two surfaces have been formed. Adsorption of oxygen to the surface in some arrangement, X, from some oxygen source of chemical potential μ_O results in a new surface formation energy that can be written as the sum of the clean slab and the changes due to adsorption:

$$\gamma_X(T, \mu_O) = \gamma_{clean}(T) + \Delta\gamma_X(T, \mu_O) \tag{6.12}$$

Writing the surface formation energy of the oxygen containing surface this way is convenient because it makes explicit that surface phase stability can be studied by examining

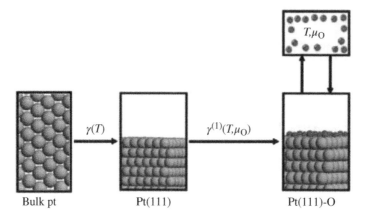

FIGURE 6.5 Schematic representation of the process of surface oxide formation. Beginning with (i) bulk cleavage and followed by (ii) oxygen adsorption and reconstruction. Schematic from reference [52]. Reproduced with permission from Getman et al. © American Chemical Society.

only the relative stability of the surface states. When relative stabilities of surfaces are considered, the factor of two in the denominator is lost as a result of the fact that we are now comparing one surface to another and no longer considering the formation of two separate surfaces. The change in surface formation energy for a fixed oxygen concentration, here written as fractional oxide coverage, Θ_X, of some phase X_s, for a given monolayer number density Γ_{ML}, may furthermore be written as

$$\Delta\gamma_X(\Theta_X, \mu_O, T) = \frac{\Theta_X F_X(T) + (1-\Theta_X)F_{clean}(T) - [\mu_O \Gamma_{ML}\Theta_X + F_{clean}(T)]}{A} \quad (6.13)$$

Writing the change in surface formation energy this way makes explicit that modification of the oxygen chemical potential may change the stable surface state and furthermore that in considering the stable surface state for a given oxygen coverage (relative to other surfaces at the same coverage), the only factor that is important is the free energy of that state $F_X(T)$. Thus, when considering the stable phase at a given coverage, it is common practice to only consider the relative and not absolute stability of phases. For the time being, the question of the most stable state as a function of oxygen chemical potential is put off and first the question of the most stable state for a given surface oxygen coverage is addressed. Because it is only the relative stability of states with respect to one another that matters, it is typical to choose two reference states, the clean surface (state 1, $\Theta_1 = 0$) and some selected known state (state 2, here $\Theta_2 = 1$) such that the change in surface energy can be written as

$$\Delta\gamma_{1:2, X}(\Theta_X) = \frac{\Theta_X F_X(T) - [(1-\Theta_X)F_1(T) + \Theta_X F_2(T)]}{A} \quad (6.14)$$

It produces curves with endpoints (states 1 and 2) at values of 0 and makes phase stability analysis facile and intuitive.

With the relative surface free energies in hand, one may perform a phase stability analysis of the surface structures. For a more detailed discussion, we refer the reader to the literature [41] but describe here the overall method. Two often referred-to concepts in constructing a phase diagram from a free energy diagram are the convex hull and the tie line. The convex hull is a representation of free energy path a mixture will take as one changes the desired exogenous variable (the emboldened line across G_0–G_2–G'_4–G_6–G_8 in Fig. 6.4). It is composed of stable phases (at a given point) and tie lines (connecting the stable phases and representing multiphase regions). Single-phase regions are formed between compositions in which there is no tie line connecting the desired endpoints or minima (G_0–G_2 and G_6–G_8 in Fig. 6.4). Two-phase regions are formed when there exists a common line connecting the desired endpoints or minima (G_2–G'_4–G_6 in Fig. 6.4). The line connecting the two stable phases in the multiphase region is known as the tie line (the emboldened line across G_2–G'_4–G_6 in Fig. 6.4). The tie line is constructed by connecting the minima of two phases with a common tangent line—a tangent being equivalent to chemical potential on a free energy diagram. When performing a first-principles analysis, if a new structure is found, which is below the tie line, it is said that the tie line is "broken" or that the new structure "breaks the tie line," indicating that the calculated structure is a newly found stable point along the free energy diagram.

In practice, it is usually the case that the complete free energy is not known and that approximations must be made to make the expression for relative stability, Equation (6.14), useful. Several factors may go into the surface free energy term: vibrational entropy, zero-point energy, configurational entropy, formation energy, and more [52]. There are certainly cases where differences in entropic contributions become important; however, it is normally assumed that the changes in surface free energy are dominated by the change in surface formation energy. This allows the replacing of the free energy terms F with a DFT calculated energy E^{DFT} and allows the approximation of the surface formation energy as:

$$\Delta \gamma_{1:2, X}(\Theta_X) \approx \frac{\Theta_X E_X^{DFT}(T) - \left[(1 - \Theta_X) E_1^{DFT}(T) + \Theta_X E_2^{DFT}(T)\right]}{A} \qquad (6.15)$$

Given this expression, all that remains is the choice of the structures that are to be evaluated using DFT. In general, configurations are chosen based on previously identified and suggested experimental observation [53, 54], previous computational work [55], and with the overall goal to iteratively minimize the change in surface formation energy [56]. More advanced search methodologies based on cluster expansion methods are also possible, but structure identification for calculation remains a major issue in this field of study.

Now, we consider the stability of a surface phase as a function of chemical potential. In writing the expression for the DFT approximation of surface stability, Equation (6.15), the formation energy F_X was approximated as E_X^{DFT}. There are many circumstances in which E_X^{DFT}, a representation of the cohesive energy of a structure, is an insufficient representation of free energy; the most common cases in which E_X^{DFT} fails to represent the free energy well are when the surface is in thermodynamic equilibrium with an environment and the environmental chemical potential is nonnegligible. In this case, additional layers of detail must be given to the free energy approximation, writing, for example,

$$F_X \approx E_X^{DFT} + \Theta_X \mu_X^{env} \qquad (6.16)$$

where μ_X^{env} may be arbitrarily written as the ideal gas contributions to free energy [57], tabulated gas fugacity data, the Nernst expression for energy contributions under an applied

TABLE 6.1   Here are commonly used approximations for free energy contributions that are not represented in the DFT determined cohesive energy, E^{DFT}

Physical property	Free energy expression
Zero-point energy	$\frac{1}{2}\sum h\nu$
Electrochemical potential	$-qU$
pH concentration-dependent entropy	$-k_BT \ \ln \ ([H^+]) = k_BT \ pH \ \ln 10$
Ideal gas contributions	$k_BT \ln \left(\frac{N}{V} \left(\frac{2\pi\hbar^2}{mk_BT} \right)^{3/2} \right)$
Vibrational/rotational entropy	$-k_BT \ \ln \ Q^{vib/rot}$

h is Planck's constant, $\hbar$ is the reduced form of Planck's constant where $\hbar = h/2\pi$, ν is the frequency of the stable modes of vibration, k_B is the Boltzmann constant, T is temperature, q is the charge transferred across the electrochemical interface, U is the applied bias, N/V is the number density of atoms in the gas phase, m is the mass of those atoms in the gas phase, and $Q^{vib/rot}$ is the partition function associated with either the vibrational or rotational motion of a species. The entropic contributions due to rotational and vibrational entropy follow the formalism of a generalized internal degree of freedom; any internal degree of freedom can be added in such a fashion given that the partition function is known. Important references for the aforementioned are given [57–60].

bias [58], or whatever appropriate experimental parameters are desirable. Including additional contributions to free energy such as zero-point energy, vibrational and rotational entropy, or others may be equally added to the DFT approximation in the form of a sum as in Equation (6.16). Commonly used free energy expressions are shown in Table 6.1. Once the selection of free energy approximation is made, it may be inserted into the generalized expression for relative surface stability, Equation (6.14), and a case-specific approximation of relative stability derived following Equation (6.15).

The goal of this work is to serve as an instructive example of how computational methods might be used to identify experimentally observed structures as a function of environment and to develop the state of knowledge within a field. With this goal in mind, here is presented the narrative of how computational tools have helped to understand oxide growth on magnesium and platinum and how this understanding has advanced the knowledge of both oxide growth and corrosion mechanisms.

6.5   CASE STUDIES

6.5.1   Magnesium

Magnesium is a material that holds a great deal of promise for lightweight structural applications. Magnesium alloys have improved strength-to-mass properties over aluminum, which if implemented could result in fuel, cost, and safety improvements for transportation vehicles and other applications [61–64]. One of the main problems preventing magnesium from widespread use has been corrosion; these corrosion mechanisms are not limited to but include metal atom dissolutionment, hydrogen embrittlement, and runaway oxide formation [62–64]. The first step in many corrosion processes is the breakdown of the passivating oxide layer making understanding its chemistry a critical problem in improving magnesium corrosion resistance.

Oxide growth on magnesium is a process that greatly contrasts with the more well-studied late transition metal oxide nucleation [53, 65–73]. "Normal" oxide nucleation over late transition metals proceeds via a four-step process. First, the O_2 dissociates over the metal surface to form chemisorbed oxygen [74–76]. These oxygen atoms remain at the surface where they typically occupy threefold hollow sites [65]. The surface oxygen atoms experience a repulsive O^*–O^* interaction, which leads to characteristic surface structures [53, 66–69]. Further addition of oxygen only adds to the chemisorbed surface oxygen until, second, a critical surface coverage is exceeded where the oxygen atoms undergo ingress [70], but remain near the surface and do not dissolve into the bulk [71, 72]. The critical coverage at which oxygen ingress occurs has been described as the result of a balance between the cost of deforming the lattice to incorporate the oxygen and the repulsive interactions between the adsorbed oxygen adatoms [70]. As such, it is possible that this ingress occurs later in the progression of oxide formation. In the third step of oxide formation over late transition metals, the subsurface oxygen atoms experience mutual attraction and combine to form the first nuclei of oxide formation [71–73]. Continued accommodation of oxygen at the surface ultimately leads to bulk oxide formation, which is the fourth and final stage [65].

Detailed experimental analysis of oxide nucleation over magnesium has shown fundamentally different oxide nucleation behavior than that which has been observed in the more well-understood "normal" transition metal oxides [53, 54, 77–80]. Just as with transition metal oxide nucleation, the first step is the dissociative chemisorption of O_2 [76]. The very first exposure of Mg(0001) to O_2 results in the formation of subsurface oxygen monomers, not surface monomers [26, 27, 47]. Rather than interacting repulsively, the oxygen monomers experience an attractive oxygen–oxygen interaction, which results in cluster formation [53, 54, 79, 81]. At higher exposures, measurements have demonstrated that approximately 90% of the oxygen atoms are on or near the surface [80]. This is precisely the opposite oxide growth mechanism that is observed in late transition metals; oxygen begins as subsurface, self-agglomerates forming oxide nuclei, and then at higher exposures oxygen surface adsorption occurs.

For insights into the underlying nature and driving forces behind the oxide evolution, we can turn to DFT modeling. In particular, we here highlight the contributions from Schröder, Kiejna, and Francis in understanding the individual structures and those of Francis in integrating the structure into broader phase behavior [55, 82–87].

Francis used DFT methods to investigate binding and structure of the very first stages of oxygen chemisorption [56]. In describing the possible surface states of oxygen, the conventional description of the binding sites was used (Fig. 6.6). DFT calculations showed that the

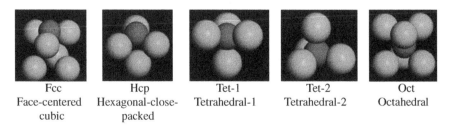

| Fcc | Hcp | Tet-1 | Tet-2 | Oct |
| Face-centered cubic | Hexagonal-close-packed | Tetrahedral-1 | Tetrahedral-2 | Octahedral |

FIGURE 6.6 Surface and subsurface sites of oxygen on the Mg hcp(0001) surface—not including atop and bridge. The emboldened text is the short name for the sites and the longer, unemboldened text is the full name. Reproduced with permission from Francis and Taylor [56]. © American Physical Society.

minimum energy monomer O^* site was the tetrahedral-1 subsurface position. Not only was it demonstrated that the tetrahedral-1 position was the most stable but that surface sites were unstable compared to this site and that there was no activation to oxygen ingress (Fig. 6.7). DFT calculations demonstrated that the magnesium-mediated oxygen–oxygen interactions were attractive. In investigating which of the sites were the most stable under cluster formation, not only was it the case that the preference for tetrahedral–1 persisted but that tetrahedral-1 clusters uniquely demonstrated a reduced work function, a work function trend that had been observed in experiment. This corroboration of calculated work function behavior with experiment allowed the positive identification of the O^* monomer as tetrahedral-1. Investigation into the structure of the oxygen clusters yielded interesting results. Coadsorption of oxygen resulted in the stabilization of surface states (Fig. 6.7). This result shed light on the previous inference that oxygen monomer adsorption was subsurface while later stage oxygen adsorption was bound to the surface [26, 27, 47, 80]. Continued growth of the oxygen clusters studied by Francis showed the beginnings of oxide nuclei growth by oxygen intercalation in the plane of the magnesium atoms of which deeper investigation was performed by Schröder et al. [55, 83].

Schröder used a lattice gas Hamiltonian model to look at the intermediate exposures of magnesium surface to an oxygen gas in which many oxygen atoms coexisted on the surface [82]. Schröder demonstrated that clusters persist to nonnegligible local coverages—as high as 3 ML. Schröder joined with Fasel and Kiejna to look at the structure of the oxide film at larger exposures [55, 83]. DFT simulations were used to find the structure and binding energies of possible structures. Rather than making predictions based on the lowest-energy states, single-scattering cluster (SSC) model simulations were performed in order to understand what X-ray photoelectron diffraction (XPD) measurements might be produced from the predicted structures (Fig. 6.8). Using a reliability factor as a means to correlate simulation with experiment, the bilayer and trilayer structures were identified as either a pure or a mixed state of tetrahedral-1 and a newly predicted "flat" structure (Fig. 6.9). In the "flat" structure, the oxygen and magnesium formed a 2D structure much like the Mg(0001) in which the oxygen atoms are intercalated between the magnesium atoms (Fig. 6.9). This "flat" structure may be the higher exposure state of the clusters described by Francis [56].

Francis combined his own data and that of Schröder, Fasel, and Kiejna to perform a FPT analysis of the possible O^*/Mg(0001) structures (Fig. 6.10). Using the FPT method (Eq. (6.15)) and the endpoints of clean Mg(0001) surface and the 3 ML structures predicted by Schröder, Fasel, and Kiejna, Francis performed an analysis of the O_2/Mg(0001) interface. The relative surface formation energy of the FPT contains stable and metastable structures and formed an upside-down U-curve indicative of a possible underlying spinodal decomposition. A spinodal decomposition is a pathway by which an existing phase is vulnerable to diffusive decomposition. If, for example, beginning with a homogeneous coverage of 1 ML, a small atomic scale diffusive event resulting in the local creation of slightly enriched and slightly starved 1 ML would be thermodynamically driven. This separation process would persist until the surface was composed only of zero and high oxygen coverage phases. The FPT diagram demonstrates that this process is thermodynamically driven. This phase separation is driven for any FPT with a negative curvature ($d^2G/d\Theta^2 < 0$).

That the phase separation of the surface oxide is thermodynamically driven has important consequences for corrosion. Given a passivating MgO film on Mg, there will be a thermodynamic driving force to the separation and subsequent renewal of exposed magnesium. The renewal of magnesium will result in increased reactivity and corrosivity of the previously passivated magnesium surface; regardless of the corrosion mechanism that is active, any

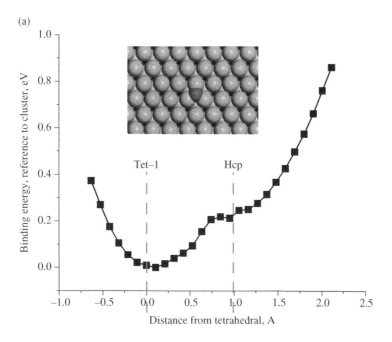

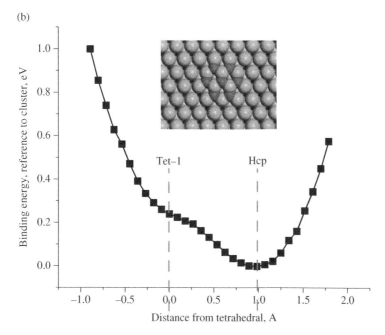

FIGURE 6.7 Oxygen binding energy as a function of relative vertical displacement. Binding energy referenced to the energy of a single oxygen atom in the tetrahedral-1 position. (a) A single oxygen monomer and (b) the central oxygen atom in a 7-atom cluster of "tetrahedral-1." Monomer is stable as tetrahedral-1, central atom of cluster "pops" onto the surface. Reproduced with permission from Francis and Taylor [56]. © American Physical Society.

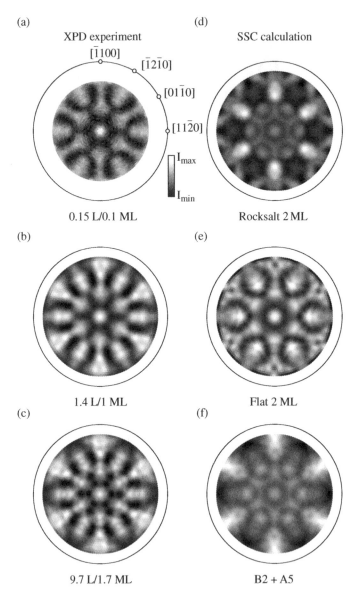

FIGURE 6.8   Comparison of (left) experimental X-ray photoelectron diffraction (XPD) (right) and simulated XPD using single-scattering cluster (SSC) [55, 83]. A reliability factor, R_{MP}, was used to compare the XPD and SSC spectra. The reliability factor demonstrated that of the SSC calculations for 1 ML or less, none satisfactorily matched the experimental XPD for any level of oxygen exposure. This reliability factor gave good agreement for ordered 2 and 3 ML oxide structures. The structures indicated by comparing the XPD and SCC were either oxides with oxygen in the interstices of the Mg(0001) surface or a graphite-like structure that was referred to as "flat" (Fig. 6.9). Reproduced with permission from Schröder [82]. © American Physical Society.

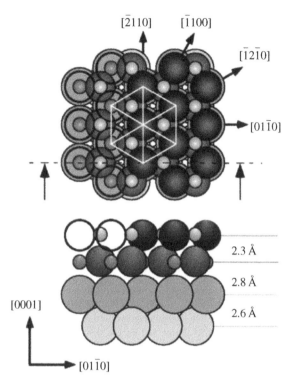

FIGURE 6.9 Representation of "flat" structure found to occur for $O_2/Mg(0001)$ when oxygen levels are multiple monolayers. This finding resulted from a combined SCC, XPD, and DFT analysis as represented in Figure 6.8 [55, 83]. Reproduced with permission from Schröder et al. [55]. © American Physical Society.

protective layer will again and again be removed, possibly resulting in the observed large vulnerability to corrosion. The case study of $O^*/Mg(0001)$ demonstrates the utility of first-principles modeling methods in understanding the precise structure and broader driving forces that may control corrosion.

6.5.2 Platinum

Platinum has a myriad of practical uses, especially in the field of electrochemistry where it is used as a catalyst and as a reference electrode. In particular, platinum is the most active known pure metal for the oxygen reduction reaction (ORR) in which O_2 is split and combined with protons to form H_2O. This is an important step in low-temperature fuel cells (polymer electrolyte fuel cells, direct methanol fuel cells, etc.) as it often is what limits the total fuel cell efficiency. Furthermore, platinum is rather expensive with the materials cost of its use in fuel cells being roughly half of the total fuel cell cost. Consequently, a great deal of effort is made in order to optimize its use.

One way platinum use is optimized is through the use of Pt nanoparticles, acting to increase the amount of Pt surface (where the ORR takes place) per amount of Pt used.

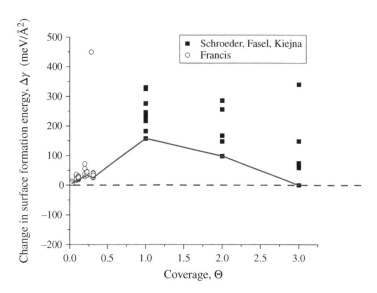

FIGURE 6.10 First-principles thermodynamic analysis of the structures occurring at the O_2/Mg (0001) interface between oxygen coverage of O and 3 ML [56]. Change in surface formation energy versus surface coverage, an example free energy diagram used for surface analysis. Hollow data points by Francis and filled in by Shroeder, Fasel, and Kiejna. Absence of a free energy point beneath the line connecting 0 and 3 ML indicates a driving force to separation of the intermediate oxide coverages. Reproduced with permission from Francis and Taylor [56]. © American Physical Society.

The use of nanoparticles, however, comes at the cost of stability as smaller particles are less stable than larger ones, particularly under fuel cell environments that are highly acidic and under varied electrostatic potentials. This stability is strongly related to the onset of oxide formation on the particle surface. In this section, we will discuss the literature pertaining to the O^*/Pt(111) surface phase diagram from first principles, Pourbaix diagrams of Pt and other metals from first principles, and the nanoparticle Pt Pourbaix diagram.

6.5.2.1 O^*/Pt(111) Surface Phase Diagram from First Principles and Place Exchange

Oxide formation on Pt surfaces in electrolyte environments has been studied for nearly a century [9] yet still much is being learned about the oxide formation and growth on these surfaces. This is in part due to the complexity of the surface/electrolyte interactions under electrostatic potential as well as the need for accurate approaches to study such systems. The drastic improvement in DFT over the past two decades coupled with the improved surface/electrolyte models have made it such that theoretical tools can be applied to these systems with an acceptable degree of accuracy. We will here discuss some recent studies on the Pt(111) surface using DFT and some of the implications of the DFT findings.

Experimental studies of oxygen on Pt(111) surfaces show that at high enough oxygen coverage (dictated by oxygen partial pressure and electrostatic potential, Eq. (6.13)), oxygen atoms go below the Pt surface due to the repulsive oxygen–oxygen interaction (oxygen ingress). The mechanism through which this occurs has been called place exchange in the Pt literature and had previously been attributed to dipole–dipole interactions [89].

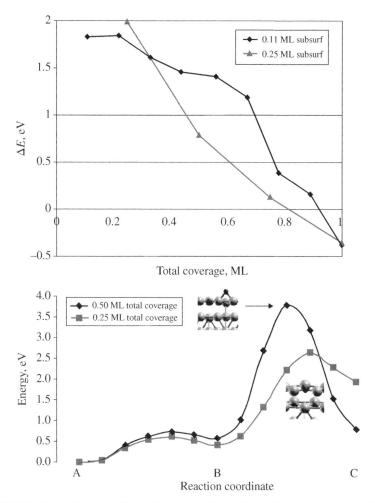

FIGURE 6.11 Plots of energy difference between surface and subsurface O adsorption as a function of oxygen coverage (top) and potential energy surface for O going subsurface through a threefold Pt site (bottom) from reference [88]. Note that subsurface occupation becomes thermodynamically stable at total coverages of approximately 0.8 ML but that a large approximately 2 eV kinetic barrier exists, limiting the likelihood of such phenomena occurring during typical CV timescales. Reproduced with permission from Gu and Balbuena [88]. © American Chemical Society.

Recently, DFT studies have suggested a more comprehensive understanding of how oxygen occupies the Pt subsurface and how an initial monolayer of oxide is formed.

DFT studies by Gu and Balbuena [88] focused on the stability of Pt(111) surfaces with oxygen occupying both hollow sites on top of the Pt layer and tetrahedral sites below the top Pt layer at varying concentrations (Fig. 6.11). It was found that at a coverage of approximately 0.8 ML the subsurface positions became more stable than those on the Pt surface. A kinetic pathway for the subsurface diffusion of O to the subsurface through the Pt hollow site produced a very large kinetic barrier (several eV), which would mean that simple

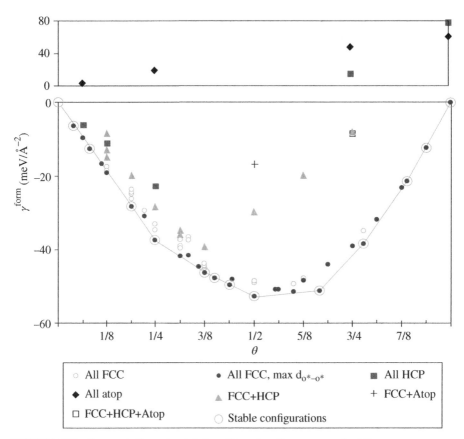

FIGURE 6.12   Convex hull constructed from Pt surface free energy calculation versus oxygen coverage from reference [52]. Reproduced with permission from Getman et al. [52]. © American Chemical Society.

oxygen diffusion into the subsurface is not a likely explanation for the place-exchange mechanism. This mechanism is believed to be seen during cyclic voltammetry (CV) with sweep times of less than 1 min. At the temperatures and potentials of these CV experiments, such a large kinetic barrier would preclude observing place exchange.

Getman et al. [52] used DFT to look at a large number of oxygen orderings on the Pt surface in order to create a O*/Pt(111) surface phase diagram up to 1 ML (Fig. 6.12). The methodology for surface phase diagram/convex hull creation presented in the work by Getman [52] is incredibly valuable and is strongly encouraged reading for any researcher interested in first-principles studies of surface adsorption. In this work, the researchers find 13 oxygen-on-surface arrangements between 0 and 1 ML that fall on the convex hull. Using the thermodynamic data, they also generated a plot of stable surface phase versus oxygen chemical potential and an Ellingham diagram that shows phase stability as a function of temperature. Unfortunately, subsurface oxygen was not considered in this study, and therefore, the higher oxygen coverage surface phases (where subsurface oxygen is likely to play a role) are not exhaustively treated. This point relates to a general issue related to surface phase diagram generation; there is a near infinite number of possible surface atom and adsorbate configurations and an exhaustive search for all of them is computationally prohibitive (without using some advanced techniques beyond DFT and simple thermodynamic models). Consequently, experimental guidance and more advanced search techniques are required to improve on phase diagrams that often evolve in the literature.

Hawkins et al. [69] also studied the $O^*/Pt(111)$ surface, including subsurface O, using DFT and thermodynamic theory. Their findings represented a shift in thinking about these surfaces. They considered not only the typical oxygen hollow site occupation but also the experimentally observed effect of Pt buckling [90]. In the buckling phenomenon, Pt is displaced from its native surface site and forms linear chains with Pt between oxygen. The Pt acts to screen the oxygen–oxygen interaction, thus leading to more stable phases at higher oxygen coverages (above 0.5 ML). When considering that the Pt lattice is no longer fixed at higher coverages, a host of other potential atomic orderings becomes possible.

Holby et al. [91] built on the concept of buckling and again, using DFT and thermodynamic theory, suggested a phase diagram that included a hybrid buckled/place-exchanged phase. Their findings showed that the phase diagram consisted of low coverage surface phases with O occupying FCC hollow sites up to 0.25 ML, a 0.75 ML buckled phase (suggested by Hawkins et al. [69]), and this new hybrid phase at 1 ML. Their explanation for this evolution with increased coverage is as follows (Fig. 6.13).

At low coverages, oxygen–oxygen repulsion is still relatively weak and so simple surface orderings to move O adsorbates apart minimize surface energy. Though phases at 0.5 ML are found to be metastable, it is at approximately this coverage at which the energy to reconstruct (buckle the Pt) is equal to that of the destabilization energy due to oxygen–oxygen repulsion. Thus, at 0.75 ML, a single O–Pt–O chain along with a complementary FCC occupied stripe is formed. Formation of a second O–Pt–O string would require a close packing of chains. In the 1 ML hybrid phase of Holby et al. this is accommodated through a rotation of one chain such that one chain is roughly parallel to the surface and one is perpendicular (Fig. 6.13). As a consequence of this rotation, oxygen ends up below the Pt (in a place-exchange-like configuration). This phase was found to be significantly more stable than previously reported 1 ML phases, including a 2D PtO_2-like phase previously assumed in the literature. The kinetics of such a rotation are likely significantly lower than the several eV barrier of direct O penetration in the undisturbed surface and so provide an energetically realistic path for oxygen to go subsurface.

The aforementioned interpretation acts to join the observation of Pt buckling and O–Pt–O chain formation with that of the traditional place exchange. In addition, this interpretation fits well with experimentally obtained CV data in which kinetic barriers for buckling/place exchange lead to hysteresis in which oxygen removal occurs at a lower voltage (as the more stable but more slowly forming buckled/place-exchanged phases are removed) than oxygen formation (where metastable phases form first due to kinetic inhibition to buckling). This again serves as an example of how the near infinite search space of surface/adsorbate configurations complicates phase diagram calculation and how increased understanding comes through incremental refinement and experimental comparison (in this case to CV and place-exchange data). Additionally, these examples show how during oxide formation, the underlying metal can become displaced from its native lattice site. Under certain cyclic conditions that cause oxide to form and be removed, this is likely to roughen the surface that can lead to a destabilization of the metal, leaving it more active for corrosion processes (dissolution in particular). These findings also demonstrate how surface phases, distinct from bulk phases, can affect the onset of oxide formation, similar to the case of $O^*/Pd(111)$.

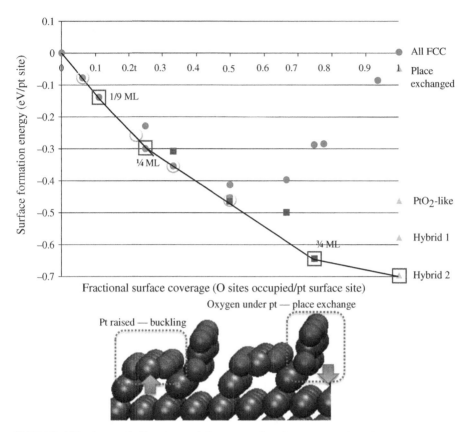

FIGURE 6.13 Convex hull constructed from Pt surface free energy calculation versus oxygen coverage from reference [91] along with the stable 1 ML "hybrid" phase that unifies place-exchange and buckling mechanisms. Reproduced with permission from Holby et al. [91]. © American Chemical Society.

6.5.2.2 *Surface Pourbaix Diagram for $O^*/Pt(111)$ from First Principles*

Phase diagrams, from a general perspective, show the thermodynamically stable phase of some combination of materials as a function of some set of environmental conditions. Input from DFT can greatly aid in the calculation of such diagrams, but as shown previously, thermodynamic models are needed to fully incorporate effects such as concentrations, pH, and electrostatic potential. In particular, the use of electrostatic reference state models aids in determining stable phases under a variety of electrochemical environmental conditions.

Named after corrosion chemist Marcel Pourbaix, the phase diagram as a function of pH and electrostatic potential is referred to as a Pourbaix diagram. The first such diagram for the Pt system is shown in Fig. 6.14 [92]. Assumed concentrations of ions are typically 1 M, though the diagrams can be made for any assumed concentration. The Nernst equation, along with the standard potential (which can be modeled using DFT), then gives the regions of metal stability (immunity), oxide stability (passivity), and susceptibility (metal dissolution). During oxide formation, we are often concerned with electrochemical reactions that, due to production or consumption of electrons and protons, are dependent on electrostatic

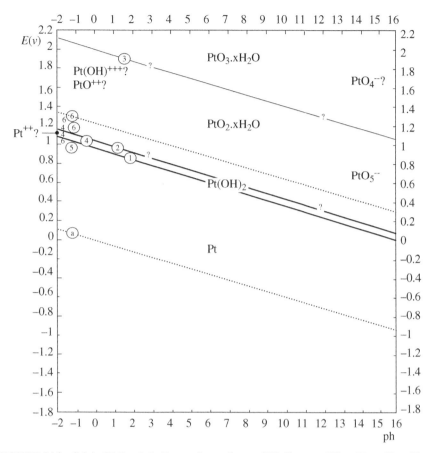

FIGURE 6.14 Original Pt Pourbaix diagram from reference [92]. Shows stability of Pt and Pt oxides as a function of pH and potential ($E(v)$). Dashed lines represent the stability region for liquid water. Solid lines represent solid state phase equilibria. Numbers represent reference equilibrium reactions. Question marks represent ambiguity regarding phase or dissolved ion. Further information may be found in the original text. Reproduced with permission from Pourbaix [92]. © National Association of Corrosion Engineers.

potential and pH, respectively. As such, for corrosion science and oxide formation in particular, Pourbaix diagrams are one of the most important phase diagrams. More information on these diagrams can be found in a number of textbooks, but our focus here is on examples of how such diagrams are calculated for metal surfaces using DFT.

Hansen et al. studied the surface Pourbaix diagrams of Pt, Ag, and Ni(111) surfaces using DFT in order to better understand ORR on these surfaces [93]. By implementing a theoretical standard hydrogen electrode, a thermodynamic equivalence between protons and H_2 gas at standard conditions sets the pH (pH = 0) and electrostatic potential (U = 0V) for the system. Deviations from these states are calculated according to the energy shifts discussed in Table 6.1. Using DFT to get energies for H_2O, the pristine metal surface, the surface with OH and O adsorbed on the surface (which, as shown previously, varies based on coverage and structural arrangement), and H_2 gas, free energies of reaction can be written for OH and O adsorption at those conditions. This negates the need to find the energy of the solvated

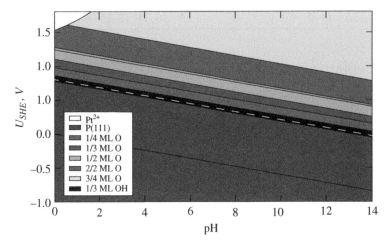

FIGURE 6.15   DFT calculated Pt(111) surface Pourbaix diagram from reference [93]. This example shows thermodynamic stability of different surface phases based on pH and electrostatic potential. Reproduced with permission from Hansen et al. [93]. © Royal Society of Chemistry.

proton using DFT (a difficult task [94]) and doesn't require any pH or electrostatic potential contributions, providing a baseline free energy for the reactions. The researchers also add in contributions due to zero-point energy and entropy (which, due to the 0 K nature of DFT, are absent and require some model for their appropriate addition for nonzero temperatures). Electrostatic and pH effects are then easy to add to the calculated free energy based on the number of electrons/electrostatic potential versus the theoretical SHE and the solution pH using the Nernst equation and the free energy/electromotive force relation. Using this technique and a sampling of surface OH and O coverages/configurations, the Pourbaix diagrams were calculated. One example for the now familiar Pt case is given in Figure 6.15.

In this study, a small window for Pt corrosion is found at very high potentials and very low pHs, which fits with Pt being a very noble metal. The unoxidized surface is stable at all pHs at moderate potentials versus SHE. The thermodynamic pathway of oxide formation at a given pH as electrostatic potential is increased is predicted to be a 1/3 ML coverage of OH, followed by 1/4 ML, 1/3 ML, 1/2 ML, 2/3 ML, and finally 3/4 ML of O, unless at very low pH in which case, Pt would dissolve after the 2/3 ML phase is formed above 1.5 V versus SHE. This is at odds with the aforementioned section discussing $O^*/Pt(111)$ oxide thermodynamics, but this study included only a handful of surface coverages/orderings and did not take into account buckling and place-exchange effects. As such, it still serves as a useful study on how to apply DFT data and thermodynamic models to surface oxide formation to produce surface Pourbaix diagrams. We will next discuss a similar model but applied to a Pt nanoparticle instead of a metal surface.

6.5.2.3   *Pt Nanoparticle Pourbaix Diagram from First Principles*   One of the main motivations for studying oxide formation on Pt is the stability of Pt nanoparticles under fuel cell conditions (large electrostatic potential and very low pH in the case of a polymer electrolyte fuel cell). The study of surfaces can serve as a less computationally intensive proxy for the nanoparticles, but increases in computational power now allow for calculations on realistically sized catalyst nanoparticles comprised of tens to hundreds of metal atoms.

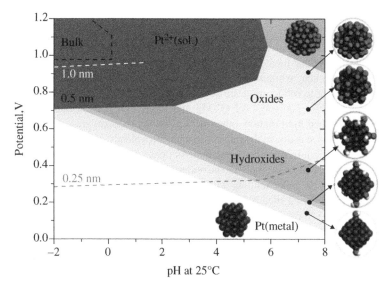

FIGURE 6.16 Pt nanoparticle (radius =0.5 nm) Pourbaix diagram from reference [95]. The orange line shows how a particle with radius of 0.25 nm does not form oxides or hydroxides at low pH, and so if potential is increased, particles go directly to the susceptible regime where they will dissolve without any oxide protection. Reproduced with permission from Tang et al. [95]. © American Chemical Society.

These particles differ from bulk in that they have multiple surfaces, as well as edge and corner atoms, which, due to their lower coordination, should oxidize at a lower thermodynamic driving force than the bulk surface. These sites are less stable, which affects the overall stability of the nanoparticle as well (Fig. 6.16).

Tang et al. studied these effects using DFT (with a complementary *in situ* STM study) [95]. They found similar trends for the Pt nanoparticles studied compared to the bulk in relation to metal/hydroxide/oxide stability but found a very strong dependence on particle size. For the smallest particle size studied (radius of 0.25 nm), at low pH, there is a direct dissolution pathway and hydroxides and oxides are not predicted to form prior to dissolution. Consequently, it can be surmised that under fuel cell conditions, there is a crossover in particle size below which the particles do not develop a passivating oxide layer and instead are subject to direct dissolution, thus decreasing their already compromised stability. This study again only looks at surface occupation of OH and O without surface/particle reconstruction but provides very useful information for the stability of Pt nanoparticles and the conditions that lead to oxide formation at the nanoscale. In addition, it shows how DFT can not only give information on phase stability but also point to corrosion mechanisms under different conditions.

The aforementioned examples of O^*/Mg and O^*/Pt surfaces illustrate the usefulness of DFT studies on oxide film formation and demonstrate some of the capabilities that first-principles modeling holds for the field. These examples are far from exhaustive but hopefully help to guide the reader in their understanding of DFT and its ability to contribute directly to applications of relevance within the corrosion community. With that, we turn our focus to the future and where we see the future of first-principles modeling of passive film formation.

6.6   THE FUTURE

We have shown how DFT can be incredibly useful for understanding oxide formation at the atomic scale, but we believe that there are still significant contributions yet to be made using these techniques. In particular, while theories for bulk oxide growth have been produced, general theories pertaining to oxide onset are still lacking. For instance, we have demonstrated different oxide onset pathways for O^*/Mg and O^*/Pt, but the reasons for these differences (which may shed light on how other metals initially oxidize) are yet unknown. It is likely that this difference is due to the relative iconicity/covalency of the metal–oxygen bond but further work is required for confirmation. As phase diagram behaviors of more $O^*/Metal$ surfaces are understood, we feel that the underlying principles of oxide layer onset will become clearer. Such an understanding will certainly be useful for understanding the corrosion mechanisms of varied metals under different environments in any case where these early onset layers are formed or removed.

There is also reason to believe that DFT calculations will become faster and more accurate as the field advances. This will come about not only due to improved algorithms and XC functionals but also through improvements in computational speed. As computation speed doubles every 18 months or so (according to Moore's law [96]), larger calculations that were previously thought to be intractable start to become possible with more computational power. Such an increase allows for more accurate modeling of these systems with less need for potentially inaccurate assumptions.

In summary, DFT is a powerful tool for gaining an atomic scale understanding of how oxides form on metals. It allows for the generation of phase diagrams and for direct comparison with experiments. This has allowed for an integrated theory/experimental pathway forward for understanding oxide formation and will ultimately aid in the understanding (and mitigation) of corrosion phenomena.

REFERENCES

1. K. Reuter, in *Nanocatalysis, edited by U. Heiz and U. Landman (Springer, Berlin, 2006), pp. 343376.*

2. G. Tammann, Z. Anorg. *Zeitschrift für Anorganische und Allgemeine Chemie* **111**, 78 (1920).

3. N. B. Pilling and R. E. Bedworth, *Journal of the Institute of Metals* **29**, 529 (1923).

4. K. R. Lawless, *Reports on Progress in Physics* **37**, 231 (1974).

5. C. Wagner, *Corrosion Science* **9**, 91 (1969).

6. D. E. Coates and A. D. Dalvi, *Oxidation of Metals* **2**, 331 (1970).

7. N. Cabrera and N. F. Mott, *Reports on Progress in Physics* **12**, 163 (1949).

8. F. P. Fehlner and N. F. Mott, *Oxidation of Metals* **2** (1), 5999 (1970).

9. B. E. Conway, *Progress in Surface Science* **49** (4), 331–452 (1995).

10. E. Lundgren, C. Kresse, M. Borg, J. N. Andersen, M. De Santis, Y. Gauthier, C. Konvicka, M. Schmid and P. Varga, *Physical Review Letters* **88** (24), 246103 (2002).

11. J. Tersoff and D. R. Hamann, *Physical Review B* **31**, 805 (1985).

12. J. Klikovits, E. Napetschnig, M. Schmid, N. Seriani, O. Dubay, G. Kresse and P. Varga, *Physical Review B* **76**, 9 (2007).

13. J. Klikovits, E. Napetschnig, M. Schmid, N. Seriani, O. Dubay, G. Kresse and P. Varga, *Physical Review B* **76** (4), 045405 (2007).

14. D. Sholl and J. A. Steckel, *Density Functional Theory: A Practical Introduction* *(Wiley-Interscience, Hoboken, NJ, 2009)*.

15. R. M. Martin, *Electronic Structure: Basic Theory and Practical Methods (Cambridge University Press, Cambridge, UK, 2004)*.

16. W. Kohn, *Reviews of Modern Physics* **71** (5), 1253–1266 (1999).

17. M. Born and J. R. Oppenheimer, *Annalen der Physik* **84**, 457 (1927).

18. P. Hohenberg and W. Kohn, *Physical Review B* **136** (3B), B864–B871 (1964).

19. W. Kohn and J. Sham, *Physical Review* **140** (4A), A1133–A1138 (1965).

20. J. P. Perdew and K. Schmidt, *AIP Conference Proceedings* **577** (1), 1–20 (2001).

21. P. Soderlind, G. Kotliar, K. Haule, P. M. Oppeneer and D. Guillaumont, *MRS Bulletin* **35**, 883 (2010).

22. W. Kohn and L. J. Sham, *Physical Review* **140** (4A), A1133–A1138 (1965).

23. J. P. Merrick, D. Moran and L. Radom, *Journal of Physical Chemistry A* **111**, 11683 (2007).

24. P. J. Feibelman, B. Hammer, J. K. Norskov, F. Wagner, M. Scheffler, R. Stumpf, R. Watwe and J. Dumesic, *The Journal of Physical Chemistry B* **105**, 4018 (2001).

25. G. Kresse, A. Gil and P. Sautet, *Physical Review B* **68** (2003).

26. S. E. Mason, I. Grinberg and A. M. Rappe, *Physical Review B* **69**, 161401 (2004).

27. Y. Wang, S. de Gironcoli, N. S. Hush and J. R. Reimers, *Journal of the American Chemical Society* **129** (34), 10402–10407 (2007).

28. S. Grimme, *Journal of Computational Chemistry* **25** (12), 1463–1473 (2004).

29. M. Dion, H. Rydberg, E. Schröder, D. C. Langreth and B. I. Lundqvist, *Physical Review Letters* **92**, 246401 (2004).

30. J. Klimes, D. R. Bowler and A. Michaelides, *Journal of Physics. Condensed Matter* **22**, 022201 (2010).

31. K. Lee, E. D. Murray, L. Kong, B. I. Lundqvist and D. C. Langreth, *Physical Review B* **82**, 081101 (2010).

32. I. D. Prodan, G. E. Scuseria and R. L. Martin, *Physical Review B* **76** (3), 033101 (2007).

33. L. Wang, T. Maxisch and G. Ceder, *Physical Review B* **73**, 195107 (2006).

34. S. L. Dudarev, G. A. Botton, S. Y. Savrasov, C. J. Humphreys and A. P. Sutton, *Physical Review B* **57** (3), 1505–1509 (1998).

35. V. I. Anisimov, J. Zaanen and O. K. Andersen, *Physical Review B* **44**, 943 (1991).

36. M. Hutchinson and M. Widom, *Computer Physics Communications* **183** (7), 1422–1426 (2012).

37. L. Kaufman and H. Bernstein. Computer Calculation of Phase Diagrams with Special Reference to Refractory Metals (Academic Press, New York, 1970).

38. H. Gaye and C. H. P. Lupis, *Metallurgical and Materials Transactions A: Physical Metallurgy and Materials Science* **6**, 1049–1056 (1975).

39. H. Gaye and C. H. P. Lupis, *Metallurgical and Materials Transactions A: Physical Metallurgy and Materials Science* **6**, 1057–1064 (1975).

40. D.D. Lee, J.H. Chay and J. K. Lee, *Materials Week (Cincinnati, OH, 1992)*.

41. D. D. Lee, J. H. Choy and J. K. Lee, *Journal of Phase Equilibria* **13** (4), 7 (1992).

42. M. Hillert, "Methods Of Calculating Phase Diagrams", in *Calculation of Phase Diagrams and Thermochemistry of Alloys*, Y. A. Chang and J. F. Smith, Eds., Proc. Conf. AIME Fall Meeting, Sept. 17–18, 1979, Milwaukee, 1979, The Metallurgical Society, pp. 1–13 (1979).

43. D. Furman, S. Dattagupta and R. B. Griffiths, *Physical Review B* **15**, 441–464 (1977).

44. R.M. Aikin Jr. and J. K. Lee, *Bulletin of Alloy Phase Diagrams* **4**, 131–134 (1983).

45. J.F. Counsell, E.B. Lees and P. J. Spencer, *Metal Science Journal* **5**, 210–213 (1971).

46. Y. Zhang, V. Blum and K. Reuter, *Physical Review B* **75** (23), 235406 (2007).

47. C. Stampfl, H. J. Kreuzer, S. H. Payne, H. Pfnuer and M. Scheffler, *Physical Review Letters* **83** (15), 14 (1999).

48. J. Schnadt, A. Michaelides, J. Knudsen, R. T. Vang, K. Reuter, E. Laegsgaard, M. Scheffler and F. Besenbacher, *Physical Review Letters* **96** (14), 4 (2006).

49. M. Schmid, A. Reicho, A. Stierle, I. Costina, J. Klikovits, P. Kostelnik, O. Dubay, G. Kresse, J. Gustafson, E. Lundgren, J. N. Andersen, H. Dosch and P. Varga, *Physical Review Letters* **96** (14), 4 (2006).

50. H. Tang, A. V. d. Ven and B. L. Trout, *Physical Review B: Condensed Matter and Materials Physics* **70** (4), 10 (2004).

51. A. Michaelides, K. Reuter and M. Scheffler, *Journal of Vacuum Science & Technology A* **23** (6), 11 (2005).

52. R. B. Getman, Y. Xu and W. F. Schneider, *Journal of Physical Chemistry C* **112**, 14 (2008).

53. A. F. Carley, P. R. Davies, K. R. Harikumar, R. V. Jones and M. W. Roberts, *Topics in Catalysis* **24** (1), 51–59 (2003).

54. A. F. Carley, P. R. Davies, R. V. Jones, K. R. Harikumar and M. W. Roberts, *Chemical Communications* **2002**(18), 2020–2021 (2002).

55. E. Schröder, R. Fasel and A. Kiejna, *Physical Review B: Condensed Matter and Materials Physics* **69**(4), 193405 (2004).

56. M. F. Francis and C. D. Taylor, *Physical Review B* **87** 075450 (2013).

57. C. Kittel and H. Kroemer, *Thermal Physics, 2nd edn. (W.H. Freeman and Company, New York, 1980).*

58. J. K. Nørskov, J. Rossmeisl, A. Logadottir, L. Lindqvist, J. R. Kitchin, T. Bligaard and H. Jónsson, *The Journal of Physical Chemistry B: Condensed matter, materials, surfaces, interfaces and biophysical* **108** (46), 7 (2004).

59. R. K. Pathria, *Statistical Mechanics, 2nd edn. (Butterworth Heinemann, 1996).*

60. C. D. Taylor, S. A. Wasileski, J.-S. Filhol and M. Neurock, *Physical Review B: Condensed Matter and Materials Physics* **73** (16), 165402–165416 (2006).

61. L. Duffy, *Materials World* **4** (3), 127–130 (1996).

62. H. N. J. Ghijsen, P.A. Thiry, J. J. Pireaux and P. Caudano, *Applied Surface Science* **8**, 397 (1981).

63. C. R. McGall, M. A. Hill and R. S. Lillard, *Corrosion Engineering, Science and Technology* **40** (3), 7 (2005).

64. G. Song and A. Atrens, *Advanced Engineering Materials* **5** (12), 837–858 (2003).

65. K. Reuter, Los Alamos National Laboratory, Preprint Archive, Condensed Matter, 1–18, arXiv: cond-mat/0409743 (2004).

66. L. Surnev, G. Rangelov and G. Bliznakov, *Surface Science* **159** (2–3), 299–310 (1985).

67. M. Lindroos, H. Pfnur, G. Held and D. Menzel, *Surface Science* **222** (2–3), 451–463 (1989).

68. H. Pfnur, G. Held, M. Lindroos and D. Menzel, *Surface Science* **220** (1), 43–58 (1989).

69. J. M. Hawkins, J. F. Weaver and A. Asthagiri, *Physical Review B* **79** (12), 125434 (2009).

70. M. Todorova, W. X. Li, M. V. Ganduglia-Pirovano, C. Stampfl, K. Reuter and M. Scheffler, *Physical Review Letters* **89** (9), 961031–961034 (2002).

71. K. Reuter, M. V. Ganduglia-Pirovano, C. Stampfl and M. Scheffler, *Physical Review B: Condensed Matter and Materials Physics* **65** (16), 165403–165410 (2002).

72. L. Wei-Xue, C. Stampfl and M. Scheffler, *Physical Review B: Condensed Matter and Materials Physics* **67** (4), 45408–45416 (2003).

73. M. V. Ganduglia-Pirovano, K. Reuter and M. Scheffler, *Physical Review B* **65** (24), 245426 (2002).

74. J. Behler, B. Delley, S. Lorenz, K. Reuter and M. Scheffler, *Physical Review Letters* **94** (3), 036104 (2005).

75. G. Katz, R. Kosloff and Y. Zeiri, *Journal of Chemical Physics* **120**, 3931 (2004).

76. C. Bungaro, C. Noguera, P. Ballone and W. Kress, *Physical Review Letters* **79** (22), 4 (1997).

77. H. Namba, J. Darville and J. M. Gilles, *Surface Science* **108** (3), 446–482 (1981).

78. B. E. Hayden, E. Schweizer, R. Kötz and A. M. Bradshaw, *Surface Science* **111** (1), 26–38 (1981).

79. A.U. Goonewardene J. Karunamuni, R.L. Kurtz, R.L. Stockbauer, *Surface Science* **501**, 102–111 (2002).

80. B. C. Mitrovic, D. J. O'Connor and Y. G. Shen, *Surface Review and Letters* **5** (2), 6 (1998).

81. A. F. Carley, P. R. Davies, K. R. Harikumar, R. V. Jones and M. W. Roberts, *Topics in Catalysis* **24** (1–4), 9 (2003).

82. E. Schröder, *Computational Materials Science* **24**, 6 (2002).

83. E. Schröder, R. Fasel and A. Kiejna, *Physical Review B: Condensed Matter and Materials Physics* **69**, 115431 (2004).

84. A. Kiejna, *Physical Review B: Condensed Matter and Materials Physics* **68** (23), 235401–235405 (2003).

85. A. Kiejna and B. I. Lundqvist, *Physical Review B: Condensed Matter and Materials Physics* **64** (4), 049901 (2001).

86. A. Kiejna and B. I. Lundqvist, *Physical Review B: Condensed Matter and Materials Physics* **63** (8), 085405–085410 (2001).

87. M. F. Francis and C. D. Taylor, *Physical Review B* **87** 075450 (2013).

88. Z. Gu and P. B. Balbuena, *The Journal of Physical Chemistry C* **111** (46), 17388–17396 (2007).

89. B. E. Conway, B. Barnett, H. Angerstein-Kozlowska and B. V. Tilak, *The Journal of Chemical Physics* **93** (11), 8361–8373 (1990).

90. S. P. Devarajan, J. A. Hinojosa Jr. and J. F. Weaver, *Surface Science* **602** (19), 3116–3124 (2008).

91. E. F. Holby, J. Greeley and D. Morgan, *The Journal of Physical Chemistry C* **116**, 9942 (2012).

92. M. Pourbaix, *Atlas of Electrochemical Equilibria in Aqueous Solutions, 2nd English edn. (National Association of Corrosion Engineers, Houston, TX, 1974)*.

93. H. A. Hansen, J. Rossmeisl and J. K. Norskov, *Physical Chemistry Chemical Physics* **10**, 3722–3730 (2008).

94. C. Knight and G. A. Voth, *Accounts of Chemical Research* **45** (1), 101–109 (2011).

95. T. Tang, B. Han, K. Persson, C. Friesen, T. He, G. Sieradzki and G. Ceder, *Journal of the American Chemical Society* **132**, 596–600 (2010).

96. G. Moore, *Electronics* **38** (8), 1–4 (1965).

7

PASSIVE FILM FORMATION AND LOCALIZED CORROSION

VINCENT MAURICE[1], ALEXIS MARKOVITS[2], CHRISTIAN MINOT[2] AND PHILIPPE MARCUS[1]

[1]*Institut de Recherche de Chimie Paris/Physical Chemistry of Surfaces, Chimie ParisTech-CNRS, Ecole Nationale Supérieure de Chimie de Paris, Paris, France*
[2]*Laboratoire de Chimie Théorique, Université Pierre et Marie Curie—CNRS, Paris, France*

7.1 INTRODUCTION

Localized corrosion of metals and alloys occurs in aggressive media (e.g., containing chloride) as a consequence of the passivity breakdown, with major impact in practical applications and on the economy. This form of corrosion is particularly insidious since a component, otherwise well protected by a well-adherent, ultrathin oxide or oxyhydroxide barrier layer (i.e., the passive film), can be perforated locally in a short time with no appreciable forewarning. Extensive studies have been conducted over the last five decades to understand localized corrosion by pitting [1–10], but the detailed mechanisms accounting for the local occurrence of passivity breakdown remain to be elucidated and combined with kinetics laws to allow reliable prediction.

Passivity breakdown is the first stage in the process leading to localized corrosion. It is categorized into three main mechanisms: (i) the adsorption-induced thinning of the passive film that involves the competitive adsorption of the aggressive ions versus hydroxyl groups at the passive film surface [2, 5, 8, 10–18], (ii) the penetration mechanism that involves the subsurface insertion and transport of the aggressive species to the metal/oxide interface [1, 2, 6, 11] as developed in the point defect model [19, 20] leading to penetration-induced

Molecular Modeling of Corrosion Processes: Scientific Development and Engineering Applications, First Edition.
Edited by Christopher D. Taylor and Philippe Marcus.

voiding [21], and (iii) the inhibited healing mechanism that involves continuous events of breakdown and repair of the passive film and repair poisoning by the aggressive species [22–26]. All these models consider the passive layer as a simple uniform and homogeneous oxide (or hydroxide) film that blocks the migration of ions through the passive film.

Oxide passive films grown on metal and alloy surfaces have been the subject of recent reviews [6, 8, 27–31]. Recent results obtained with structure-sensitive methods (scanning tunneling microscopy, atomic force microscopy, X-ray diffraction at grazing incidence with synchrotron radiation) have shown that almost all passive films consist of crystalline grains of nanometer dimensions with intergranular boundaries. A model for passivity breakdown explicitly including the role of the passive film nanostructure and the presence of grain boundaries has been proposed [32]. Since oxide grain boundaries have ionic conductivity many orders of magnitude higher than oxide grains [33], the model postulates that grain boundaries in the passive film are less resistant to ion transfer, and thus, the potential drop within the oxide barrier layer in the metal/oxide/electrolyte system is locally smaller at grain boundaries [32]. As a direct consequence, the potential drops at the metal/oxide and/or at the oxide/electrolyte interfaces are locally larger and thus favor the interfacial reaction leading to passivity breakdown at these intergranular boundaries. Depending on the interface (metal/oxide or oxide/solution) at which the increased potential drop is located, different mechanisms were proposed leading to (i) local thinning, (ii) voiding and collapse of the passive film, and (iii) stress-induced fracture.

Useful atomic and subatomic scale information on hydroxylated oxide surfaces and their interaction with aggressive ions (e.g., Cl^-) can be provided by theoretical chemistry, whose application to corrosion-related issues has been developed in the context of the metal/liquid interfaces [34–49]. The application of ab initio density functional theory (DFT) and other atomistic methods to the problem of passivity breakdown is, however, limited by the complexity of the systems that must include three phases, metal(alloy)/oxide/electrolyte, their interfaces, electric field, and temperature effects for a realistic description. Besides, the description of the oxide layer must take into account its orientation, the presence of surface defects and bulk point defects, and that of nanostructural defects that are key actors for the reactivity. Nevertheless, these methods can be applied to test mechanistic hypotheses.

The objective of this chapter is to present the reader with recent modeling work using DFT (mostly) and molecular dynamics (MD) and describing the passive film surface interacting with aqueous environments and aggressive species in the context of passivity breakdown. The chapter includes a short introduction to DFT followed by a section on the modeling of oxide surfaces. Then the interaction of oxide surfaces with water and their hydroxylation is presented. This is followed by a section on the interaction with halides (including chlorides) where the implications for passivity breakdown are discussed.

7.2   DFT: A SHORT INTRODUCTION

7.2.1   The Dirac Challenge: The Limitation of Traditional Approaches

In 1929, theoretical physicist Paul Dirac announced: "The general theory of quantum mechanics is now complete …. The underlying physical laws necessary for the mathematical theory of a large part of physics and the whole of chemistry are thus completely known." The enthusiasm has nearly vanished a century after. The mathematical theory, the Schrödinger equation [50], is too complex to be solved for useful systems. Treatments require approximations. The most frequent is the orbital approximation that states that

FIGURE 7.1   An average electron distribution does not allow taking into account the influence of the real position of one electron on another one, just like the soldiers in the figure.

one can solve the Schrödinger equation for a single electron in the field of all the others. This reduces the many-body problem of N electrons with 3N spatial coordinates to only 3 spatial coordinates for a single electron. Considering a one-electron problem and assuming a known distribution for all the other electrons, the Hartree–Fock (HF) method [51] thus provides the best solution within a mathematical scheme (a set of atomic orbitals) for a system containing several electrons. The many-body problem remains the superposition of individual solutions for each electron (molecular orbital) without coupling. One can reach a limit (self-consistent field (SCF) convergence) by improving the mathematical description of each atomic orbital. Nevertheless, this approach suffers from its original sin: electrons are correlated. The position of an electron, its charge or spin interaction with another one, is sensitive to the real relative position of the two electrons that is not described by an average distribution. It is obvious that the two soldiers in Figure 7.1 do not ignore one another; so do the electrons.

In order to take into account the coupling of electrons (correlation), theoreticians perform calculations *beyond HF*. They calculate the configuration interaction (CI), either rigorously, which requires a huge computational effort, or with approximations, which is less costly but still requires computational efforts. To obtain quantitative results, correlation cannot be neglected. Correlation in systems as simple as C_2 represents a factor of six (in computer time). In solids where many electrons fluctuate, the effect is larger. Correlation is also important in ionic oxides since electrons in ions are strongly correlated. Methods like valence bond (VB) that do not make the orbital approximation (treating several electrons simultaneously) are equivalent to SCF+CI approaches and do not help, except when they allow justifying a choice of approximations.

Hence, the Dirac statement finds its limitation caused by technical problems as soon as the number of electrons is large. Traditional treatment of a many-body problem of N electrons is in practical very difficult, requiring severe approximations. DFT is an efficient way to include correlation, preserving the simplicity of the treatment of a single electron. This is the main reason for its success. Its advantages are that it requires much less computation than the IC or VB methods and that it is adapted to solids and metal–metal bonds and easy to use. The disadvantages are that it is less reliable than IC or VB methods and is not ab initio in a strict sense since an approximate (fitted) term is introduced in the Hamiltonian. It also does not allow comparison of results obtained using different functionals.

7.2.2   DFT

DFT is based on two theorems by Hohenberg and Kohn (H–K) [52]. The first H–K theorem demonstrates that the ground-state properties of a many-electron system are uniquely determined by an electron density. This theorem has been extended to excited states as well. The first remark is that this lays the groundwork for reducing the many-body problem of N electrons with 3N spatial coordinates to only 3 spatial coordinates, through the use of functional of the electron density. As stated earlier, this is the key of its efficiency.

What is a functional? A function of another function. In mathematics, a functional is traditionally a map from a vector space to the field underlying the vector space, which is usually the real numbers. In other words, it is a *function that takes a vector as its argument or input and returns a scalar*. For example, Thomas and Fermi have shown in 1929 that the kinetic energy for an electron gas may be represented as a functional of the density. The first H–K theorem demonstrates that the electronic energy of a system is functional of a single electronic density only.

The second H–K theorem defines the energy functional for the system and proves that the correct ground-state electron density minimizes this energy functional. This is important since it allows using the variational principle as is done for the HF method. If $\rho(r)$ is the exact density, the energy functional $E[\rho(r)]$ is minimum, and we search for ρ by minimizing $E[\rho(r)]$ with $\int \rho(r)dr = N$ (N is the total number of electrons). $\rho(r)$ is a priori unknown.

To implement this theorem, Kohn and Sham proposed three equations [53]:

1. The first equation reintroduces orbitals: the density is defined from the square of the amplitudes. This is needed to calculate the kinetic energy.
2. The second equation defines an effective potential that has a one-body expression. The intractable many-body problem of interacting electrons in a static external potential is reduced to a tractable problem of noninteracting electrons moving in an effective potential.
3. The third equation is the writing of an effective single-particle potential. This effective potential includes the external potential and the effects of the Coulomb interactions between the electrons, for example, the exchange that is already present in HF methodology and *correlation* interactions. The exchange is a functional of the density (LDA method). DFT developments make this term also depend on derivatives of the density: $E_{XC} = E_{XC}(\rho, \nabla\rho)$ for generalized gradient approximation (GGA) and $E_{XC} = E_{XC}(\rho, \nabla\rho, \Delta\rho)$ for meta GGA. GGA methods differ by the expression of the exchange and correlation terms. Hybrid methods, B3LYP (the most popular), PBE0, and HSE03, modify the expression of E_{XC} by incorporating a portion of exact exchange from HF theory with exchange and correlation from other sources.

DFT is good for evaluating total energy, ionization potentials, bond energies, distances, electron density, forces, vibrational frequencies, phonon frequencies, dipole moments, and polarizabilities. However, DFT cannot rigorously predict excited state energies, band gap, and band structures. Hybrid functionals are a way of improving the width of the band gap; indeed, as HF methods overestimate the gap, incorporating a portion of HF results in an improvement. The introduction of Hubbard repulsion (DFT+U methods) on metal atoms, a penalty for localizing electrons at the same place, also enlarges the gap. In an oxide, the conduction band (CB), localized on the metal atoms, is shifted up more than the valence band (VB), localized on the oxygen atoms. The problem remains of the evaluation of the U term. There is no univocal way of imposing U, and U is not transferable from one property to another. For TiO_2, a large value (8 eV) is necessary to fix the band gap of the bulk; this value is excessive to describe the electronic states of a slab. A value of 4 eV is much more adapted to describe the surface states in the gap and does not modify other properties such as binding energies.

For solids, calculation methods taking into account periodicity are much more appropriate than cluster methods that do not. With periodic methods, one has to define a unit cell that is repeated in the space by translations (defined by one to three cell vectors). As an example, two atoms and three cell vectors are enough to generate the MgO bulk. A MgO cluster, whatever its size, is limited, and self-consistent calculations take a lot of time to describe the unsaturated atoms, whereas bulk properties are described by the core atoms of the cluster. This problem remains for surfaces when we are interested in an active site. Calculations take time improving the positions of atoms that are far from this site and represented by drastic approximations such as point charges. For corrosion, we are also interested in surfaces. Four atoms (2 Mg and 2 O) and two cell vectors generate an MgO slab composed of two layers. A cube of width twice the number of atoms (4Mg and 4O) is the smallest reasonable cluster. Not only it is twice as large, but the coordination of each atom is lower; each vertex has three neighbors, while in the double layer, there are five.

The main codes for periodic DFT calculations differ by the choice of basis sets. There are three possibilities: atomic orbitals (CRYSTAL, SIESTA, GAUSSIAN codes), plane waves (Vienna ab initio simulation package (VASP), CASTEP, DACAPO, CPMD, Quantum-ESPRESSO codes), or both with a partition of the space between atomic spheres and intermediate regions (APW, LMTO, KKR, WIEN2K codes). We will shortly describe the advantage and disadvantage of the two extreme choices taking CRYSTAL and VASP as representative codes.

CRYSTAL [54] uses atomic orbitals and this has many advantages. All the expertise of theoreticians, analyzing properties in terms of atomic orbitals, such as Mulliken charges and overlap populations, is preserved. Periodicity may be absent (molecular calculation) or present with any dimension: 1D (polymers), 2D (surfaces), or 3D (bulk materials). Empty space between one slab and its image created by periodicity does not complicate calculation since atomic orbitals are localized. A slab calculation defined as a 3D calculation with a large vacuum between the slabs (see the following for VASP calculations) differs from the 2D calculations even when the vacuum is very large. All the Hamiltonians are available. HF is the most natural choice using atomic orbitals. This means that hybrid functionals are easily available since they include a HF contribution. The inconveniences are mainly of two kinds. One is the difficulty to treat sophisticated problems. It may take much effort to get convergence, and one has to start by HF and small basis sets before using hydride DFT and larger basis sets. Metallic systems are almost impossible to handle. The other is related to the so-called basis set superposition error, BSSE. To describe a molecule with accuracy, one has to use diffuse orbitals in addition to localized atomic orbitals. When these diffuse orbitals superpose, there are numerical artifacts. In a periodic calculation, especially using a small unit cell, the diffuse orbitals of neighboring cells overlap yielding too diffuse crystalline orbitals. The solution is to truncate the usual basis sets used for molecular calculations. This may be seen as an advantage since we thus eliminate redundant orbitals and obtain implicitly very diffuse orbitals from the overlap of others less diffuse. This has however two drawbacks. Calculations strongly depend on the basis sets, and BSSE may be important. This is also the case for systems which include both periodic and molecular components, such as adsorption. The evaluation of adsorption energy requires the calculation of the adsorbate alone (the molecule) and that of the surface with and without the adsorbate (periodic slabs). If the molecule is calculated with a truncated basis set, it will not be accurately described. One has to calculate the molecule with the complete set of orbitals of the extended system.

VASP [55, 56] is a periodic DFT program in three dimensions using plane waves and pseudopotentials. The choice of plane waves is natural for a solid because they are delocalized in the whole space and have long been used by physicists. Plane waves constitute a universal basis set since they do not need to be defined for species in different chemical environments. The mathematical manipulations are simple and easy to implement, but the number of plane waves necessary to describe correctly the wave function is huge. Plane waves must fill the unit cell, imposing 3D. Surfaces must then be described using slabs repeated in the normal direction and introducing a vacuum region between them large enough to prevent any interaction. Nevertheless, the vacuum region must be filled by plane waves, and this contributes to the computational cost. Electrons of the inner shells of an atom, localized in a very small region close to the nuclei, are the most difficult to describe with plane waves; indeed localization is obtained by the superposition of numerous plane waves. One avoids calculating them by using pseudopotentials as soft as possible or the projector-augmented wave (PAW) method. The latter mimics the atomic orbitals and allows calculating HF contributions. Hybrid DFT methods are thus available; nevertheless, pure DFT is more natural using plane waves. The advantage of VASP is a fast and robust electronic convergence that renders its use easy. Geometry optimization methods are very efficient. VASP avoids the problem of choice of basis set and of the BSSE even though a cutoff has to be chosen for the number of plane waves. The disadvantages are the lack of easy information derived from orbital analysis (Mulliken charges and overlap populations are not available) and the difficulty to have accurate reference energies for atoms and molecules. Plane waves are not appropriate for their description. In the case of evaluation of the adsorption energy of an atom, this is the difficult part. One has to define as unit cell a huge asymmetric box avoiding interaction between atoms equivalent by translation and avoid broadening techniques that are usually present in the calculation to compensate the limited sampling of spaces.

7.3   MODELING OF OXIDE SURFACES

The most important task for a theoretician is the choice of a model. It may seem a paradox to emphasize that taking the largest model may not be recommended, even after a presentation of the codes with the implicit desire to be able to calculate the largest possible systems. The main contribution of a theoretician is to help analysis, which is by nature a simplification rather than to produce numbers. In this regard, a model does not have to be complete and sophisticated; it should be constructed so as to explore the strict necessary causes of a phenomenon. Ockham's razor is a principle of parsimony that recommends selecting the fewest number of assumptions that provides the correct answer. Nevertheless, if one wants to reach the limit of the state of the art, the improvement of the model should prevail over that of the technique. Physics should not depend on the technique provided that mistakes are avoided. In the following, we want to outline some constraints in the choice of models for oxides.

7.3.1   Respect of Stoichiometry

The respect of stoichiometry is the first requirement for modeling an oxide. Oxidation states must be those that we want to describe. A metal oxide in which the cation is in the highest oxidation state has a set of crystal orbitals completely occupied, the VB. These are predominantly localized on O atoms with the oxidation number −2 even though electrons are partially delocalized on the cations. The other crystal orbitals are empty and mainly localized on the cations, forming the CB. Chemistry (adsorption, defects) acts to maintain this feature or to restore it, if it is lost. A deviation of stoichiometry does not allow the ideal filling; there are states in the gap, population of the CB or depopulation of the VB.

A cluster model of one atom, supposed to model the reaction site, its first neighbors (ions of opposite charges) and next-nearest neighbors (ions of the same charge), and the following shells will never comply with the requirement of correct stoichiometry. Even in slabs and periodic calculations, the same problem may appear. Let us consider the ABO_3 perovskite with the $(11\bar{2}4)$ orientation in which AO and BO_2 layers alternate. A slab model is generated by two cleavages. Each cleavage between successive layers generates equal amounts of AO and BO_2 surfaces and even if one surface is more stabilized by relaxation than the other, the cleavage associates the existence of the surfaces. In the case of $BaTiO_3$, the layers BaO and TiO_2 are neutral, if we consider formal oxidation states. In the case of $LaMnO_3$, the corresponding layers LaO^+ and MnO_2^- are charged. However, this orientation imposes less charge separation than the (100) orientation, and the stacking is of the lowest energy.

To avoid the presence of a dipole moment perpendicular to the surface, it is tempting to build a symmetric model. In this case, the slab cannot respect the stoichiometry, and the filling of the bands is incorrect. Chromium oxide, a very important metal oxide for corrosion, has the same corundum structure as aluminum oxide. Perpendicular to the basal plane, this structure alternates A and B hexagonal compact planes of O^{2-} anions. The Cr^{3+} cations fill interstitial octahedral sites following three planes, a, b, and c. Hence, the plane stacking is A-a-b-B-c-a-A-b-c-B-a-b-A-c-a-B-b-c-A- and so on. Since this stacking is that of the real crystal, it becomes easy to define a cleavage respecting the stoichiometry, for example, between two successive chromium planes a-b and b-c, thus forming a stoichiometric slab b-B-c-a-A-b.

7.3.2 Electroneutrality

Electroneutrality is the second requirement. It is not possible to introduce a global charge in a periodic system in a straightforward way because the repetition would generate unrealistic macroscopic charges. As a consequence, a chlorine anion, for example, must be associated with a counter ion. Thus, in order to model a chlorine anion, either we have to replace an OH^- ion by Cl^- (a substitution leaving the global charge unaffected) or we have to add HCl or NaCl (an addition with the charge of the cation compensating that of an ion). O vacancies in MgO are neutral (F centers with two electron in the cavity) or charged (F^+ centers with only one electron in the cavity). Removal of any atoms from a stoichiometric MgO slab creates neutral defective entities. A charged system (F^+ centers) necessitates the introduction of a charged ion. The hydroxylation of a surface cation (addition of OH^-) allows localizing one electron there, leaving only one electron instead of two in the cavity. To be complete, VASP and CRYSTAL allow adjusting, in a punctual calculation, the number of electrons to simulate the desired charge state. This procedure thus yields a supercell with an overall charge, which is compensated by introduction of a uniform jellium background charge to avoid divergence of the Coulomb term. This treatment does not allow comparing different systems, unless careful corrections are applied [57–59].

There are mainly four defects that can have an important role on several properties of an oxide: oxygen vacancy, metal vacancy, interstitial oxygen, and a Frenkel defect (interstitial chromium coupled to a chromium vacancy in the case of chromium oxide). Some properties depend on the experimental method of synthesis and so can it be for defects. Some charged defects induce a polaron made of electrons and holes and a polarization field. The local structural deformation extends to 3 Å around the defect. For chromium oxide, a neutral chromium or oxygen vacancy or Frenkel defect yields an electronic state in the band gap and modifies the magnetic state near the defect. Migration of these defects is a very important aspect of this oxide. Nevertheless, stoichiometry is the rule: defects are difficult to model.

Redox reactions often lead to open shell system with a spin state. Imposing the correct spin state to the model is often a requirement for a realistic description of the oxide and to maintain low-energy states.

7.3.3   Inclusion of Temperature and Pressure

Studies on perfect surfaces are most often model studies that do not express the complexity of the real systems. *In situ* conditions imply the consideration of a reaction medium that cannot be simplified to a few individual molecules. An adsorbate is never one single molecule except in ultrahigh vacuum (UHV) conditions. Pressure and medium (i.e., solvent) have to be taken into account.

A first modeling consists in the introduction of numbers extracted from periodic calculations of thermodynamic formula. Trying to bridge the gap between zero-temperature/zero-pressure techniques, the appropriate thermodynamic potential to consider is the Gibbs free energy G(T,p). If DFT total energies are entered in a suitable way into a calculation of G(T,p) for a material surface, ab initio thermodynamics is the result, and the predictive power of the first-principles technique can be extended to a more relevant temperature and pressure range [60].

Let us take the example of water adsorption. The (de)hydration process on a surface is described by equilibrium:

$$\text{Surface} + n_{\text{ads}}\text{H}_2\text{O}_{\text{gas}} \leftrightarrow (n_{\text{ads}}\text{H}_2\text{O})/\text{surface} \qquad (7.1)$$

where n_{ads} stands for the number of adsorbed water molecules per surface unit cell. The Gibbs free energy associated is approximated as

$$G = (E(n_{\text{ads}}\text{H}_2\text{O})/\text{surface}) - E(\text{surface}) + n_{\text{ads}}E(\text{H}_2\text{O gas}) - n_{\text{ads}}\mu\text{H}_2\text{O} \qquad (7.2)$$

where the term

$$\Delta E = (E(n_{\text{ads}}\text{H}_2\text{O})/\text{surface}) - E(\text{surface}) \qquad (7.3)$$

is the internal energy of the surface at 0 K as calculated by a periodic DFT code. The Gibbs energy depends as linear expression on $\mu\text{H}_2\text{O}$, the chemical potential of water in the gas phase depending on T and P, (PV-TS). It is next introduced as a parameter in an expression including the chemical potential of the gas and thus P and T.

7.3.4   Main Features of Adsorption on Oxides

Adsorption of molecules on stoichiometric oxides occurs via acid–base mechanisms maintaining the electronic gap associated with the stoichiometry. In UHV and dry conditions, semiconducting oxides, such as TiO_2, have smaller gaps than insulating materials such as MgO and are more reactive, allowing dissociation of molecules on the surfaces. This dissociation is formally heterolytic, the fragments adsorbed on surfaces being ions of opposite charges. Surface hydroxylation may change this picture, affecting all cationic and anionic sites and leaving only Brønsted sites with weaker interactions.

The adsorption of radicals or the creation of surface defects (vacancies) does not allow maintaining the gap associated with the stoichiometry. However, this may be achieved by pairing them. For example, 2H will give $H^+ + H^-$ as it happens on surfaces of irreducible oxides such as MgO [61] for which H_2 adsorption produces $H^+ + H^-$, one being absorbed on Mg and the other on O. Reactivity is often controlled by the possibility of maintaining the electron count of the stoichiometric compound, that is, maintaining the oxidation states of the atoms and the presence of a gap between a fully occupied VB and a completely empty CB. If the initial conditions have introduced a deviation (the presence of defects or that of adsorbates that have reduced the surface), the reactivity will tend to restore the optimal conditions. On a pristine surface, an O atom adsorbs on a surface oxygen atom to form the peroxide O_2^{2-}, whereas on a reduced surface, it will adsorb inside a defect or on a cation to restore the stoichiometry. The radical Cl (an acceptor-like atomic adsorbate) adsorbs weakly on a perfect surface, and there is no large difference between the adsorption sites, for example, the Ti or O sites of the stoichiometric TiO_2 surface [62]. This is confirmed for the Cl/MgO(100) system [36]. However, on the reduced TiO_2 surface, Cl adsorption occurs preferentially at the O vacancy, forming two Ti–Cl bonds and replacing the missing oxygen [36]. The formal Cl_2 dissociation takes place with formation of Cl^-/Ti^{4+} and $Cl^-/$vacancy entities. The reduced surface is here more reactive than the clean nondefective one. The gap is restored and this is a low spin state. Since the O vacancy costs more energy to create on MgO than on TiO_2, the restoration returns a larger energy gain with the former [63].

Metal oxides differ by the reducibility of the cation. Some metal cations presenting different oxidation states and being in the highest oxidation state (e.g., Ti^{4+} in TiO_2) can be reduced (to Ti^{3+} in Ti_2O_3). CeO_2, Nb_2O_5, V_2O_5, and Ta_2O_5 are reducible oxides. This possibility decreases the cost of a radical adsorption on the O atom. An electropositive atom (hydrogen or metal) can thus become a cation adsorbed at the O site, each adsorbate being associated with the reduction of one cation (Ti^{4+} to Ti^{3+} in the case of TiO_2). Hydrogen atoms are adsorbed as H^+ on the bridging O^{2-} surface ions of TiO_2, building only OH bonds that are stronger than MH bonds [63, 64]. For metal oxides in which the cation is in the highest oxidation state, defects are O vacancies and then only reduction may occur. Defect (O vacancy) formation is easier to perform on reducible oxides. Metal adsorption and binding of atoms (H) to a supported metal on oxide depend on the number of electrons available for a transfer. Metal oxides in which the cations are not in the highest oxidation state are already reduced. In this case, cation vacancies may appear as a reoxidation process.

Chromium oxide has a rather complicated electronic structure. It is antiferromagnetic. The so-called spinwaves are difficult to take into account in periodic calculations. The band gap, approximately 3.4 eV, is a bit larger than the one of TiO_2. The very complex magnetic properties make the choice of the functional for DFT calculations very delicate.

7.4 INTERACTION WITH WATER AND SURFACE HYDROXYLATION

7.4.1 Adsorption on MgO(100)

MgO(100) represents a controversial case with regard to molecular versus dissociative water adsorption [65]. The controversy is mainly between experiment and theory. Experimentally, high hydroxyl coverage is observed [66]. However, most of the quantum calculations [64, 67–78] agree in predicting that the (100) faces do not dissociate water even at very low coverage [79]. The presence of defects can change the adsorption leading to dissociation [80, 81].

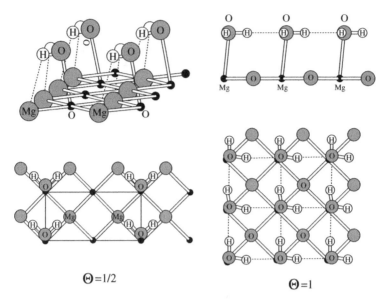

$\Theta = 1/2$ $\Theta = 1$

FIGURE 7.2   Side view (top row) and top view (bottom row) models of the molecular adsorption of water on the MgO(100) surface. At $\Theta=1/2$ (left column), the OH bonds, parallel to the surface, are oriented along the Mg–O rows. At $\Theta=1$ (right column), the interaction of the H atoms between the adsorbed species forces the molecule to rotate with the OH bonds oriented along the Mg–Mg rows. Reprinted from Ahdjoudj et al. [78], with permission from Elsevier.

A few recent calculations [82, 83] have found a mixed molecular and dissociative adsorption. Dissociation is supposed to occur on the low-coordinated surface atoms [68, 79, 84–88], on steps [75, 76, 86], and when surfaces are reconstructed [89], eventually leading to the presence of facets with other orientations [90].

One main argument supporting that isolated water molecule is adsorbed without dissociation on perfect flat MgO(100) surfaces is the weak basicity of the surface oxygen ions that are strongly stabilized by the Madelung field. At moderate coverage ($\theta = 1/2$), the electron lone pair of the oxygen atom interacts with one surface Mg^{2+} Lewis acidic site (Fig. 7.2). The water molecule is oriented roughly parallel to the surface. This orientation allows forming hydrogen bonds with the surface O^{2-} basic sites.

The orientation of the molecule changes with the coverage (Fig. 7.2) The sequence of the H atoms between the adsorbed species at $\theta = 1$ forces the molecule to rotate with the OH bonds oriented along the Mg–Mg rows with respect to the orientation for $\theta = 1/2$. Obviously, increasing the coverage leads to water molecules interacting one with the other, simulating a solvent effect. An ordered layer of water molecules is formed, resembling ice [78, 91, 92]. According to both CRYSTAL [78, 93] and VASP calculations [63], the molecular adsorption mode is more stable than the dissociative one. Indeed, the proton and the hydroxyl of the dissociated molecule recombine to form water. The weak binding also explains the adsorption mode on MgO(100), roughly parallel to the surface. The *frontier orbitals* are not very close in energy and the orbital control is modified. A stronger interaction is that maximizing the overlap rather than that controlled by the smaller energy difference. The reactive electron pair of the water is thus the p pair and the orientation of the water is parallel to the surface plane [68].

7.4.2   Adsorption on TiO$_2$ (110)

The interaction of water with stoichiometric rutile TiO$_2$ (110) surface has been extensively studied. It is both a probe system for fundamental research and a system of great interest for many technological issues. Experiments show the existence of a stable phase of molecular adsorbed water at low temperature [94]. Dissociation occurs at higher temperature and is mediated by surface defects. Most theoretical studies conclude to facile dissociation on the perfect surface [95, 96] or to a mixed mode with both molecular and dissociated molecules [97].

This comparison of the molecular adsorption mode versus the dissociative one is not trivial. The choice of the method of calculation as well as the slab thickness is difficult. The water molecule, once adsorbed, may have many orientations. For coverage of one monolayer, the molecular mode may be stabilized, the adsorption energy being 1.090 eV versus 1.051 eV for the dissociated mode [98]. A clear picture is thus difficult to get. For rutile TiO$_2$, the dissociative mode is closer in energy to the molecular one than for MgO. This may be explained by the stronger basicity of surface oxygen atoms in rutile TiO$_2$ than in MgO.

Anatase is another interesting phase of TiO$_2$. The natural crystallographic faces are the (101) and (100) orientations. The acid–base adsorption mechanism is very similar to the one of the metal/oxide already discussed. On the (101) surface, molecular adsorption is exothermic (16.6 kcal/mol for full coverage) [99]. In contrast, water dissociates on the (001) plane, which is more reactive [100]. It is worth noticing that the surface morphology has a very important effect on the hydroxyl coverage [100].

7.4.3   Water–Oxide Interface

The previous section deals with adsorption of individual water molecules in ultravacuum conditions. Are these considerations still valid at coverage higher than the monolayer, taking into account the solvent effect? Water is obviously a solvent that has been widely studied, the structure and chemistry at the water/metal interface being critical for the properties in many biological, chemical, and materials systems. Understanding the complex structure of the water–ion–adsorbate/metal interface and its dependence on potential has posed a major theoretical challenge for over 100 years [101].

We have mentioned that in VASP, the slabs have to be repeated in the direction perpendicular to the surface, leaving a large interstitial vacuum region. This region may be used to mimic situations that differ from UHV conditions and filled by *the medium*. This approach allows us to represent explicitly the molecules of the medium that influence the adsorbate/substrate interaction. Filling the empty space by water is the first step. The structure of water is taken from the ice structure and the width of the vacuum region is adapted to have an epitaxy on both interfacial regions.

For modeling the water/MgO interface, first, pure water was inserted as a solvent [39]. Since quantum calculations are performed at 0 K, the water structure at the interface was found to be proton ordered and to possess the structure of ice.

Solvents may promote heterolytic cleavage by stabilization of the ions formed, and dissociation of a water molecule may be facilitated by the surrounding water molecules if the uncovered surface is not reactive enough. It is thus natural to expect that the water solvent phase is necessary to allow dissociation of individual water molecules at the MgO(100) surface. Modeling was performed with several layers of ice in epitaxy with the MgO slab above

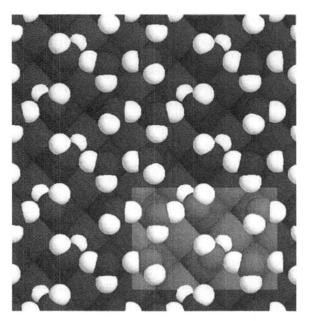

FIGURE 7.3 Top view model of the ordered structure of water on MgO(100) at 1 monolayer coverage. The rectangular highlighted c(3 × 2) unit cell contains 6 adsorbed water molecules: 2, nearly perpendicular to the surface plane, are dissociated, and 4, lying parallel to the surface, are nondissociated. Reprinted from Odelius [83], with permission from the American Physical Society.

and below [102]. Cubic ice was chosen to match the MgO(100) symmetry. Similarly, hexagonal ice has been introduced in the interspace of the nickel slab with (111) orientation, which better matches the hexagonal symmetry [103]. Optimizations were run for different interspaces, and it was found that partial dissociation occurred for a high local water density allowing full coverage. The first water layer was found to partially dissociate in agreement with experimental data that also show partial dissociation of a physisorbed water layer on a nondefective MgO surface (Fig. 7.3) [82, 83].

Miyamoto et al. [104] performed MD simulation of a NaCl solution on MgO(001). It was found that the nearest water molecules to the surface dissociate into H^+ and OH^-. H^+ quickly adsorbs on the surface, and the nondissociated water molecules remain well ordered.

Taylor et al. have investigated the changes in the structure of water at the metal interface as a function of potential for Cu(111) [105] and Ni(111) [103] surfaces. A method was introduced for tuning the electrochemical potential of a half cell using periodic plane wave DFT and a homogeneous countercharge, which explicitly models the countercharge by a plane of ions. The method uses two reference potentials, one related to the potential of the free electron in vacuum and the other related to the potential of H_2O species far from the electrode. This approach allows modeling electrochemical systems [57, 58]. The results show that the applied potential modifies the atomic structure of the water/metal interface as shown on Figure 7.4 for the Pd(111) surface. An interfacial phase diagram was proposed with changes in the reactivity of the interface. One of the key features for simulation was found to be the modeling of the polarizability of the interface. The calculations were able to estimate the potentials of both the electroreduction and electrooxidation of water. It was shown that

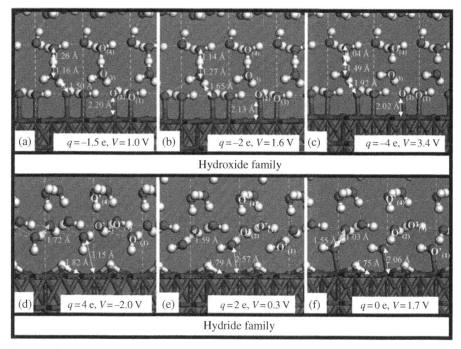

FIGURE 7.4 Side view models showing the variation of the structure of the electrode/water interface as a function of applied potential for (a–c) Pd(111)/hydroxide and (d–f) Pd(111)/hydride surface terminations. Reprinted from Filhol and Neurock [57], with permission from Wiley.

an electron transfer from the metal to the water layer yields an adsorbed hydride and an aqueous hydroxide ion via a proton transfer.

More specifically, Filhol and Doublet, with a new correction of the method, have investigated the structure and stability of a water monolayer over a Pd(111) surface in electrochemical environment with simulation of the potential [59]. They have analyzed water monolayers and found that reactivity depends on the orientation of the water molecules. The water molecules pointing toward the vacuum were found to be mostly electrochemically inactive. On the contrary, those lying parallel to the surface underwent oxidative adsorption under oxidizing conditions, and those pointing toward the surface showed reductive adsorption under reducing conditions. These authors have also proposed an electrochemical diagram for the stability of the water molecules layers. A phase transformation occurs from a positively charge H-up to a negatively charged H-down phase at a given potential. This can be considered as a kind of disproportionation with different charged domains.

VASP is also able to simulate the Daniell battery [106]. The model is a bimetallic slab consisting of two parts in epitaxy, one made of copper and the other of zinc. The interspace between the successive slabs is filled by four layers of water, originally in a hexagonal ice arrangement, and the optimization is run. One layer of water decomposes, the OH⁻ being adsorbed on Cu and the H⁺ on the Zn. There is a relaxation of the metal surfaces at the interfaces, some Zn atoms moving significantly outward. A limited periodic model accounts then for a typical electrochemical reaction.

7.5 INTERACTION WITH AGGRESSIVE SPECIES AND IMPLICATIONS FOR PASSIVE FILM BREAKDOWN

In this section, we discuss atomistic modeling studies of the interaction of model passivated surfaces with halide ions. First, the study using conventional DFT of chloride adsorption and subsurface penetration on defect-free hydroxylated nickel oxide surfaces characteristic of passivated nickel surfaces is presented. Then the implications of using DFT+U are discussed as well as the interaction with other halides. This is followed by one example on the effect of implementing surface defects characteristic of those observed experimentally on a passivated nickel surface. Finally, the application of reactive MD modeling to more complex systems including a substrate metal (copper) covered by a passive film (copper oxide) in interaction with a chloride-containing aqueous solution is presented.

7.5.1 Interaction of Defect-Free Hydroxylated NiO Surfaces with Cl Atoms Modeled by DFT

Periodic conventional DFT was used in this study [47]. The calculations were performed using VASP [55, 56] with GGA [107] and using an energy cutoff of 270 eV (defined from plane wave convergence test performed on bulk NiO) and ultrasoft pseudopotentials [108, 109]. The Brillouin zone integrations have been performed using a Monkhorst–Pack grid [110] of 4×4×1 for the (2×2) supercell used as derived from the 8×8×8 grid with which accurate bulk parameters were obtained with good compromise between calculation accuracy (<0.01 eV) and computational cost but without simulating the magnetic properties of NiO (see the following).

The defect-free hydroxylated NiO(111) structure was modeled by a slab of three NiO bilayers covered by an outermost layer of hydroxyl groups (Fig. 7.5) [34, 47]. The surface is unreconstructed in agreement with the experimental results on the passivation of Ni(111) surface in aqueous acid solution [111–119] and on the hydroxylated surface of thermal NiO (111) oxide surfaces [120, 121].

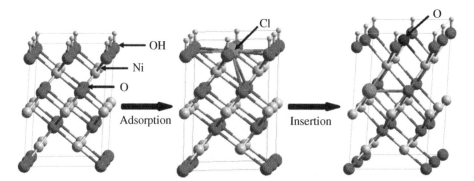

FIGURE 7.5 Side view models for the adsorption and insertion of chloride ions on the hydroxylated NiO(111) surface showing the Cl-free hydroxylated surface, adsorption of chloride ions by substitution of hydroxide ions, and subsurface insertion by exchange between oxygen and chloride ions. A (2×2) supercell with a Cl coverage of 25% is shown. Reprinted from Bouzoubaa et al. [47], with permission from Elsevier.

Figure 7.5 shows the protocol adopted to simulate the reactions of adsorption and subsurface insertion of chlorides. The adsorption reaction was selected as a model for the competitive adsorption of the aggressive ions with hydroxyl groups at the surface of the passive film and is characteristic of the adsorption-induced local thinning mechanism of passivity breakdown. The reaction of subsurface insertion was also modeled because it is the initial step of the penetration-induced voiding mechanism [5, 8, 10]. Adsorption was modeled by substituting the surface OH groups by Cl atoms at coverages of 25, 50, 75, and 100% made possible by using a (2×2) supercell. Subsurface insertion was modeled by exchanging one adsorbed Cl of the topmost anionic layer with one O atom of the first inner anionic layer of the oxide at Cl coverages of 25, 50, 75, and 100. All structures mapping the configuration space were optimized using a conjugated gradient algorithm. The structural optimization was performed at 0 K. The atomic positions of the three lower atomic layers of the oxide slab were frozen to mimic the bulk. The four upper layers of the adsorbed (Ni–O–Ni–OH(Cl)) and inserted (Ni–O(Cl)–Ni–OH(Cl)) structures were allowed to relax in the x-, y-, and z-directions.

The adsorption and insertion energies were calculated assuming the following reaction (X stands for Cl):

$$NiO-(OH)_4 + nHX \rightarrow NiO-(OH)_{4-n}-X_n + nH_2O \qquad (7.4)$$

and using the following expression:

$$\Delta E_{subst/insert} = \left[E\left(NiO-(OH)_{4-n}-nX\right) + nE(H_2O) - E\left(NiO-(OH)_{4-n}\right) - nE(HX)\right]/n$$
$$(7.5)$$

where $E\left(NiO-(OH)_{4-n}-nX\right)$ and $E\left(NiO-(OH)_{4-n}\right)$ are the total electronic energies of the hydroxylated NiO surface with and without X, respectively, obtained after separate geometry optimizations and $E(HX)$ and $E(H_2O)$ are the total electronic energies of HX and H_2O optimized separately, as isolated molecules. The number of X atoms in the slab is $0 \leq n \leq 4$. A negative value of the energy indicates an exothermic process.

Figure 7.6 shows the optimized adsorbed structures for a Cl surface coverage increasing up to surface saturation. The adsorbed Cl structures tend to form a O–Ni–OH/Cl atomic trilayer (marked by a rectangle) characteristic of the building trilayers encountered in the layered crystalline structure of the nickel hydroxychloride (Ni(OH)Cl) bulk compound. However, strong relaxation is observed at low coverage (25%), splitting the mixed OH/Cl topmost anionic layer. It decreases with increasing Cl coverage and vanishes at surface saturation. Increasing the Cl surface coverage increases the repulsive interactions in the topmost layer constrained by the lattice parameter of the oxide. The adsorption energy increases by approximately 1.1 eV for a Cl coverage increasing from 25 to 100%, going from exothermic at low coverage to endothermic at high coverage as also shown in Figure 7.6. Thus, these conventional DFT calculations indicate that Cl adsorption is energetically favorable at low coverage on the nondefective hydroxylated NiO(111) surface but would not be at high surface coverage. In addition, the geometry optimization suggests that dissolution is not promoted by adsorption on this surface, since no bonding is lost between the Ni atoms of the topmost (Ni–OH(Cl)) bilayer and the O atoms of the underlying oxide.

The optimized Cl-inserted structures are also shown in Figure 7.6 for Cl coverages of 25, 50, 75, and 100%. The subsurface insertion of a Cl atom into the first anionic plane of the

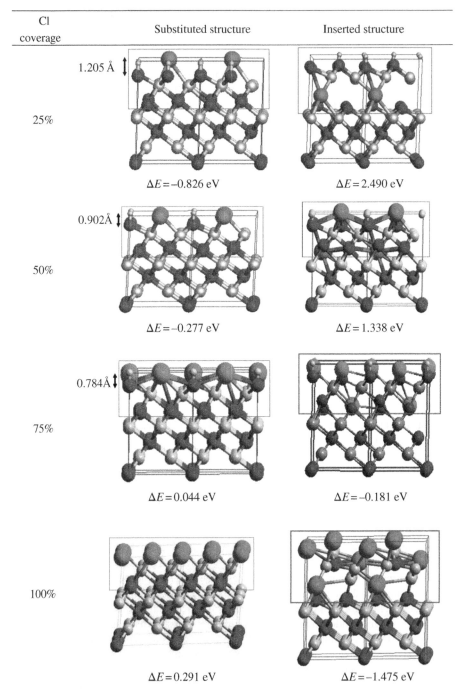

Cl coverage	Substituted structure	Inserted structure
25%	1.205 Å $\Delta E = -0.826$ eV	$\Delta E = 2.490$ eV
50%	0.902Å $\Delta E = -0.277$ eV	$\Delta E = 1.338$ eV
75%	0.784Å $\Delta E = 0.044$ eV	$\Delta E = -0.181$ eV
100%	$\Delta E = 0.291$ eV	$\Delta E = -1.475$ eV

FIGURE 7.6 Side view models of the hydroxylated NiO(111) surface showing Cl-substituted structures and Cl-inserted structures after geometry optimization at adsorption coverages of 25, 50, 75, and 100%. The substitution/insertion energies are given. Adapted from Bouzoubaa et al. [47], with permission from Elsevier.

oxide leads to surface relaxations that depend on the Cl surface coverage. At 25%, little relaxation is observed in the topmost Ni–OH bilayer, but bonding is lost between the Ni atoms and O atoms of the underlying oxide, giving a highly unstable surface structure (high positive value of ΔE) likely to dissolve. This suggests a hybrid mechanism of local thinning of the passive film in which dissolution and thus thinning of the passive film could be promoted not only by Cl adsorption in the topmost plane of the surface but would require penetration below the topmost surface. At 50 and 75%, relaxations are observed in the surface O/Cl–Ni–OH/Cl trilayer (Fig. 7.6), but the bonding with the underlying oxide increases with coverage stabilizing the inserted structures (decreasing positive value of ΔE). At 100%, subsurface insertion leads to reconstruction characterized by a strong interlayer mixing occurring in response to the increase of the electrostatic repulsion in the topmost layer. The reconstruction allows maintaining bonding with the underlying oxide, and the saturated adsorbed structure becomes stable (negative value of ΔE).

7.5.2 Interaction with Cl⁻ and Other Halides Using Periodic DFT+U

Spin polarized, periodic DFT+U calculations have also been used to simulate the interaction of defect-free hydroxylated NiO(111) structure with chlorides and with other halides [49]. In this approach, the effects of strong intra-atomic electronic correlations are taken into account by adding an on-site Coulomb repulsion, namely, a Hubbard term (U) to the DFT Hamiltonian [122–129], allowing a more adequate reproduction of band gaps and magnetic properties for transition metal oxides. NiO is an antiferromagnetic insulator with planes of collinear spin parallel to (111). With the use of a U term of 6.3 eV, a band gap of 3 eV was obtained in this study, much closer to the range of 3.6–4.3 eV found experimentally [130, 131] than the value of 0.3 eV calculated with conventional DFT. An antiferromagnetic state, a magnetic moment of 1.67 µ/Ni atom, and a cell parameter of 0.41636 nm were found. The Brillouin zone integrations were performed using a Monkhorst–Pack grid of 6×6×6 k-points for bulk simulation.

The defect-free hydroxylated NiO(111) structure was modeled by a NiO slab covered by an outmost layer of hydroxyl groups similar to that described earlier (Fig. 7.5), except that it contained four NiO bilayers instead of three in order to take into account the +–+– antiferromagnetic state of (111)-oriented NiO. A (2×2) supercell allowed varying the halide (X) coverage from 25 to 100%. Adsorption and subsurface insertion were modeled using the same protocol as described earlier (Fig. 7.5). The structural optimization was performed at 0 K. The atomic positions of the two upper NiO bilayers were allowed to relax in the x-, y-, and z-directions, and those of the two lower bilayers were frozen to mimic the bulk. The cell parameters were frozen to the bulk values. The energies of substitution/insertion of the halides were calculated using Equation (7.2).

Figure 7.7 shows the variation of the substitution energies with increasing halide coverage for Cl, F, Br, and I and that of the insertion energies for Cl and F. The values for OH substitution by Cl vary from −0.465 to −0.108, 0.146, and 0.343 eV for 25, 50, 75, and 100% Cl coverages, respectively, showing the same trend as that calculated with conventional DFT (Fig. 7.6). The optimized adsorbed structures (not shown) also show the splitting of the mixed OH/Cl topmost anionic layer that decreases with increasing Cl coverage and vanishes at surface saturation. The increasing anion–anion lateral interaction explains the decreasing substitution energy that becomes positive at 75 and 100% coverage. When an F ion substitutes the OH group, splitting in the mixed topmost anionic layer vanishes due to the smaller size of F compared to Cl. Accordingly, the substitution energy increases, and the process

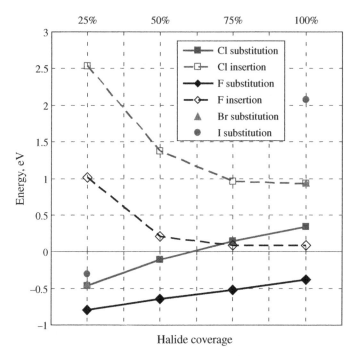

FIGURE 7.7 Substitution/insertion energies as a function of the increasing halide coverage: Cl substitution, F substitution, Br substitution, I substitution, Cl insertion, and F insertion. Adapted from Bouzoubaa et al. [49], with permission from Elsevier.

remains exothermic (negative values) up to 100% coverage despite the decrease observed with increasing coverage (Fig. 7.7). For bigger anions such as Br and I, Figure 7.7 shows that the substitution energy decreases with increasing anion size for a given coverage as calculated at 25 and 100%. These data show that OH substitution by halides at the defect-free hydroxylated NiO(111) structure is favored both by a decreasing size halide size and a decreasing surface coverage. Among the halides, F is the most strongly adsorbed with an F/OH substitution that remains exothermic at all coverages. This calculated tendency for halide adsorption is in line with the tendency reported for halide complexation with metal ions that also show that fluorides form the most stable complexes [10], which has been proposed as an explanation for the stronger increase of the passive current and faster and more general depassivation observed in the presence of fluorides.

Figure 7.7 also shows the variation of the insertion energies with increasing halide coverage for Cl and F. The same trend is found for Cl as that calculated with conventional DFT, except at high coverage for which the energies remain positive. This is because the surface reconstruction observed at the highest coverage and stabilizing the Cl-inserted structure is not reproduced with the DFT+U calculation. The subsurface Cl insertion remains an endothermic process at all coverages. This endothermic behavior relates to strong constraints imposed by the insertion of Cl in the NiO lattice. The same tendency is observed when inserting F in the subsurface anionic layer: the F insertion energy is endothermic. The endothermic behavior is however less marked than for Cl, which relates to the NiO structure less distorted by insertion of F compared to Cl. These data show that the substitution of OH by halides is favored over the halide insertion in the subsurface anionic layer of the model passive film surface. They confirm and extend to all halides the insight obtained by the conventional DFT calculations that the larger the halide is, the more endothermic the subsurface insertion is. It is also confirmed that the subsurface insertion energy decreases with increasing OH substitution.

7.5.3 Effect of Implementing Surface Defects in the Hydroxylated Surface Structure

A suitable model for investigating the surface reactivity of the NiO barrier layer of the passivated Ni surface must include (111)-oriented terraces separated by monoatomic step edges according to the experimental observations [111–119]. In addition, the terrace width and the slab thickness must be similar to those used for the investigation of the nondefective surface in order to assess the step edge effect by a relevant comparison. A periodic model satisfying these criteria was obtained from a cut of the NiO bulk lattice oriented parallel to the (533) crystallographic plane (Fig. 7.8) [48].

The terrace and step edge orientations are (111) and (010), respectively. The slab cell includes 3 NiO bilayers as for the nondefective surface, and the surface was saturated by hydroxyl groups (i.e., fully hydroxylated) [34, 47]. Like on the defect-free surface, Cl adsorption was modeled by substituting the surface OH groups by Cl atoms at 25, 50, 75, and 100% coverages, and subsurface insertion was modeled by exchanging one adsorbed Cl of the topmost anionic layer with one O atom of the first inner anionic layer of the oxide (Fig. 7.8). Conventional DFT was also applied to optimize the geometry and to calculate the substitution/insertion energies as described earlier. The (2×1) supercell adopted for the

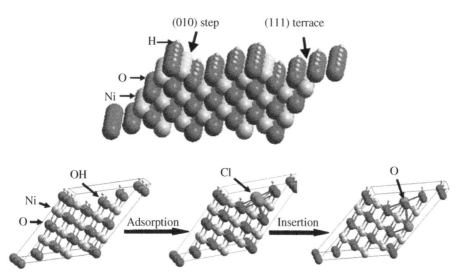

FIGURE 7.8 Side view models of the hydroxylated NiO(533) surface with the (010) step edges and the (111) terraces and showing the adsorption of chloride ions on the Cl-free hydroxylated surface by substitution of hydroxide ions and their subsurface insertion by exchange between oxygen and chloride ions. A (1×2) supercell with a Cl coverage of 50% is shown. Adapted from Bouzoubaa et al. [48], with permission from Elsevier.

hydroxylated NiO(533) surface allowed varying the surface coverage of adsorbed chlorides with finite values of 25, 50, 75, and 100%.

A noticeable effect of the presence of the step edges on the surface reconstruction of the Cl-substituted structures is observed. Substructures of Ni(OH)$_2$, Ni(OH)Cl, or Ni(Cl)$_2$ composition are formed and detached from the step edges, as shown in Figure 7.9. Due to repulsive interactions in the topmost layer constrained by the lattice parameter of the oxide, the adsorption energy increases by 0.4 eV for a Cl coverage increasing from 25 to 100%, but less markedly than on the nondefective surface, and remains exothermic with this conventional DFT simulation (Fig. 7.9). Unlike the substituted structures, the reconstruction of the subsurface inserted structures does not lead to the formation of substructures of Ni(OH)Cl or Ni(OH)$_2$ type and to clear detachment from the step edges (Fig. 7.9). This indicates that after

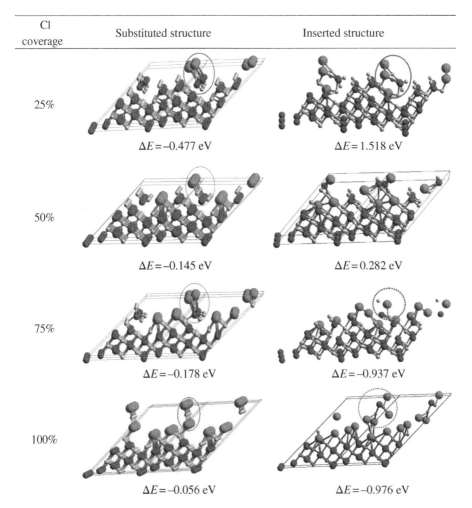

FIGURE 7.9 Side view models of the hydroxylated NiO(533) surface showing the (010) step edges, the (111) terraces, and Cl-substituted structures and Cl-inserted structures after geometry optimization at adsorption coverages of 25, 50, 75, and 100%. The substitution/insertion energies are given. Adapted from Bouzoubaa et al. [48], with permission from Elsevier.

subsurface insertion of the Cl atoms, film dissolution is not promoted. At surface saturation, the subsurface inserted structures become more stable than the adsorbed structures (Fig. 7.9), indicating a possible bifurcation from the Cl adsorption-induced oxide thinning mechanism to the penetration-induced mechanism of passivity breakdown.

Although the boundaries between crystalline oxide grains, which play a key role in passivity breakdown [32], were not modeled, nor were the oxygen vacancies postulated in the point defect model of growth of passive films [6, 19–21], the aforementioned summarized data have strong implications on the corrosion behavior of passivated surfaces. These implications apply to passive films grown on Ni surfaces in Cl-free electrolyte and subsequently exposed to chloride, for which the Cl content is significantly below a coverage of one equivalent monolayer [132] and the terrace structure of which is unchanged [117, 133]. They do not apply directly to passive films grown in Cl-containing electrolytes, which have a markedly higher Cl content in the film and at the surface [132] and the crystallization of which is blocked at high Cl concentration [134].

The aforementioned data point to the determining role of the step edges in the surface reactivity. Substructures of $Ni(OH)_2$, $Ni(OH)Cl$, and $Ni(Cl)_2$ composition are formed at the step edges, depending on the Cl coverage of the substituted structures, and their possible continuous detachment suggests that the dissolution of the oxide may lead to local thinning. The calculations of the energy of detachment of the substructures from the step edges also indicate a promoting effect on their dissolution, due to the presence of Cl atoms in the substructures in agreement with the catalytic effect of chlorides at the basis of the adsorption-induced thinning mechanism of passivity breakdown. Thus, it appears that although the dissolution of the oxide cannot be induced by Cl adsorption on the extended nondefective (111) terraces, this becomes possible in the presence of step edges at the surface. Quite interestingly, the simulation data also suggest that, in the case of terraces that are more extended than those modeled by the (533) surface, the global Cl coverage of the oxide surface may remain low and that it is only necessary to maintain a near saturation at or in the immediate vicinity of the step edges to ensure a continuous dissolution of the oxide film.

The orientation and height of the step edges is also expected to significantly influence their stability and contribute to inhibit or promote dissolution. Monoatomic step edges including kinks are expected to be less stable and to dissolve more rapidly than (010) step edges, as indicated by the experimental observations [117]. Multiatomic step edges (e.g., (010) nanofacets) are expected to be more stable than monoatomic step edges and would inhibit the oxide dissolution at the surface of the oxide grains. Further insight into these structural aspects could be obtained with atomic-scale modeling.

The subsurface inserted structures suggest a possible bifurcation from the Cl adsorption-induced oxide thinning mechanism to the penetration-induced mechanism of passivity breakdown at saturation coverage of the surface. Again, this would not require the saturation of the extended (111) terraces but only that of the step edges and their immediate vicinity. Indeed, the simulated structure and the calculated insertion energy show that at 100% Cl coverage, the subsurface insertion of Cl atoms becomes favorable, whereas the detachment energy of the step edge substructure markedly increases, showing that the subsurface inserted structure will form and that the dissolution at the step edge will be inhibited. This is an atomic-scale evidence supporting a possible route for the penetration of Cl atoms into the oxide lattice, as postulated by the Cl penetration mechanism of passivity breakdown. Further discussion of the validity of this mechanism would require simulating O vacancies, transport processes through the oxide, and voiding mechanisms, as well as taking into account the role of the grain boundaries.

7.5.4   Reactive MD Modeling of Cl Interaction with Passivated Copper Surfaces

For many atomistic-scale simulations, conventional MD is used due to its capability to handle large numbers of particles ($>1 \times 10^5$). However, the method is poorly adapted to electrochemical reactions due to fixed charge implementation. To overcome this problem, reactive methods implementing variable charge or charge transfer and temporal evolution of charge states have been developed. In the study discussed in the following [135], the only one so far addressing passive film breakdown, reactive force field (ReaxFF) [136], a bond order-based empirical force field method simulating bond breaking and formation during MD simulation [137–139], has been used. ReaxFF combines a bond order/bond distance concept with a polarizable, geometry-dependent charge calculation method [140] and has been developed to include metal oxides and chlorides [141, 142]. Characteristics of quantum chemistry effect are employed in multiple components of particle interactions such as bond energy, over-/undercoordination, lone-pair energy, valence angle, torsion, hydrogen bond, and van der Waals and Coulomb interactions. Dynamic, self-consistent charges of cations/anions are also implemented using electronegativity equalization method. For a detailed description, the reader is directed to the literature of the original authors [135, 136, 141, 142].

Figure 7.10 shows the system constructed as well as the protocol used for simulation of the interaction of passivated copper with chloride-containing water. (111)-oriented copper forms the base metal substrate. The Cu slab (2×504 Cu atoms) is reunited by the periodic boundary conditions along the z-direction. Two 0.6–0.7 nm thick Cu_2O films, each composed of 768 and 384 Cu and O atoms, respectively, are constructed on each side of the metal slab. Each Cu_2O film exposes a different termination, O enriched (bottom) or O deficient (top), so as to simulate different corrosion resistances of the oxide. We note that this construction does not take into account the actual orientation ($Cu_2O(111)$) nor the actual thickness somehow larger (up to 1.2 nm) reported for the Cu(I) passive film grown on Cu(111) [143–146]. As an initial thermal perturbation, this system was relaxed in a $3.063 \times 3.095 \times 6.5 \text{ nm}^3$ unit cell for 10 ps at 300 K (Fig. 7.10, left) prior to filling the empty space with aqueous solution (1269 H_2O molecules) of 10 M (228 Cl^- ions) or 20 M (457 Cl^- ions) concentration in chloride ions (Fig. 7.10, middle). The unrealistic high Cl concentration, even for stable pits, was argued by the authors to compensate for the limited timescales (300 ps) of MD simulations in order to make accessible passive film breakdown by accelerated corrosion kinetics. The system relaxed for 250 ps is shown in Figure 7.10 (right). Surface hydroxylation of the oxide surface, an essential feature of passive film surfaces, is not discussed by the authors.

The resulting structure shows that Cl adsorption is promoted on the O-deficient termination of the oxide film as shown by much more adsorbed chloride ions at both 10 and 20 M concentrations. Examining the density distribution of copper along the z-axis of the unit cell, the authors conclude that Cl adsorption enables preferential removal of Cu atoms from the O-deficient copper oxide. This would proceed by formation of nonuniform Cu–Cl complexes that subsequently dissolve, thus leading to oxide thinning. After reaction, the copper atoms attain higher charge as shown by the temporal evolution of their average charge, the effect being more marked on the O-deficient termination of the oxide and increasing with 20 M Cl^- concentration. The authors speculate that, after dissolution, this may increase nonuniform adsorption by attracting nearby chloride ions by Coulomb interactions based on the increasing intensity of the Cu–Cl pair distribution function with time. The authors also conclude to the absence of passivity breakdown of the passive film in these conditions of simulation since no perforation of the passive film by chloride ions is attained.

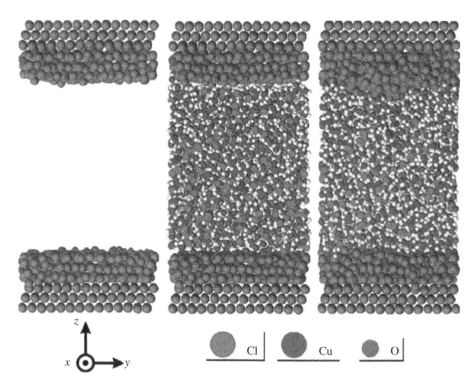

FIGURE 7.10 Side view models showing the system and protocol adopted for the reactive molecular dynamics simulation of the interaction of chloride ions with passivated copper surfaces. Left: Cu(111) slab covered by Cu_2O thin films with O-deficient (top) and O-enriched (bottom) terminations after thermal relaxation at 300 K. Middle: filling the gap with 20 M Cl^- aqueous solution (pH 7). Right: complete system after relaxation for 250 ps at 300 K showing preferential interaction of the chlorides ions with the O-deficient surface. Periodic boundary conditions apply along the x-, y-, and z-directions.Adapted from Jeon et al. [135], 1229, with permission from the American Chemical Society.

Implementation of a surface defect was performed in the same study to facilitate chloride adsorption and promote passive film breakdown within the limited timescales of the MD simulations. Figure 7.11 shows the system constructed for simulation. The unit cell now larger ($8.168 \times 8.843 \times 9$ nm^3) includes 12960 and 2640 Cu and O atoms to describe the copper substrate covered by the passive film. It has been thermally relaxed at 300 K for 10 ps. A circular defect has been implemented on the O-enriched surface of the passive film by removing all O atoms within a radius of 0.8 nm. This means introducing O vacancies exposing the Cu atoms of the passive films to the aggressive ions. It is however quite unlikely that O vacancies would cluster to form such a large ad-island at the surface of the passive film in contact with an electrolyte without being hydroxylated. The void space has been filled with 15708 H_2O molecules and 2827/5655 Cl^- ions, equivalent to 10/20 M Cl concentration conditions. Each concentration set has been relaxed for 300 ps at 373 K, assuming boiling temperature to accelerate corrosion in the time frame of the simulations. Figure 7.11 (right) shows the resulting structure for the 20 M concentration. The simulated structures show that adsorption is favored in the present case on and near the defective site of the passive film,

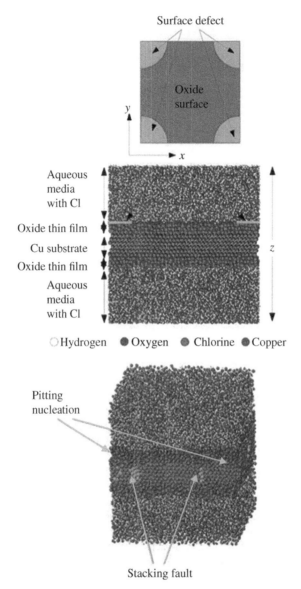

FIGURE 7.11 Implementation of a defect exposing Cu atoms on the O-enriched passive film for copper in interaction with a 20 M Cl⁻ aqueous solution (pH 7). Left: schematic top view of the unit cell at the passive film surface. The circular defect (radius of 0.8 nm) is configured using periodic boundary conditions. Middle: side view of the unit cell for the complete system. Right: side view after 300 ps relaxation showing Cl adsorption and pit nucleation at the implemented defect and defects generated in the bulk substrate around the defects site. Adapted from Jeon et al. [135], 1229, with permission from the American Chemical Society.

which confirms that Cl adsorption is promoted on the O-deficient ad-island on the passive film. The authors conclude pit nucleation on the basis that the passive film is invaded by chloride ions and perforated at the defect site with chloride ions reaching the interface between oxide and substrate. Based on the calculation of the number of opposing nearest pair atoms, the authors also conclude the presence of stacking faults in the substrate, a possible artifact of the ReaxFF simulation caused by difficulties in the potentials being able to discriminate between fcc and hcp stacking. The thin metal substrate combined with the high amount of adsorbed chloride would result in high pressure being produced from the developing Cu–Cl clusters, leading to mechanical failure observed near the perforated site.

The data analysis points to the faster reaction at the top surface containing the defect site based on the temporal evolution of the average charge of the Cu atoms, observed at both 10 and 20 M concentrations. Maps of the charge distribution of the Cu atoms of the top passive film and their temporal evolution confirm that corrosion evolves preferentially at/near the defect site of initiation. Cross-sectional distributions along the z-axis of the system of the Cu and Cl atomic density and of the average charge on Cl are shown in Figure 7.12 at 1 and 300 ps. It is shown that the Cu atomic density is smeared at the top/bottom sides of the passivated slab, while chloride infiltrates the passive film and the topmost substrate planes, yielding copper chlorides clusters as confirmed by charge on the Cl ions and confirming the perforation of the passive film. The formation of copper chlorides clusters and their steady increase are confirmed by the temporal evolution of the Cu–Cl pair distribution function (not shown). We note that these cross sections evidence the penetration of chloride ions into the passive films independently of the presence of the defect site at the surface, only its intensity increases if the defect site is present. The Cl distribution also shows locally depleted regions in the aqueous media due to the consumption by the corrosion reaction, suggesting that Cl⁻ ions from far field would diffuse to sustain further corrosion in the locally depleted sites.

The authors did not specifically discuss the breakdown mechanism of the passive film from the point of view of the existing models of passivity breakdown in their report. However, they emphasize oxide thinning based on the results that Cu atoms are displaced from the passive film into the first layers in the aqueous media and that their charge is consistent

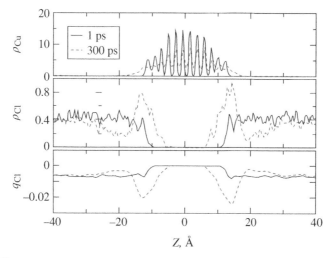

FIGURE 7.12 Cross-sectional distributions of the Cu and Cl atomic density and of the average charge on Cl along the z-axis of the 20 M concentrated system relaxed for 1 and 300 ps. Reprinted from Jeon et al. [135], 1229, with permission from the American Chemical Society.

with the formation of copper chloride clusters. This would be in support of the adsorption-induced thinning mechanism of passivity breakdown. However, it is not clear from the data and their report if the observed changes are indeed caused by the dissolution of the passive oxide (displacement of the corrosion products in the bulk electrolyte is not shown) or by the nucleation and growth of copper chlorides at the surface of the passive film by a solid-state reaction mechanism. The observed changes could indeed be caused by the volume increases accompanying the formation of these compounds due to a Pilling–Bedworth ratio superior to unity.

The penetration of chlorine atoms into the passive films is suggested by close examination of the relaxed structures in Figures 7.10 and 7.11. It seems to occur independently of the O-enriched or O-deficient nature of the films and of the implemented defect site. However, this aspect was not addressed by the authors in their study and thus cannot be further discussed. Detailed studies relevant for testing the penetration-induced voiding mechanism of passivity breakdown would require implementing O vacancies as point defects not only at the surface but also in the bulk of the passive films of appropriate crystalline structure. Implementation of field-assisted transport in the passive film and at its interfaces would also be required.

Despite its current limitations, reactive MD appears powerful to perform large-scale simulations and thus to test systems approaching the complexity of passivated metal or alloy surfaces interacting with aqueous environments containing aggressive ions. Future work should include more realistic simulations of passivated surfaces based on the experimental knowledge of their atomic structure, their defects, and their nanostructural features. Potential-driven atomic transport and pH should be implemented. Testing of the existing model of passivity breakdown would also require implementing realistically their characteristic features.

7.6 CONCLUSION

The application of atomistic modeling, mostly using DFT but also using reactive MD, to oxide surfaces, characteristic of passivated metallic surfaces, and their interaction with water and aggressive ions has been presented in this chapter and discussed in the context of passivity breakdown. The modeling of oxide surfaces and the reviewed DFT data on MgO and TiO_2 surfaces and their hydroxylation can be summarized as follows:

- The choice of the model is crucial.
- Stoichiometry is required and is sometimes difficult to obtain. The quality of the description of the electronic structure depends also on this requirement.
- A straightforward modeling follows the electroneutrality rule. Attention must be paid when substitution or charged defects is under scrutiny.
- Acid–base mechanisms allow maintaining the electronic gap with surface ions in their highest oxidation states.
- Surface hydroxylation may induce surface reduction.
- At low coverage, water dissociation is more likely to occur on TiO_2 than on MgO surfaces.
- At high coverage, the structure of the adsorbed water layer has a very important role and may change the picture.

- Recent DFT modeling allows us to explore complex water structures in electrochemical environment.

Due to the complexity of the real passivated systems, atomistic modeling of passivity breakdown is only nascent. Only a few existing DFT studies have addressed the interaction of hydroxylated oxide surface with chlorides and other halides. It is shown that:

- The OH substitution by halides is favored on defect-free hydroxylated NiO surfaces both by a decreasing halide size and a decreasing surface coverage, F being the most strongly adsorbed. F/OH substitution is exothermic at all substitution proportions, and Cl/OH only at relatively low proportions. The substitution of OH by X is favored over the X insertion in the subsurface layers. The larger the halide, the more endothermic the insertion is. Subsurface energy decreases with increasing OH substitution proportion.
- With increasing Cl/OH substitution proportion, substructures of $Ni(OH)_2$, $Ni(OH)Cl$, or $Ni(Cl)_2$ composition are formed and detached from the surface at step edges, suggesting a major role of these surface complexes in the Cl adsorption-induced thinning mechanism of the passive film. At saturation of adsorbed Cl at step edges, the results suggest a possible bifurcation from the Cl adsorption-induced oxide thinning mechanism to the penetration-induced mechanism of passivity breakdown.

Reactive MD can be applied to simulate systems including the base metal, the passive film, and the chloride ion-containing aqueous environment as shown by one study on passivated copper. This study has shown that interaction with chloride ions is favored by an oxygen-deficient termination of the passive film, either local on a defect site or homogeneous on the surface. The interaction leads to the formation of copper chloride clusters making the oxide layer thinner by consumption of copper but with no dissolution in the bulk electrolyte. Perforation occurs faster on passive films with surface defects exposing Cu atoms, suggesting a promoting effect of surface oxygen (or hydroxyl) vacancies on breakdown by an adsorption-induced thinning mechanism.

Future work should address more realistic conditions for the passivated surfaces, including the complete system: metal(alloy)/oxide/electrolyte, its interfaces, electric field, and temperature effects. The experimental knowledge of the atomic structure of passivated surfaces, their defects, and their nanostructural features needs to be faithfully input when available. Potential-driven atomic transport and pH should be implemented. Testing of the existing models of passivity breakdown also requires a realistic implementation of their characteristic features. It is foreseen that DFT will be applied to test more accurately specific steps of the complex pathways leading to passivity breakdown, while MD simulations will be developed to test the complete reaction pathways.

REFERENCES

1. T. P. Hoar, D. C. Mears, G. P. Rothwell, *Corros. Sci.* 5 (1965) 279.
2. T. P. Hoar, W. R. Jacob, *Nature* 216 (1967) 1299.
3. G. S. Frankel, R. C. Newman (Eds.), *Critical Factors in Localized Corrosion, PV 92–9*, The Electrochemical Society Proceedings Series, Pennington, NJ, 1992.
4. G. S. Frankel, *Mater. Sci. Forum* 1 (1997) 247.

5. G. S. Frankel, *J. Electrochem. Soc.* 145 (1998) 2186.

6. D. D. Macdonald, *Pure Appl. Chem.* 71 (1999) 951.

7. M. Schütze (Ed.), *Corrosion and Environmental Degradation*, Wiley-VCH, Weinheim, 2000.

8. H.-H. Strehblow, Passivity of Metals, in *Advances in Electrochemical Science and Engineering*, Vol. 8, R. C. Alkire, D. M. Kolb (Eds.), Wiley-VCH, Weinheim, 2003, pp. 271–374.

9. P. Marcus (Ed.), *Corrosion Mechanisms in Theory and Practice*, 3rd Edition, CRC Press, Taylor and Francis, New York, 2011.

10. H.-H. Strehblow, P. Marcus, Mechanisms of pitting corrosion, in *Corrosion Mechanisms in Theory and Practice*, 3rd Edition, P. Marcus (Ed.), CRC Press, Taylor and Francis, New York, 2011, pp. 349–393.

11. Y. J. Kolotyrkin, *Corrosion* 19 (1964) 261.

12. H. Böhni, H. H. Uhlig, *J. Electrochem. Soc.* 116 (1969) 906.

13. H. H. Strehblow, *Werkst. Korros.* 27 (1976) 792.

14. B. MacDougall, *J. Electrochem. Soc.* 124 (1977) 1185.

15. H. H. Strehblow, B. Titze, B. P. Loechel, *Corros. Sci.* 19 (1979) 1047.

16. B. P. Loechel, H. H. Strehblow, *J. Electrochem. Soc.* 131 (1984) 522.

17. B. P. Loechel, H. H. Strehblow, *J. Electrochem. Soc.* 131 (1984) 713.

18. C. J. Boxley, H. S. White, *J. Electrochem. Soc.* 151 (2004) B256.

19. L. F. Lin, C. Y. Chao, D. D. MacDonald, *J. Electrochem. Soc.* 128 (1981) 1187.

20. C. Y. Chao, L. F. Lin, D. D. MacDonald, *J. Electrochem. Soc.* 128 (1981) 1194.

21. M. Urquidi, D. D. Macdonald, *J. Electrochem. Soc.* 132 (1985) 555.

22. J. A. Richardson, G. C. Wood, *Corros. Sci.* 10 (1970) 313.

23. K. J. Vetter, H. H. Strehblow, *Ber. Bunsenges. Phys. Chem.* 74 (1970) 1024.

24. N. Sato, *Electrochim. Acta* 16 (1971) 1683.

25. N. Sato, K. Kudo, T. Noda, *Electrochim. Acta* 16 (1971) 1909.

26. B. MacDougall, *J. Electrochem. Soc.* 126 (1979) 919.

27. P. Marcus, V. Maurice, Passivity of metals and alloys, in *Corrosion and Environmental Degradation*, Vol. 19, M. Schütze, (Ed.), Wiley-VCH, Weinheim, 2000, pp. 131–169.

28. J. W. Schultze, M. M. Lohrengel, *Electrochim. Acta* 45 (2000) 2499.

29. C.-O. A. Olsson, D. Landolt, *Electrochim. Acta* 48 (2003) 1093.

30. H.-H. Strehblow, V. Maurice, P. Marcus, Passivity of metals, in *Corrosion Mechanisms in Theory and Practice*, 3rd Edition, P. Marcus (Ed.), CRC Press, Taylor and Francis, New York, 2011, pp. 235–326.

31. P. Marcus, V. Maurice, Oxide passive films and corrosion protection, in *Oxide Ultrathin Films. Science and Technology*, G. Pacchioni and S. Valeri (Eds.), Wiley-VCH Verlag GmbH & Co. KGaA, Weinheim, 2012. pp. 119–144.

32. P. Marcus, V. Maurice, H. H. Strehblow, *Corros. Sci.* 50 (2008) 2698.

33. J. H. Harding, K. J. Atkinson, R. W. Grimes, *J. Am. Ceram. Soc.* 86 (2003) 554.

34. N. Pineau, C. Minot, V. Maurice, P. Marcus, *Electrochem. Solid-State Lett.* 6 (2003) B47.

35. O. Blajiev, A. Hubin, *Electrochim. Acta* 49 (2004) 2761.

36. M. Menetrey, A. Markovits, C. Minot, *Surf. Sci.* 566 (2004) 693.

37. K. Segerdahl, J. E. Svensson, M. H. I. Panas, L. G. Johansson, *Mater. High Temp.* 22 (2005) 69.

38. N. C. Hendy, N. J. Laycock, M. P. Ryan, *J. Electrochem. Soc.* 152 (2005) B271.

39. F. Tielens, C. Minot, *Surf. Sci.* 600 (2006) 357.

40. A. Markmann, J. L. Gavartin, A. L. Shluger, *Phys. Chem. Chem. Phys.* 8 (2006) 4359.

41. J. Kelber, N. Magtoto, C. Vamala, M. Jain, D. R. Jennison, P. A. Schultz, *Surf. Sci.* 601 (2007) 3464.

42. C. D. Taylor, M. Neurock, J. R. Scully, *J. Electrochem. Soc.* 154 (2007) F55.

43. C. D. Taylor, R. G. Kelly, M. Neurock, *J. Electroanal. Chem.* 607 (2007) 167.

44. C. D. Taylor, M. Neurock, J. R. Scully, *J. Electrochem. Soc.* 155 (2008) C407.

45. B. Ingham, N. C. Hendy, N. J. Laycock, M. P. Ryan, *Electrochem. Solid-State Lett.* 10 (2007) C57.

46. M. M. Islam, B. Diawara, V. Maurice, P. Marcus, *Surf. Sci.* 603 (2009) 2087.

47. A. Bouzoubaa, B. Diawara, C. Minot, V. Maurice, P. Marcus, *Corros. Sci.* 51 (2009) 941.

48. A. Bouzoubaa, B. Diawara, V. Maurice, C. Minot, P. Marcus, *Corros. Sci.* 51 (2009) 2174.

49. A. Bouzoubaa, D. Costa, B. Diawara, N. Audiffen, P. Marcus, *Corros. Sci.* 52 (2010) 2643.

50. E. Schrödinger, *Phys. Rev.* 28 (1926) 1049.

51. D. R. Hartree, *Math. Proc. Camb. Philos. Soc.* 24 (1928) 89.

52. P. Hohenberg, W. Kohn, *Phys. Rev. B* 136 (1965) B864.

53. W. Kohn, L. J. Sham, *Phys. Rev. A* 140 (1965) 1133.

54. R. Dovesi, V. R. Saunders, C. Roetti, R. Orlando, C. M. Zicovich-Wilson, F. Pascale, B. Civalleri, K. Doll, N. M. Harrison, I. J. Bush, Ph. D'Arco, M. Llunell, *CRYSTAL09 User's Manual*, University of Torino, Torino, 2009

55. G. Kresse, J. Hafner, *Phys. Rev. B* 47 (1993) 558.

56. G. Kresse, J. Furthmuller, *Comput. Mater. Sci.* 6 (1996) 15.

57. J.-S. Filhol, M. Neurock, *Angew. Chem. Int. Ed.* 45 (2006) 402.

58. M. Mamatkulov, J. S. Filhol, *Phys. Chem. Chem. Phys.* 13 (2011) 7675.

59. J.-S. Filhol, M.-L. Doublet, *Catal. Today* 202 (2013) 87.

60. K. Reuter, M. Scheffler, *Phys. Rev. B* 65 (2001) 035406.

61. J. Leconte, A. Markovits, M. K. Skalli, C. Minot, A. Belmajoub, *Surf. Sci.* 497 (2002) 194.

62. U. Diebold, W. Hebenstreit, G. Leonardelli, M. Schmid, P. Varga, *Phys. Rev. Lett.* 81 (1998) 405.

63. M. Calatayud, A. Markovits, M. Menetrey, B. Mguig, C. Minot, *Catal. Today* 85 (2003) 125.

64. M. Calatayud, A. Markovits, C. Minot, *Catal. Today* 89 (2004) 269.

65. M. A. Henderson, *Surf. Sci. Rep.* 46 (2002) 1.

66. D. Abriou, J. Jupille, *Surf. Sci.* 430 (1999) L527.

67. A. Markovits, J. Ahdjoudj, C. Minot, *Mol. Eng.* 7 (1997) 245.

68. M. R. Chacon-Taylor, M. I. McCarthy, *J. Phys. Chem.* 7610 (1996). 100.

69. C. A. Scamehorn, A. C. Hess, M. I. McCarthy, *J. Chem. Phys.* 99 (1993) 2786.

70. J. Goniakowski, C. Noguera, *Surf. Sci.* 330 (1995) 337.

71. J. Goniakowski, S. Bouette-Russo, C. Noguera, *Surf. Sci.* 284 (1993) 315.

72. D. Ferry, S. Picaud, P. N. M. Hoang, C. Girardet, L. Giordano, B. Demirdjan, J. Suzanne, *Surf. Sci.* 409 (1998) 101.

73. S. Russo, C. Noguera, *Surf. Sci.* 262 (1992) 245.

74. S. Russo, C. Noguera, *Surf. Sci.* 262 (1992) 259.

75. W. Langel, M. Parrinello, *Phys. Rev. Lett.* 73 (1994) 504.

76. W. Langel, M. Parrinello, *J. Chem. Phys.* 103 (1995) 3240.

77. W. Langel, *Surf. Sci.* 496 (2002) 141.

78. J. Ahdjoudj, A. Markovits, C. Minot, *Catal. Today* 50 (1999) 541.

79. F. Finocchi, J. Goniakowski, *Phys. Rev. B* 64 (2001) 125426.

80. C. Chizallet, G. Costentin, M. Che, F. Delbecq, P. Sautet, *J. Phys. Chem. B*, 110 (2006) 15878.

81. D. Costa, C. Chizallet, E. Ealet, J. Goniakowski, F. Finocchi, *J. Chem. Phys.* 125 (2006) 054702.

82. L. Giordano, J. Goniakowski, J. Suzanne, *Phys. Rev. Lett.* 81 (1998) 1271.

83. M. Odelius, *Phys. Rev. Lett.* 82 (1999) 3919.

84. C. Duriez, C. Chapon, C. R. Henry, J. Rickard, *Surf. Sci.* 230 (1990) 123.

85. M. J. Stirniman, C. Huang, R. C. Smith, S. A. Joyce, B. D. Kay, *J. Chem. Phys.* 105 (1996) 1295.

86. J. Günster, G. Liu, J. Stultz, S. Krischok, D. W. Goodman, *J. Chem. Phys. B* 104 (2000) 5738.

87. R. Schaub, P. Thostrup, N. Lopez, E. Laegsgaard, I. Stensgaard, J. K. Nørskov, F. Besenbacher, *Phys. Rev. Lett.* 87 (2001) 26104.

88. M. Menetrey, A. Markovits, C. Minot, *Surf. Sci.* 524 (2003) 49.

89. N. H. de Leeuw, G. W. Watson, S. C. Parker, *J. Chem. Phys.* 99 (1995) 17219.

90. K. Refson, A. Wogelius, D. G. Fraser, M. C. Payne, M. H. Lee, M. Milman, *Phys. Rev. B* 52 (1995) 10823.

91. S. Picaud, C. Girardet, *Chem. Phys. Lett.* 209 (1993) 340.

92. T. Bredow, K. Jug, *Surf. Sci.* 327 (1995) 398.

93. A. Fahmi, C. Minot, *Surf. Sci.* 304 (1994) 343.

94. F. Allegretti, S. O'Brien, M. Polcik, D. I. Sayago, D. P. Woodruff, *Phys. Rev. Lett.* 95 (2005) 226104.

95. A. V. Bandura, D. G. Sykes, V. Shapovalov, T. N. Troung, J. D. Kubicki, R. A. Evarestov, *J. Phys. Chem. B* 108 (2004) 7844.

96. J. Goniakowki, M. J. Gillan, *Surf. Sci.* 350 (1996) 145.

97. P. J. D. Lindan, N. M. Harrison, M. J. Gillan, *Phys. Rev. Lett.* 80 (1998) 762.

98. L. A. Harris, A. A. Quong, *Phys. Rev. Lett.* 93 (2004) 086105.

99. A. Vitatadini, A. Selloni, F. P. Rotzinger, M. Grätzel, *Phys. Rev. Lett.* 81 (1998) 2954.

100. C. Arrouvel, M. Digne, M. Breysse, H. Toulhoat, P. Raybaud, *J. Catal.* 222 (2004) 152.

101. J. O. M. Bockris, K. T. Jeng, *Adv. Colloid Interface Sci.* 33 (1990) 1.

102. C. Minot, *Surf. Sci.* 562 (2004) 237.

103. C. Taylor, R. Kelly, M. Neurock, Phase Transitions Involving Dissociated States of Water at the Electrochemical Ni(111)/H2O Interface, DOE Technical Report LM-06K041, USA, (2006).

104. U. Mart, C. Jung, M. Koyama, M. Kubo, A. Miyamoto, *Appl. Surf. Sci.* 244 (2005) 640.

105. C. D. Taylor, S. A. Wasileski, J.-S. Filhol, M. Neurock, *Phys. Rev. B* 73 (2006) 165402.

106. M. Calatayud, A. Markovits, C. Minot, Periodic DFT studies on adsorption and reactivity on metal and metal oxide surfaces, in *Quantum Chemical Calculations of Surfaces and Interfaces of Materials*, V. A. Basiuk, P. Ugliengo (Eds.), California 91: American Scientific Publisher, 2009, pp. 183–210.

107. J. Perdew, A. Zunger, *Phys. Rev. B* 23 (1981) 5048.

108. D. Vanderbilt, *Phys. Rev. B* 41 (1990) 7892.

109. G. Kresse, J. Hafner, *J. Phys. Condens. Matter* 6 (1994) 8245.

110. H. J. Monkhorst, J. D. Pack, *Phys. Rev. B* 13 (1976) 5188.

111. V. Maurice, H. Talah, P. Marcus, *Surf. Sci.* 284 (1993) L431.

112. V. Maurice, H. Talah, P. Marcus, *Surf. Sci.* 304 (1994) 98.

113. T. Suzuki, T. Yamada, K. Itaya, *J. Phys. Chem.* 100 (1996) 8954.

114. D. Zuili, V. Maurice, P. Marcus, *In situ* investigation by ECSTM of the structure of the passive film formed on Ni(111) single-crystal surfaces, in *Passivity and Its Breakdown*, P. Natishan, H. S. Isaacs, M. Janik-Czachor, V. A. Macagno, P. Marcus, M. Seo (Eds.), PV 97–26, The Electrochemical Society Proceedings Series, Pennington, 1997, p. 1013.

115. D. Zuili, V. Maurice, P. Marcus, *J. Electrochem. Soc.* 147 (2000) 1393.

116. O. M. Magnussen, J. Scherer, B. M. Ocko, R. J. Behm, *J. Phys. Chem. B* 104 (2000) 1222.

117. V. Maurice, L. H. Klein, P. Marcus, *Surf. Interface Anal.* 34 (2002) 139.

118. N. Hirai, H. Okada, S. Hara, *Mater. Trans. JIM* 44 (2003) 727.

119. J. Scherer, B. M. Ocko, O. M. Magnussen, *Electrochim. Acta* 48 (2003) 1169.

120. F. Rohr, K. Wirth, J. Libuda, D. Cappus, M. Baumer, H-J. Freund, *Surf. Sci.* 315 (1994) L977.

121. N. Kitakatsu, V. Maurice, C. Hinen, P. Marcus, *Surf. Sci.* 407 (1998) 36.

122. A. Rohrbach, J. Hafner, G. Kresse, *J. Phys. Condens. Matter* 15 (2003) 979–996.

123. A. Rohrbach, J. Hafner, G. Kresse, *Phys. Rev. B* 70 (2004) 125426 (17 pp).

124. G. Rollmann, A. Rohrbach, P. Entel, J. Hafner, *Phys. Rev. B* 69 (2004) 165107 (12 pp).

125. A. Rohrbach, J. Hafner, *Phys. Rev. B* 71 (2005) 045405 (7 pp).

126. M. Gajdos, J. Hafner, *Surf. Sci.* 590 (2005) 117.

127. M. Petersen, J. Hafner, M. Marsman, *Phys. Condens. Matter* 18 (2006) 7021.

128. J. Hafner, *J. Comput. Chem.* 29 (2008) 2044.

129. A. Rohrbach, J. Hafner, G. Kresse, *Phys. Rev. B* 69 (2004) 075413 (13 pp).

130. M. Zöllner, S. Kipp, K. D. Becker, *Cryst. Res. Technol.* 35 (2000) 299.

131. G. A. Sawatzky, J. W. Allen, *Phys. Rev. Lett.* 53 (1984) 2339.

132. P. Marcus, J.-M. Herbelin, *Corros. Sci.* 34 (1993) 1123.

133. V. Maurice, L. H. Klein, P. Marcus, *Electrochem. Solid-State Lett.* 4 (2001) B1.

134. A. Seyeux, V. Maurice, L. H. Klein, P. Marcus, *J. Electrochem. Soc.* 153 (2006) B453.

135. B. Jeon, S. K. R. S. Sankaranarayanan, A. C. T. Van Duin, S. Ramanathan, *Appl. Mater. Interfaces.* 4 (2012) 1225.

136. A. C. T. van Duin, S. Dasgupta, F. Lorant, W. A. Goddard, *J. Phys. Chem. A* 105 (2001) 9396

137. G. C. Abell, *Phys. Rev. B* 31 (1985) 6184.

138. J. Tersoff, *Phys. Rev. Lett.* 61 (1988) 2879.

139. D. W. Brenner, *Phys. Rev. B* 42 (1990) 9458.

140. W. J. Mortier, S. K. Ghosh, S. Shankar, *J. Am. Chem. Soc.* 108 (1986) 4315.

141. A. C. T. van Duin, V. S. Bryantsev, M. S. Diallo, W. A. Goddard, O. Rahaman, D. J. Doren, D. Raymand, K. J. Hermansson, *Phys. Chem. A* 114 (2010) 9507.

142. O. Rahaman, A. C. T. van Duin, V. S. Bryantsev, J. E. Mueller, S. D. Solares, W. A. Goddard, D. J. J. Doren, *Phys. Chem. A* 114 (2010) 3556.

143. H.-H. Strehblow, B. Titze, *Electrochim. Acta* 25 (1980) 839–850.

144. J. Kunze, V. Maurice, L. H. Klein, H.-H. Strehblow, P. Marcus, *J. Phys. Chem. B* 105 (2001) 4263.

145. J. Kunze, V. Maurice, L. H. Klein, H.-H. Strehblow, P. Marcus, *Electrochim. Acta* 48 (2003) 1157.

146. J. Kunze, V. Maurice, L. H. Klein, H.-H. Strehblow, P. Marcus, *J. Electroanal. Chem.* 554–555 (2003) 113.

8

MULTISCALE MODELING OF HYDROGEN EMBRITTLEMENT

Xu Zhang and Gang Lu

Department of Physics and Astronomy, California State University Northridge, Northridge, CA, USA

8.1 INTRODUCTION

Hydrogen is a major reactant with solids as a result of its strong chemical activity, high lattice mobility, and wide occurrence as H_2 molecule and a constituent of molecular gases and liquids (e.g., moisture in air) [1]. As a consequence, interactions of H with lattice defects, such as vacancies, dislocations, grain boundaries, and cracks, are crucial in determining the influence of this impurity on the properties of solids. These interactions are the ultimate culprit of H embrittlement, which is one of the most important problems in materials science and engineering as almost all metals and their alloys suffer to some extent of H-induced brittleness [2]. Since the early 1960s, H-induced cracking has been responsible for many, if not most, service failures in numerous applications where components and structures come into contact with natural or technological environment—whether aqueous solution, gas, elevated temperature, or irradiation. Moreover, the problem of H-induced cracking directly affects the safety and reliability of engineering systems such as aircraft and aerospace structures, nuclear and fossil fuel power plants, oil and gas pipelines, field equipment, chemical plants, and marine structures, which, if they fail, can cause serious human, environmental, and financial losses. The economic and humanitarian aspects of H-induced cracking failures have led to considerable scientific and engineering efforts directed at understanding and preventing such failures. More recently, materials problems for emerging H economy have attracted great attentions, such as materials for fuel cells, hydrogen storage and production, and so on. All these applications require fundamental understanding of H–metal interactions. In particular, the mechanical properties of the materials are crucial for the success of the relevant applications.

Molecular Modeling of Corrosion Processes: Scientific Development and Engineering Applications, First Edition.
Edited by Christopher D. Taylor and Philippe Marcus.
© 2015 John Wiley & Sons, Inc. Published 2015 by John Wiley & Sons, Inc.

In the past four decades, intense research on H embrittlement has been carried out, and three general mechanisms of H embrittlement have been put forward, including (i) stress-induced hydride formation and cleavage of the brittle phase [3–5], (ii) H-enhanced local plasticity (HELP) [6–9], and (iii) H-induced decohesion and grain boundary weakening [10–12]. However, despite the impressive progress, the underlying atomic processes and the relative importance of the three mechanisms remain uncertain and controversial. It is fair to say that a complete mechanistic understanding of H embrittlement still eludes us [1].

Some of the most fascinating problems in all fields of science involve multiple spatial and/or temporal scales: processes that occur at a certain scale govern the behavior of the system across several (usually larger) scales [13]. In many problems of materials science, this notion arises quite naturally: the ultimate microscopic constituents of materials are ions and electrons; interactions among them at the atomic level (of order nanometers and femtoseconds) determine the behavior of the material at the macroscopic scale (of order centimeters and milliseconds and beyond), the latter being the scale of interest for technological applications [14]. The idea of performing simulations of materials across several characteristic length and timescales has therefore obvious appeal as a tool of potentially great impact on technological innovation. The advent of ever more powerful computers that can handle such simulations provides further argument that such an approach can address realistic situations and can be a worthy partner to the traditional approaches of theory and experiment [15–17]. At each length and timescale, well-established and efficient computational approaches have been developed over the years to handle the relevant phenomena. To treat electrons explicitly and accurately at the atomic scale, methods based on the density functional theory (DFT) [18] and local density approximation [19] can be readily applied to systems containing several hundred atoms. For materials properties at the microscopic scale, molecular dynamics or statics simulations are usually performed employing classical interatomic potentials. Although not as accurate as the DFT methods, the classical simulations are able to provide insight into atomic processes involving considerably larger systems, reaching up to 10^9 atoms [20]. Finally, for the macroscopic scale, finite element (FE) methods are routinely used to examine the large-scale properties of materials considered as an elastic continuum [21].

The challenge in modern simulations of materials science and engineering is that real materials usually exhibit phenomena at one scale that require a very accurate and computationally expensive description and phenomena at another scale for which a coarser description is satisfactory and in fact necessary to avoid prohibitively large computations. Since none of the methods above alone would suffice to describe the entire system, the goal becomes to develop models that combine different methods specialized at different scales, effectively distributing the computational power where it is needed most. It is the hope that a multiscale approach is the answer to such a quest. Sometimes, a full-blown brute force calculation could shadow the crucial physics owing to its sheer amount of information or complexity. On the other hand, effective theories and well-constructed multiscale models could capture essential physics without the distraction of less important details [13, 14]. Overall, multiscale modeling is a vibrant enterprise of multidisciplinary nature. It combines the skills of physicists, materials scientists, chemists, mechanical and chemical engineers, applied mathematicians, and computer scientists. The marriage of disciplines and the concomitant dissolution of traditional barriers between them represent the true power and embody the great promise of multiscale approaches for enhancing our understanding of and our ability to control complex physical phenomena [14].

Modeling H embrittlement is an incredibly challenging task that requires sophisticated multiscale modeling. For example, H-assisted cracking is at the heart of H embrittlement.

To model it, one has to involve *ab initio*-based multiscale approaches. This is because on one hand, a crack propagates by breaking atomic bonds at the crack tip; thus, the atomic process of H attacking metallic bonds has to be captured quantum mechanically. On the other hand, the crack is loaded remotely from exterior surfaces that are far away from the crack tip. Hence, the processes that transmit and magnify the applied load to the crack tip have to be accounted for, which can be modeled most effectively by continuum mechanics. Similarly, the interactions between H and other extended defects such as dislocations, grain boundaries, voids, and so on, inevitably require *ab initio*-based multiscale modeling. Therefore, owing to the inherent multiscale nature of H embrittlement, *ab initio*-based multiscale approaches are the most promising theoretical means to address outstanding experimental and theoretical problems in H embrittlement. In this chapter, we present several *ab initio* multiscale modeling approaches that we have developed recently and have been used to study crucial problems in H embrittlement of metals.

In the following, we first provide an introduction to the multiscale methods, including the Peierls–Nabarro (P-N) model for dislocations, quantum mechanical/molecular mechanical (QM/MM) methods, and quasicontinuum density functional theory (QCDFT) method. We then apply these methods to study some key phenomena and processes in H embrittlement of metals, ranging from HELP, H-assisted cracking, and crucial role of vacancies to H diffusion in extended defects. Finally, we propose a tentative mechanism for H embrittlement and speculate on one future research direction in this area that is interesting to us. The materials presented here primarily reflect our own research interests in multiscale modeling of H embrittlement, and they are by no means exhaustive. Nevertheless, we hope that they offer a glimpse of the current modeling effort of the field and perhaps can serve as an inspiration for future work in this area.

8.2 MULTISCALE MODELING APPROACHES

Conceptually, two categories of multiscale modeling approaches can be envisioned—sequential and concurrent approaches [14]. The sequential modeling approaches attempt to piece together a hierarchy of computational methods in which large-scale models use the coarse-grained representations with information obtained from more detailed, smaller-scale models. The sequential modeling approaches have proven effective in systems in which the different scales are only weakly coupled. The concurrent approaches on the other hand attempt to link models appropriate at each scale together in a combined theory, where the different scales of the system are considered concurrently and communicate with some type of handshaking procedure. These approaches are necessary for systems that are inherently multiscale; that is, systems whose behavior at each scale depends strongly on what happens at the other scales. One important advantage of the concurrent approaches is that they do not require a priori knowledge of the system of interest; neither do they rely on phenomenological models. Thus, the concurrent approaches are particularly useful to explore problems about which little is known at the atomistic level and its connection to larger scales and to discover new phenomena.

8.2.1 P-N Model of Dislocations

The P-N model was first proposed by Peierls [22] and Nabarro [23] to incorporate the details of a discrete dislocation core into a framework that is essentially a continuum. More specifically, the atomistic scale information of the P-N model is contained in the form of the so-called generalized stacking fault energy surface (also referred to as the γ-surface), and the higher scale information is described by a phenomenological continuum description. In fact, the former can be determined quantum mechanically by *ab initio* calculations. The P-N model serves as a link between atomistic and continuum approaches, by providing a means to incorporate information obtained from atomistic calculations (*ab initio* or empirical) directly into continuum models. The resultant approach can then be applied to problems that neither atomistic nor conventional continuum models could handle separately. The simplicity of the P-N model makes it an attractive alternative to direct atomistic simulations of dislocation properties [24–36]. It provides a rapid and inexpensive route to determine dislocation core structure and mobility. Combined with *ab initio* determined GSF energy surface, the P-N model could give rather reliable quantitative predictions for various dislocation properties.

Consider a solid with an edge dislocation in the middle as shown in Figure 8.1: the solid containing this dislocation is represented by two elastic half-spaces joined by atomic-level forces across their common interface, known as the glide plane (---). The goal of the P-N model is to determine the slip distribution on the glide plane, which minimizes the total energy. The dislocation is characterized by the slip (relative displacement) distribution

$$\mathbf{f}(x) = \mathbf{u}(x,0^+) - \mathbf{u}(x,0^-) \tag{8.1}$$

which is a measure of the misfit across the glide plane; $\mathbf{u}(x,0^+)$ and $\mathbf{u}(x,0^-)$ are the displacement of the half-spaces at position x immediately above and below the glide plane. The total energy of the dislocated solid includes two contributions: the nonlinear potential energy resulting from the atomistic interaction across the glide plane, and the elastic energy stored in the two half-spaces associated with the presence of the dislocation. Both energies are functionals of the slip distribution $\mathbf{f}(x)$. Specifically, the nonlinear misfit energy can be written as

$$U_{\text{misfit}} = \int_{-\infty}^{\infty} \gamma[\mathbf{f}(x)]dx, \tag{8.2}$$

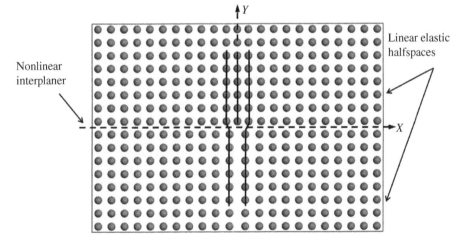

FIGURE 8.1 A schematic illustration showing an edge dislocation in a lattice. The partition of the dislocated lattice into a linear elastic region and a nonlinear atomistic region allows a multiscale treatment of the problem.

where $\gamma(\mathbf{f})$ is the generalized stacking fault energy surface introduced by Vitek [37]. The nonlinear interplanar γ-surface can be determined from *ab initio* calculations. The elastic energy of the dislocation can be calculated reasonably from elasticity theory: the dislocation may be thought of as a continuous distribution of infinitesimal dislocations whose Burgers vectors integrate to that of the original dislocation [38]. Therefore, the elastic energy of the original dislocation is just the sum of the elastic energy caused by all the infinitesimal dislocations (from the superposition principle of linear elasticity theory), which can be written as

$$U_{\text{elastic}} = \frac{\mu}{2\pi(1-\nu)} \int dx \int dx' \ln \frac{L}{|x-x'|} \frac{d\mathbf{f}(x)}{dx} \frac{d\mathbf{f}(x')}{dx'}, \tag{8.3}$$

where μ and ν are the shear modulus and Poisson's ratio, respectively. The variable L is an inconsequential constant introduced as a large-distance cutoff for the computation of the logarithmic interaction energy [39]. The gradient of $\mathbf{f}(x)$ is called dislocation (misfit) density, denoted by $\rho(x)$. The successful application of the P-N models depends on the reliability of both γ-surface and the underlying elasticity theory, which is the basis for the formulation of the phenomenological theory.

The total energy of the dislocation is a function of misfit distribution $\mathbf{f}(x)$ or, equivalently, $\rho(x)$, and it is invariant with respect to arbitrary translation of $\rho(x)$ and $\mathbf{f}(x)$. To regain the lattice discreteness, the integration of the γ-energy in Equation (8.2) was discretized and replaced by a lattice sum in the original P-N formulation

$$U_{\text{misfit}} = \sum_{i=-\infty}^{\infty} \gamma[\mathbf{f}(x_i)]\Delta x, \tag{8.4}$$

with x_i the reference position and Δx the average spacing of the atomic rows in the lattice. This procedure, however, is inconsistent with evaluation of elastic energy in Equation (8.3) as a continuous integral. Therefore, the total energy is not variational. Furthermore, in the original P-N model, the shape of the solution $\mathbf{f}(x)$ is assumed to be invariant during dislocation translation, a problem that is also associated with the nonvariational formulation of the total energy.

To resolve these problems, a so-called semidiscrete variational P-N (SVPN) model was developed [31] that allows the study of narrow dislocations, a situation that the standard P-N model cannot handle. Within this approach, the equilibrium structure of a dislocation is obtained by minimizing the dislocation energy functional

$$U_{\text{disl}} = U_{\text{elastic}} + U_{\text{misfit}} + U_{\text{stress}} + Kb^2 \ln L, \tag{8.5}$$

where

$$U_{\text{elastic}} = \sum_{i,j} \frac{1}{2} \chi_{ij} \left[K_e \left(\rho_i^{(1)} \rho_j^{(1)} + \rho_i^{(2)} \rho_j^{(2)} \right) + K_s \rho_i^{(3)} \rho_j^{(3)} \right], \tag{8.6}$$

$$U_{\text{misfit}} = \sum_{i} \gamma_3 [\mathbf{f}(x_i)]\Delta x, \tag{8.7}$$

$$U_{stress} = -\sum_{i,l} \frac{x_i^2 - x_{i-1}^2}{2} \rho_i^{(l)} \tau_i^{(l)}, \qquad (8.8)$$

with respect to the dislocation misfit density. Here, $\rho_i^{(1)}$, $\rho_i^{(2)}$, and $\rho_i^{(3)}$ are the edge, vertical, and screw components of the general interplanar misfit density at the ith nodal point, respectively, and $\gamma_3(\mathbf{f})$ is the corresponding three-dimensional γ-surface. The components of the applied stress are $\tau^{(1)} = \sigma_{21}$, $\tau^{(2)} = \sigma_{22}$, and $\tau^{(3)} = \sigma_{23}$, respectively. The variables K, K_e, and K_s are the prelogarithmic elastic energy factors, related to the shear modulus, Poisson's ratio, and the dislocation character [39]. The dislocation density at the ith nodal point is $\rho_i = (f_i - f_{i-1})/(x_i - x_{i-1})$, and χ_{ij} is the elastic energy kernel [31].

The first term in the energy functional $U_{elastic}$ is now discretized to be consistent with the discretized misfit energy, which makes the total energy functional variational. Another modification in this approach is that the nonlinear misfit potential in the energy functional U_{misfit} is a function of all three components of the nodal misfit $\mathbf{f}(x_i)$. Namely, in addition to the misfit along the Burgers vector, lateral and even vertical misfits across the glide plane are also included. This allows for the treatment of straight dislocations of arbitrary orientation in arbitrary glide planes [32, 33]. Furthermore, because the misfit vector $\mathbf{f}(x_i)$ is allowed to change during the process of dislocation translation, the energy barrier (referred to as the Peierls barrier) can be significantly lowered compared to the corresponding value taken from a rigid translation. The response of a dislocation to an applied stress is achieved by minimization of the energy functional with respect to ρ_i at the given value of the applied stress $\tau_i^{(l)}$. An instability is reached when an optimal solution for ρ_i no longer exists, which is manifested numerically by the failure of the minimization procedure to converge. The Peierls stress is defined as the critical value of the applied stress that gives rise to this instability. The SVPN model has been applied to study various interesting materials problems related to dislocation phenomena [32–36].

8.2.2 Quantum Mechanics/Molecular Mechanics Method

In atomistic modeling of materials, quantum mechanics (QM) is necessary for a proper treatment of phenomena such as bond breaking, charge transfer, electronic and optical excitations and magnetism, and so on; however, owing to its computational demand, the application of QM has to be restricted to relatively small systems consisting of up to a few hundreds of atoms. On the other hand, atomistic simulations based on empirical interatomic potentials are often capable of describing small-amplitude vibrations and torsions, elastic deformation, electrostatic interactions, and so on, in many materials. Termed as molecular mechanical (MM) methods, these empirical atomistic approaches can treat millions of atoms or more. Therefore, algorithms that combine quantum mechanics and molecular mechanics (QM/MM) poise to offer a promising solution to the computational challenge in atomistic simulations of materials [40–42].

For many molecular systems that are of interest in chemistry and biochemistry, one can partition the QM/MM system by cutting chemical bonds linking the QM and MM parts and then saturate the dangling bonds at the boundary of QM region by the so-called link atoms [41, 43, 44]. This procedure and the similar ones can be justified because of the presence of well-defined and localized chemical bonds in such molecular systems. Unfortunately, for metallic materials, the procedure is no longer valid owing to the delocalized nature of metallic bonding; it becomes impractical to cut and saturate bonds. In fact, the very concept of chemical bonds becomes less appropriate than the band picture for metals. Therefore, more sophisticated ideas have to be developed to deal with metallic cohesion represented by the delocalized electron states across the QM/MM boundary. As a result, far fewer QM/MM-like simulations have been attempted in metallic systems [42, 45–49]. Here, we present some recent development of such ideas for modeling metallic materials [42, 50–53].

In a QM/MM simulation, the computational domain is partitioned into two regions: region I is the primary region where QM simulations are performed, and region II is the surrounding region where classical atomistic simulations are carried out. Although different levels of QM simulations could be employed in region I, here, we focus on the Kohn–Sham density functional theory (KS-DFT) [19]. Similarly, although many empirical potentials could be used in region II, we choose the embedded atom method (EAM) [54] as an example for MM calculations. The total energy of a system including the energy of the region I, the energy for the region II, and the interaction energy between them can be expressed as

$$E_{\text{tot}}[\text{I} + \text{II}] = E_{\text{DFT}}[\text{I}] + E^{\text{int}}[\text{I}, \text{II}] + E_{\text{EAM}}[\text{II}]. \tag{8.9}$$

The interaction between regions I and II is at the heart of any QM/MM method. Depending on the formulation of the interaction energy, the QM/MM methods can be divided into two categories: mechanical coupling and quantum coupling. In the quantum coupling, the interaction energy $E^{\text{int}}[\text{I}, \text{II}]$ is formulated by DFT; the quantum mechanical calculation for region I is carried out in the presence of region II; and a single-particle embedding potential that represents the quantum mechanical effects of region II enters the QM Hamiltonian of region I. On the other hand, in the mechanical coupling, one performs quantum mechanical simulation of region I in the absence of region II and treats the interaction energy $E^{\text{int}}[\text{I}, \text{II}]$ at the MM level.

8.2.2.1 Quantum Coupling The self-consistent determination of the interaction energy at a QM level is the hallmark of the quantum coupling approach. The spatial partition of the entire system is shown in Figure 8.2. In specific, the total energy can be expressed as:

$$E_{\text{tot}}[\rho^{\text{tot}}; \mathbf{R}^{\text{tot}}] = \min_{\rho^{\text{I}}} \left\{ E_{\text{KS}}[\rho^{\text{I}}; \mathbf{R}^{\text{I}}] + E_{\text{OF}}^{\text{int}}[\rho^{\text{I}}, \rho^{\text{II}}; \mathbf{R}^{\text{I}}, \mathbf{R}^{\text{II}}] \right\} + E_{\text{MM}}[\mathbf{R}^{\text{II}}], \tag{8.10}$$

where $\mathbf{R}^{\text{tot}} \equiv \mathbf{R}^{\text{I}} \cup \mathbf{R}^{\text{II}}$ and $\mathbf{R}^{\text{I}}$ and $\mathbf{R}^{\text{II}}$ denote atomic coordinates in region I and II, respectively. The charge density of region I, ρ^{I}, which is the degree of the freedom of the problem, is determined self-consistently by minimizing the total energy functional (Eq. (8.10)). We associate each MM atom in region II with an atomic-centered electron density (ρ^{at}) and a pseudopotential; both of them are constructed a priori. The charge density of region II, ρ^{II}, is defined as a superposition of atomic-centered charge densities ρ^{at} via $\rho^{\text{II}}(\mathbf{r}) = \sum_{i \in \text{II}} \rho^{\text{at}}(\mathbf{r} - \mathbf{R}_i)$, which only changes upon the relaxation of region II ions. The total charge density ρ^{tot} is given by $\rho^{\text{tot}} = \rho^{\text{I}} + \rho^{\text{II}}$. The interaction energy between regions I and II, $E_{\text{OF}}^{\text{int}}$, formulated by orbital-free density functional theory (OF-DFT) [55–57] is defined as follows:

$$E_{\text{OF}}^{\text{int}}[\rho^{\text{I}}, \rho^{\text{II}}; \mathbf{R}^{\text{I}}, \mathbf{R}^{\text{II}}] = E_{\text{OF}}[\rho^{\text{tot}}; \mathbf{R}^{\text{tot}}] - E_{\text{OF}}[\rho^{\text{I}}; \mathbf{R}^{\text{I}}] - E_{\text{OF}}[\rho^{\text{II}}; \mathbf{R}^{\text{II}}] \tag{8.11}$$

The unique feature of OF-DFT is that it allows a QM calculation of energetics by knowing *only* the charge density. The accuracy of OF-DFT is in between KS-DFT and EAM, which is consistent to its usage in the QM/MM method. Finally, the EAM method was developed

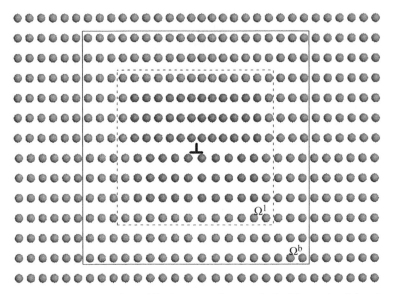

FIGURE 8.2 Schematic domain partition in the QM/MM method with an edge dislocation as an example. The dashed box represents Ω^I, and the solid box represents the periodic box Ω^b. The charge density of region I ρ^I is confined within Ω^I, and the periodic boundary conditions are imposed over Ω^b. From Zhang et al. [53]. © American Physical Society.

specifically to treat metallic systems. $\mu_{emb}[\rho^I, \rho^{II}]$, termed as embedding potential, is of crucial importance to the QM/MM method and is defined as a functional derivative of the interaction energy E_{OF}^{int} with respect to ρ^I:

$$\mu_{emb}(\mathbf{r}) \equiv \frac{\delta E_{OF}^{int}[\rho^I, \rho^{II}; \mathbf{R}^I, \mathbf{R}^{II}]}{\delta \rho^I} \tag{8.12}$$

$\mu_{emb}(\mathbf{r})$ represents the effective single-particle potential that the region I electrons feel due to the presence of the region II atoms (both electrons and ions). We have recently developed a quantum coupling approach in which the interaction energy is calculated by KS-DFT, thus overcoming the limitations of OF-DFT. Based on the constrained DFT, the method is more general and is particularly useful for magnetic materials [58].

8.2.2.2 Mechanical Coupling In the mechanical coupling, the interaction energy E^{int} is evaluated with EAM calculations:

$$E^{int}[I,II] = E_{EAM}[I+II] - E_{EAM}[I] - E_{EAM}[II], \tag{8.13}$$

and thus, Equation (8.9) becomes

$$E_{tot}[I+II] = E_{DFT}[I] + E_{EAM}[I+II] - E_{EAM}[I] \tag{8.14}$$

The advantage of the mechanical coupling is simplicity [59]. It demands nothing beyond what is required for a DFT cluster calculation and two classical MM calculations (one for the entire system and the other for a cluster of region I). On the other hand, the quantum coupling allows a more accurate description, in particular, for the electronic structure of region I. For example, the fictitious surface states due to the cluster calculation of region I in the mechanical coupling can be eliminated in the quantum coupling [52]. Of course, the quantum coupling is more complicated than the mechanical coupling in general.

8.2.3 QCDFT

A widely used and successful concurrent multiscale method is the so-called quasicontinuum (QC) method originally proposed by Tadmor et al. [60]. The QC method combines atomistic models with continuum theories and thus offers an advantage over conventional atomistic simulations in terms of computational efficiency. The idea underlying the QC method is that atomistic processes of interest often occur in very small spatial domains (e.g., crack tip), while the vast majority of atoms in the material behave according to well-established continuum theories. To exploit this fact, the QC method retains atomic resolution only where necessary and coarsens to a continuum FE description elsewhere. This is achieved by replacing the full set of N atoms with a small subset of N_r *representative atoms* or *repatoms* ($N_r \ll N$) that approximate the total energy through appropriate weighting. The energies of individual repatoms are computed in two different ways depending on the deformation in their immediate vicinity. Atoms experiencing large deformation gradients on an atomic scale are computed in the same way as in a standard fully atomistic method. In QC, these atoms are called *nonlocal* atoms. In contrast, the energies of atoms experiencing a smooth deformation field on the atomic scale are computed based on the deformation gradient in their vicinity as befitting a continuum model. These atoms are called *local* atoms.

The total energy E_{tot} (which for a classical system can be written as $E_{tot} = \sum_{i=1}^{N} E_i$, with E_i the energy of atom i) is approximated as

$$E_{tot}^{QC} = \sum_{i=1}^{N_{nl}} E_i(\{\mathbf{q}\}) + \sum_{j=1}^{N_{loc}} n_j E_j^{loc}(\{\mathbf{F}\}). \qquad (8.15)$$

The total energy has been divided into two parts: an atomistic region of N_{nl} nonlocal atoms and a continuum region of N_{loc} local atoms ($N_{nl} + N_{loc} = N_r$).

The original formulation of QC was limited to classical potentials for describing interactions between atoms. However, since many materials properties depend crucially on the behavior of electrons, such as bond breaking/forming at crack tips or defect cores, chemical reactions with impurities, surface reactions and reconstructions, and magnetism, it is desirable to incorporate appropriate quantum mechanical descriptions into the QC formalism. QCDFT is one strategy to fill this role. In specific, QCDFT combines the coarse graining idea of QC and the coupling strategy of QM/MM approaches mentioned earlier [61–63]. Therefore, QCDFT can capture the electronic structure at the defect cores (e.g., crack tip) within the accuracy of DFT and at the same time reach the length scale that is relevant to experiments.

The original QC formulation assumes that the total energy can be written as a sum over individual atom energies. This condition is not satisfied by quantum mechanical models. To address this limitation, in the present QCDFT approach, the nonlocal region is treated by either the mechanical or the quantum coupling QM/MM approaches [61, 62]. Here, for simplicity, we only discuss the mechanical coupling QCDFT in which the nonlocal

QC formulation is based on the coupling between KS-DFT and EAM with the interaction energy determined by EAM. The local region, on the other hand, is also dealt with by EAM. This makes the passage from the atomistic to the continuum seamless since the same underlying material description is used in both. This description enables the model to adapt automatically to changing circumstances (e.g., the nucleation of new defects or the migration of existing defects). The adaptability is one of the main strengths of QCDFT, which is missing in many other multiscale methods.

More specifically, in the QCDFT approach, the material of interest is partitioned into three distinct domains as shown in Figure 8.3: (i) a nonlocal quantum mechanical DFT region (region I), (ii) a nonlocal classical region where classical EAM potentials are used (region II), and (iii) a local region (region III) that employs the same EAM potentials as region II. The coupling between region II and III is achieved via the QC formulation, while the coupling between region I and II is accomplished by the QM/MM scheme [59]. The total energy of the QCDFT system is thus given by [61]

$$
\begin{aligned}
E_{\mathrm{tot}}^{\mathrm{QCDFT}} &= E^{\mathrm{nl}}[\mathrm{I+II}] + \sum_{j=1}^{N_{\mathrm{loc}}} n_j E_j^{\mathrm{loc}}(\{\mathbf{F}\}) \\
&= E_{\mathrm{DFT}}[\mathrm{I}] - E_{\mathrm{EAM}}[\mathrm{I}] + E_{\mathrm{EAM}}[\mathrm{I+II}] + \sum_{j=1}^{N_{\mathrm{loc}}} n_j E_j^{\mathrm{loc}}(\{\mathbf{F}\}),
\end{aligned}
\tag{8.16}
$$

where $E^{\mathrm{nl}}[\mathrm{I+II}]$ is the total energy of the nonlocal region (I and II combined with the assumption that region I is embedded within region II), $E_{\mathrm{DFT}}[\mathrm{I}]$ is the energy of region I in the absence of region II computed with DFT, $E_{\mathrm{EAM}}[\mathrm{II}]$ is the energy of region II in the absence of region I computed with EAM, and $E_{\mathrm{EAM}}[\mathrm{I+II}]$ is the energy of the nonlocal region computed with EAM.

Other combinations of quantum mechanical and classical atomistic methods may also be implemented in QCDFT. The advantage of the present implementation is its simplicity; it has nothing more than a DFT calculation and an EAM–QC calculation. Another advantage of the QCDFT method is that if region I contains multiple atomic species while region II contains only one atom type, there is no need to develop reliable EAM potentials to describe each species and their interactions. This is because if the various species of atoms are well within region I, the energy contributions of these atoms are canceled out in the total energy calculation. This advantage renders the method useful in dealing with impurities, which is an exceedingly difficult task for conventional empirical simulations.

The equilibrium structure of the system is obtained by minimizing the total energy in Equation (8.16) with respect to all degrees of freedom. Because the time required to evaluate $E_{\mathrm{DFT}}[\mathrm{I}]$ is considerably more than that required for computation of the other EAM terms in $E_{\mathrm{tot}}^{\mathrm{QCDFT}}$, an alternate relaxation scheme turns out to be useful. The total system can be relaxed by using conjugate gradient approach on the DFT atoms alone while fully relaxing the EAM atoms in region II and the displacement field in region III at each step. An auxiliary energy function can be defined as

$$
E'[\{\mathbf{q}^{\mathrm{I}}\}] \equiv \min_{\{\mathbf{q}^{\mathrm{II}}\},\{\mathbf{q}^{\mathrm{III}}\}} E_{\mathrm{tot}}^{\mathrm{QCDFT}}[\{\mathbf{q}\}],
\tag{8.17}
$$

which allows for the following relaxation scheme: (i) minimize $E_{\mathrm{tot}}^{\mathrm{QCDFT}}$ with respect to the atoms in regions II ($\{\mathbf{q}^{\mathrm{II}}\}$) and the atoms in region III ($\{\mathbf{q}^{\mathrm{III}}\}$) while holding the atoms in

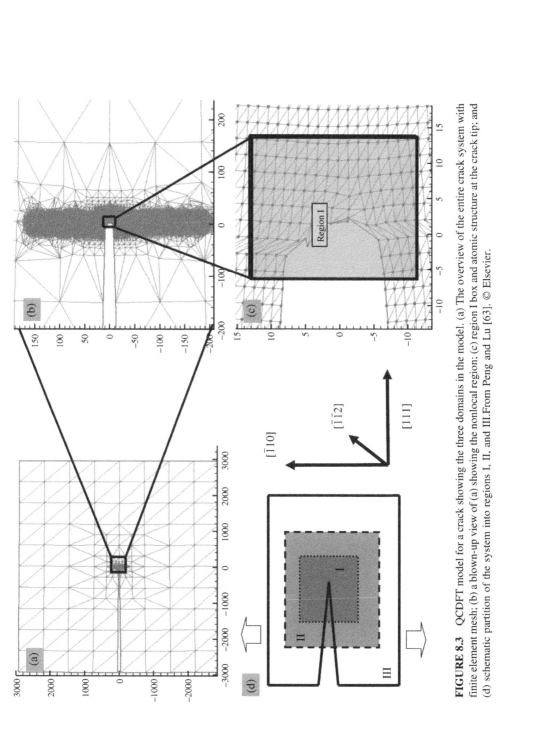

FIGURE 8.3 QCDFT model for a crack showing the three domains in the model. (a) The overview of the entire crack system with finite element mesh; (b) a blown-up view of (a) showing the nonlocal region; (c) region I box and atomic structure at the crack tip; and (d) schematic partition of the system into regions I, II, and III. From Peng and Lu [63]. © Elsevier.

region I fixed; (ii) calculate $E_{tot}^{QCDFT}[\{q\}]$ and the forces on the region I atoms; (iii) perform one step of conjugate gradient minimization of E'; and (iv) repeat until the system is relaxed. In this manner, the number of DFT calculations performed is greatly reduced, albeit at the expense of more EAM and local QC calculations. It has been shown that the total number of DFT energy calculations for the relaxation of an entire system is about the same as that required for DFT relaxation of region I alone.

8.3   MULTISCALE MODELING OF HYDROGEN EMBRITTLEMENT

In the past decade, we have used multiscale modeling to study some key problems in H embrittlement of metals by focusing on H–defect interactions, which are summarized in the following.

8.3.1   HELP

In recent years, a unified understanding of H embrittlement is emerging based on the key concept of HELP [1]. There is a large body of experimental evidence suggesting that H strongly affects plastic deformation in metals in a manner leading to enhanced fracture. One of the most convincing experimental evidence of HELP is the controlled *in situ* TEM observation that is performed in real time and at high spatial resolution [7]. Observations were made on specimens that were under stress, but the deformation processes had ceased to operate in vacuum. On adding H_2 gas to the environmental cell, dislocations began to move. Subsequently, removal of the H_2 gas caused a cessation of the dislocation motion, and the cycle could be repeated many times. This H-enhanced dislocation mobility has been observed for screw, edge, and mixed dislocations as well as for isolated dislocations and dislocation tangles. Observations have been made in a wide range of materials with different crystal structures, including Fe, Ni, steels, Al and Al alloys, Ti and Ti alloys, and Ni_3Al [5, 8, 9, 64–70]. Another important result, which is believed to be responsible for H embrittlement, is H-induced slip planarity. It has been revealed by microscopic observations that solute H atoms can inhibit dislocation cross-slip in Al, austenitic stainless steels, Ni, NiCo, and Ni-based superalloy, leading to slip planarity [71–75]. This slip planarity forces dislocations from a given source to remain localized in narrow slip bands because the relaxation of dislocation pileups via cross-slip is suppressed. As a consequence, a microcrack may be initiated in front of a pileup when the stress intensity factor reaches its critical value over there. In contrast to the vast body of experimental evidence, theoretical studies of the HELP mechanism are scarce. The SVPN and QM/MM calculations have been performed to gain an understanding of the physics behind the HELP mechanism [34, 53, 76, 77].

In the SVPN approach, *ab initio* calculations are carried out for the γ-surface of Al with H impurities placed at the interstitial sites [34]. The γ-surfaces for both pure Al and the Al + H systems are shown in Figure 8.4. Comparing the two γ-surfaces, one finds an overall reduction (up to 50%) in γ-energy in the presence of H, which is attributed to the change of atomic bonding across the glide plane, from covalent like to ionic like [78]. This reduction of γ-energy could change dislocation properties significantly. For example, one expects that dislocations would be emitted more easily from a crack tip when H is present based on the lowering of the unstable stacking fault energy along the $[1\bar{2}1]$ direction [79]. In fact, this

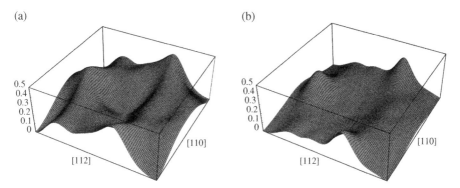

FIGURE 8.4 The γ-surface (J/m^2) for displacements along a (111) plane for (a) pure Al and (b) Al + H systems. The corners of the plane and its center correspond to identical equilibrium configurations (i.e., the ideal lattice). The two surfaces are displayed in exactly the same perspective and on the same energy scale to facilitate comparison of important features. From Lu et al. [34]. © American Physical Society.

TABLE 8.1 Peierls stress (σ_p, MPa), core energies (E_{core}, ev/Å), and binding energy (E_b, eV/ atom) for the four dislocations in the pure Al and the Al + H systems

		Screw			Edge
		0°	30°	60°	90°
σ_p	Al	254	51	97	3
	Al + H	1.7	1.2	1.4	0.3
E_{core}	Al	−0.08	−0.11	−0.17	−0.20
	Al + H	−0.14	−0.18	−0.27	−0.32
E_b		−0.06	−0.07	−0.10	−0.12

From Lu et al. [34]. © American Physical Society.

prediction is in accordance with atomistic simulations for Ni, where dislocations are found to be emitted more rapidly from a crack tip in the presence of H [80].

The core properties of four different dislocations, screw (0°, 30°, and 60°) and edge (90°), have been studied using the SVPN model. The Peierls stress for these dislocations is shown in Table 8.1. It was found that the Peierls stress, which is the minimum stress to move a stationary dislocation, is reduced by more than an order of magnitude in the presence of H. Moreover, this H-enhanced mobility is observed for screw, edge, and mixed dislocations, which is compatible with the experimental results. This result invalidates the perception that the only plausible explanation of HELP must be based on elastic interactions among dislocations and that dislocation–lattice interaction is not important [1]. On the contrary, it was found that the dislocation–lattice interaction is very important and that the dislocation core structure is responsible for the observed H-enhanced dislocation mobility. However, it does not exclude the possibility that dislocation–dislocation interaction does play a role in HELP [81]. The results of binding energy E_b in Table 8.1 shows that there is strong binding between H and the dislocation cores; that is, H is attracted (trapped) to dislocation cores, which lowers the core energies. More importantly, the binding energy was found to be a function of dislocation character, with the edge dislocation having the greatest and the screw dislocation having the lowest binding energies. For a mixed dislocation, the binding energy increases with the amount of edge component of the Burgers vector. These results indicate that in the presence of H, it costs more energy for an edge dislocation to transform into a screw dislocation to cross-slip, as the edge dislocation has almost twice the binding energy of the screw dislocation. In the same vein, it costs more energy for a mixed dislocation to transfer its edge component to a screw component for cross-slip. Therefore, the cross-slip process is suppressed because of the presence of H, and the slip is confined to the primary glide plane, exhibiting the experimentally observed slip planarity.

The interaction between the H impurities with dislocations in α-Fe has been examined by using the QM/MM approach [76]. In the QM/MM simulations, the entire dislocation is partitioned into two regions as shown in Figure 8.2: the QM region contains the dislocation core and the impurities; the MM region on the other hand contains the rest of the system including the long-range elastic field of the dislocation. The impurity–dislocation interaction is investigated by calculating the impurity–dislocation solution energy and the impurity binding energy to the dislocation cores. The impurity solution energy is calculated as follows:

$$E_{X,H}^{s} = E_{X+H} - E_X - E_H \qquad (8.18)$$

here, X could stand for bulk, edge, and screw dislocations, respectively. The three energy contributions are the energy of X in the presence of the H impurity, the energy of X in the absence of the H impurity, and the energy of the H impurity by itself. The more negative the solution energy, the stronger the binding between X and H (or the site is more preferred by the impurity). It was found that in α-Fe, H prefers the tetrahedral (T–) site over the octahedral (O–) site at the dislocation core, as well as in the bulk. We note that the observed site preference of H is materials specific and thus one could be cautious to generalize the conclusion to other materials. The solution energies of the most stable sites in the bulk, edge, and screw dislocations are −2.33, −2.80, and −2.60 eV, respectively. The impurity binding (or segregation) energy to the dislocation core can be defined as the solution energy difference between the bulk and the dislocations. The lower the binding/segregation energy, the more stable the impurity at the dislocation core; the negative value of the binding/segregation energy indicates that the impurity prefers to segregate to the dislocation core as opposed to staying in the bulk. The lowest binding energies of H to the screw and the edge dislocation are −0.27 and −0.47 eV, respectively. Therefore, the H impurity is energetically most stable in the edge dislocation, less stable in the screw dislocation and least stable in the bulk, similar to the preceding results of Al. We have also performed QM/MM calculations for Al dislocations and find a binding energy between H and the edge dislocation of −0.12 eV [53], which is identical to the previous P-N model result. However, H has a weaker binding to dislocations in Mg where the binding energies are −0.03 and −0.05 eV for the screw and edge dislocations, respectively [77].

To study the effect of H on dislocation mobility, the QM/MM method is employed to calculate the Peierls energy barrier, which is the minimum energy that a straight dislocation has to overcome in order to move in a crystal. The screw dislocation is studied here because its mobility is much lower than that of the edge dislocation in Fe; hence, its mobility is more relevant to the overall mobility of a curved dislocation or a dislocation loop. The dislocation is placed at two adjacent *easy* positions as the initial and final state, as shown in Figure 8.5a and c, respectively. The minimal energy path is found to be in the {110} plane, which is consistent with the experiments [82]. The saddle point configuration (Fig. 8.5b) has a *none-split* core structure. The none-split core is very close to the *hard* core configuration, while the split core means that the dislocation spreads into two *easy* core configurations.

(a)   (b)   (c)

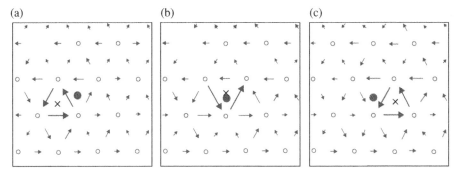

FIGURE 8.5 The differential displacement maps of the screw dislocation moving along the minimum energy path: (a) initial structure, (b) saddle point, and (c) final structure. The dislocation core is represented by the cross, and the position of the H impurity is indicated by a dot. From Zhao and Lu [76]. © Institute of Physics.

The minimal energy path profile has a single peak, that is, the Peierls barrier. The Peierls energy barrier for the pure screw dislocation is 0.06 eV/b, where b is the length of Burgers vector. In the presence of the H impurity, the Peierls barrier becomes 0.03 eV/b so that H can enhance the dislocation mobility considerably. Thus, the HELP mechanism is also operating in Fe, which is consistent with experimental observations. The reduction in the Peierls energy barriers can be explained from the γ-energy. With one H atom per $\langle 112 \rangle \times \frac{3}{2} \langle 111 \rangle$ cell (49 Å^2), we find that H can lower the γ-energy by 0.08 J/m^2. This is because in the presence of H, the covalent Fe–Fe bonding across the slip plane is significantly disrupted and Fe–H ionic bonding is formed at its expense. Since the ionic Fe–H bonding across the slip plane resembles the formation of positively and negatively charged plates, the energy cost does not depend as sensitively to the shear as for the covalent Fe–Fe bonding. Hence, the γ-energy in the presence of H is lower in Fe.

8.3.2 Hydrogen-Assisted Cracking

The presence of H is known to promote crack growth in a variety of technologically important materials [83]. A recent work on large-scale atomistic simulations of Al has suggested an interesting mechanism for ductile fracture, which depends on dislocation mobility and pinning behavior [84]. Based on EAM simulations, the authors have observed that the cleavage fracture can occur after emission of dislocations from the crack tip. The emitted dislocations can shield the crack tip, so that the macroscopic toughness exceeds the Griffith value, and hence, the failure is a combination of both *brittle-like* and *ductile-like* phenomena. To fully address this problem, QCDFT that combines *ab initio* calculations for H–metal interactions near the crack tip and large-scale QC simulations for the rest of the system is required. Here, we study H-assisted cracking in Al, which is at the heart of H embrittlement. If one understands the fundamental processes of H-assisted cracking, it might be possible to establish connections between the various mechanisms and arrive at a comprehensive picture of H embrittlement.

A semi-infinite crack (as shown in Fig. 8.3) in a single Al crystal with H atoms at the crack tip has been studied by using the QCDFT method [85]. *Ab initio* studies have showed that the Al–Al bond can be attacked by H with a significant charge transfer from the Al atoms

TABLE 8.2 Summary of K_{IC} and Δ for eight crack configurations. Symbols in the configuration reflect the adsorbed H positions, for example, F_6^B means that there are six H atoms adsorbed on the bridge or B site at the crack front surface. N_H is the total number of adsorbed H atoms at the crack tip

Case	Configuration	N_H	K_{IC} (eV/Å$^{2.5}$)	Δ (Å)
Pure	Pure	0	0.28	0.16
A	$U_2^B + D_2^B$	4	0.29	1.02
B	$U_2^T + D_2^T$	4	0.28	0.21
C	F_6^B	6	0.32	1.45
D	F_1^T	1	0.31	6.1
E	F_5^T	5	0.32	9.07
F	$F_6^B + U_2^B$	8	0.30	1.35
G	$F_6^B + U_2^B + D_2^B$	10	0.32	1.55
H	$F_5^T + U_2^T + D_2^T$	9	0.30	8.42

From Sun et al. [85]. © American Physical Society.

to H [34, 78, 86]. Hence, we consider two H adsorption sites at the crack tip in the QCDFT simulations: **B** site on the bridge center of an Al–Al bond and **T** site right on top of an Al atom. The H atom at the **B** site forms an Al–H–Al bond, whereas the H atom at the **T** site forms an Al–H bond. At the crack tip, H atoms can be adsorbed on the up surface (U), down surface (D), and front surface (F). To examine all these possibilities, we study eight different cases labeled from (A) to (H) as in Table 8.2. The primary quantities of interest here are the critical stress intensity for dislocation emission, K_{IC}, and the crack opening at the stress intensity of 0.3 eV/Å$^{2.5}$, Δ. As shown in Table 8.2, we find that H-assisted cracking occurs only when H atoms are adsorbed at the **T** site of the crack front surface; H concentration at 0.2 ML would be sufficient to lead to crack growth as shown in case D. However, H atoms adsorbed on the side surfaces (both up and down surfaces) cannot yield crack growth. On the other hand, H atoms at the **T** site and **B** site on the crack front surface could modify the pattern of dislocation nucleation from the crack tip.

The preceding results suggest the following atomic processes of H-assisted cracking. (i) At the crack front surface, H atoms adsorbed on the **T** sites attack Al–Al bonds, which ultimately leads to the fracture of the first atomic layer as shown in Figure 8.6a; the second atomic layer then becomes the new front surface. (ii) The H atoms at the *original* front surface and/or the side surfaces can diffuse to the fresh front surface. The diffusion energy barrier is about 0.2 eV, which is similar to the H diffusion barrier on Al surfaces [87]. H adsorption and diffusion are key steps in H embrittlement. (iii) With increasing loading level, H atoms on the **T** sites of the fresh front surface continue attacking the Al–Al bonds at the second atomic layer as shown in Figure 8.6b. Steps (ii) and (iii) repeat for a continuous crack growth. To confirm that *only* the H atoms on the **T** sites can contribute to step (iii), we study two contrasting cases with and without H atoms at the **B** sites of the fresh front surface. At the stress intensity of 0.35 eV/Å$^{2.5}$, the fracture of the second atomic layer occurs only when H atoms are adsorbed on the **T** sites as shown in Figure 8.6b with a significant opening of $\Delta = 3.9$ Å at the fresh front surface. On the other hand, in the two contrasting cases, the crack does not grow as indicated in Figure 8.6c and d.

To understand ductile-to-brittle transitions, the Rice criterion in which the ratio between the surface energy (Griffith fracture energy) and the unstable stacking fault energy is often

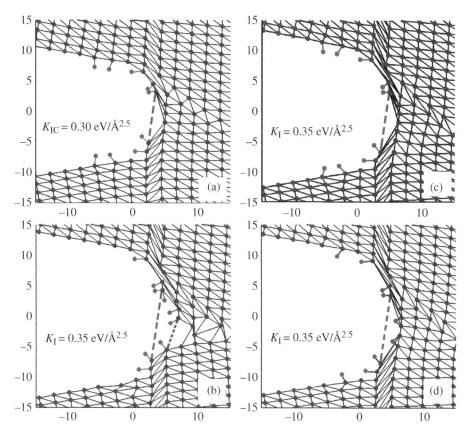

FIGURE 8.6 (a) Cleavage of the first atomic layer at the crack tip; the second atomic layer becomes the new front surface subsequently. (b) Cleavage of the second atomic layer when H atoms are adsorbed on the **T** sites of the fresh front surface. No cleavage of the second atomic layer for (c) without H atoms on the fresh front surface and (d) H atoms adsorbed on the **B** sites of the fresh front surface. The dashed and dotted lines represent the first and second atomic layers, respectively. From Sun et al. [85]. © American Physical Society.

used [79]. The lower the ratio, the stronger the tendency for a brittle fracture; the higher the ratio, the more ductile the material behaves. While this picture may be overly simplistic for a quantitative analysis, it does give a useful insight for general trends and can even lead to predictions that have been verified experimentally [88]. We have used the Rice criterion to examine H embrittlement in Mg and find that the H impurities generally lower the ratio leading to the brittle fracture of Mg [77].

8.3.3 Crucial Role of Vacancies

Vacancies, being ubiquitously present in solids and having the ability to act as impurity traps, could play a central role in H embrittlement as hinted by some recent experiments. One set of experiments has established that H could induce superabundant vacancy formation in a number of metals, such as Pd, Ni, Cr, and so on [89, 90]. The estimated vacancy concentration,

C_v, in these systems can reach a value as at 23% [90]. A conclusion drawn from these experiments is that H atoms, originally at bulk interstitials, are trapped at vacancies in multiple numbers with rather high binding energies. The consequence of H trapping is that the formation energy of a vacancy defect is lowered by a significant amount. Such reduction in the vacancy formation energy could result in a drastic increase (10^7-fold for Fe) of equilibrium vacancy concentrations [91]. The superabundant vacancy formation in turn provides more trapping sites for H impurities, effectively increasing the apparent H solubility in metals by many orders of magnitude. For example, it was observed experimentally that about 1000 at. ppm of H atoms can enter Al accompanied by vacancy formation at the surface under aggressive H charging conditions, which should be contrasted with the equilibrium solubility of H in Al of about 10^{-5} at. ppm at room temperature where the experiments were carried out [92]; this is a staggering change of eight orders of magnitude in concentration. It was further observed that the H–vacancy defects clustered and formed platelets lying on the {111} planes, which directly lead to void formation or crack nucleation on the {111} cleavage planes [92]. Because of the extremely low solubility of H in metals such as aluminum, experiments are usually difficult and results are dependent on H charging conditions; for such systems, first-principles calculations are particularly useful to complement experiments.

First-principles calculations show that the tetrahedral interstitial site for H atoms in bulk Al is slightly more favorable than the octahedral interstitial site by 0.07 eV. And the H atom prefers to occupy the vacancy site over the interstitial tetrahedral site in bulk by 0.40 eV in excellent agreement with the experimental value of 0.52 eV [1]. The lowest energy position for the H atom in the presence of a vacancy is not at the geometric center of the vacancy site, but rather at an off-center position close to a tetrahedral site adjacent to the vacancy site as shown in Figure 8.7a. The H atom is negatively charged, and the kinetic energy of the H^{-1} electrons is lowered at the vacancy site where the conduction electron density is lower. At the same time, it is energetically favorable for the H^{-1} ion to sit off center of the vacancy, to minimize the Coulomb interaction energy with the nearby Al ions. Having established the stability of a single H atom at a single vacancy in Al, the ensuing question is whether multiple H atoms, in particular, H_2 molecules would be stable at this defect. This question is relevant to H_2 bubble formation that gives rise to H embrittlement. The binding energy E_b of the H_2 unit at a vacancy site is calculated as

$$E_b = E_c(V_{Al} + H_2) + E_c(V_{Al}) - 2E_c(V_{Al}H), \qquad (8.19)$$

where $E_c(V_{Al} + H_2)$ is the cohesive energy of a system with an H_2 unit at the center of the vacancy, $E_c(V_{Al})$ is the cohesive energy of a system with a single vacancy in the absence of the H_2 unit, and $E_c(V_{Al}H)$ is the cohesive energy of a system with a single H atom at the vacancy (in the off-center tetrahedral site). This binding energy is 0.06 eV, simply stating that these 2H atoms would prefer to be trapped at two single vacancy sites *individually* rather than in the same vacancy site *as a pair*. One can conclude that if the single vacancy concentration C_v is greater than the H concentration C_H, each vacancy in equilibrium should contain no more than 1 H atom.

On the other hand, if C_H is greater than C_v, the question arises as to where will the extra H atoms be situated, at interstitial or at vacancy sites? To answer this question, one can calculate the trapping energy E_{trap} of multiple H atoms at a single vacancy site, which is defined as

$$E_{trap} = \frac{1}{n}[E_c(V_{Al} + nH) - E_c(V_{Al})] - [E_c^0(H) - E_c^0], \qquad (8.20)$$

(a)

(b)

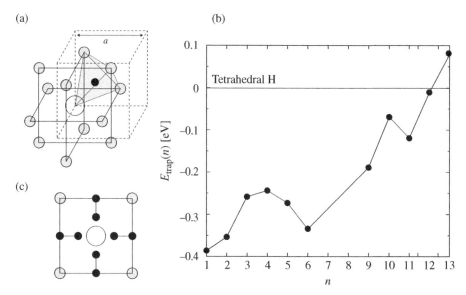

(c)

FIGURE 8.7 (a) The lowest energy site for a H atom (black circle) occupying in the vacancy. The vacancy is shown as a large open circle and its 12 nearest neighbors as smaller filled circles. (b) Trapping energy per H atom in eV as a function of the number of H atoms being trapped at a single vacancy site. (c) The arrangement of four of the six H_2 molecules surrounding the vacancy on a [93] plane. From Lu and Kaxiras [94]. © American Physical Society.

where $E_c(V_{Al} + nH)$ is the cohesive energy of a system with nH atoms each situated at a single vacancy site, $E_c^0(H)$ is the cohesive energy of bulk Al with a H atom at the tetrahedral interstitial site, and E_c^0 is the cohesive energy of the ideal bulk without H. A negative value for the trapping energy represents the energy gain when the H atoms are trapped at a single vacancy site relative to being dispersed at n different tetrahedral interstitial sites. Here, the trapping energy was calculated to estimate the maximum number of H atoms trapped at a vacancy. An entire thermodynamic description of H–vacancy interaction would depend on H chemical potential, temperature, pressure, and other parameters. A more detailed analysis of such interaction has been put forward recently by Ismer et al. [95].

The results for E_{trap} as a function of n are summarized in Figure 8.7b. It is energetically favorable for multiple H atoms to be trapped at a single vacancy site relative to being dispersed at interstitial sites as individual atoms. In fact, up to 12 H atoms can be trapped at a single vacancy in Al. The atomic arrangement of the 12 H atoms trapped at a single vacancy is indicated in Figure 8.7c. The ordered arrangement of the H atoms is necessary to minimize the electrostatic energy. Due to the trapping effect, vacancies can lead to a significant reduction in apparent lattice diffusivity of H in Al, causing a strong composition dependence and non-Arrhenius behavior of the effective diffusion coefficient [96].

To check the possibility of H-induced vacancy clustering, a number of relevant configurations have been examined including (i) two vacancies, each with 1 H atom, forming a NN divacancy with 2 H atoms trapped, and (ii) n vacancies, each with 2 H atoms, forming a complex of NN multivacancies with $2n$ H atoms trapped, for $n = 2$ and 3. The results are summarized by using the notation of chemical reactions as

$$\text{(i)}: 2V_{Al}H \rightarrow (V_{Al})_2 H_2 - 0.21 \text{ eV},$$

$$\text{(ii)}: nV_{Al}H_2 \rightarrow (V_{Al})_n H_{2n} + n \; 0.29 \text{ eV},$$

(8.21)

where the last number in each equation represents the reaction enthalpy. A positive value of enthalpy means the reaction is exothermic, that is, the process from left to right is energetically favorable. Consistent with the previous discussion, the reaction (i) is unfavorable (endothermic) because the effect of a single H atom on the covalent or metallic bonding of the NN Al atoms around the vacancy site is small and localized. On the other hand, reaction (ii) is favorable for $n = 2$ and 3, because the H_2 units can attract more conduction electrons from the nearby Al atoms, weakening the bonding among the NN Al atoms, which in turn drives the formation of multivacancies. The large energy gain in forming the trivacancy ($n = 3$) is of particular interest. First, it is consistent with the experimental observation that the single vacancy defects occupied by H atoms can coalesce to form platelets on {111} planes of Al. Although the calculations primarily concern the formation of the trivacancy, it is likely that even larger vacancy clusters can also be formed based on the same mechanism. Second, these vacancy clusters can serve as embryos of cracks and microvoids with local H concentrations much higher than the average bulk value.

In general, the fracture surface is along the active slip planes where shear localization occurs. For fcc metals, the slip planes are the {111} planes. In many cases, microvoids open up along these active slip planes in front of the crack tip; these microvoids can open and close in response to the local stress. Fracture occurs when these microcracks are joined to the crack tip, upon reaching the critical stress. The H-enriched microvoids may be created along the slip planes by the coalescence of vacancies with trapped H. These microvoids can be formed *only* in the presence of H, which produces an additional source of microcracks necessary for the H embrittlement. On the other hand, the apparent lattice mobility of H atoms is also enhanced since multiple H atoms may be trapped at a single vacancy. All these vacancy-based mechanisms contribute to the H embrittlement as they increase the rate of crack growth. Finally, the significant H trapping at vacancies provides a scenario by which drastic increase of local H concentration may occur without improbable accumulation of H at bulk interstitial sites. This new feature resolves the long-standing problem of how a sufficiently high H concentration can be realized to induce H embrittlement in materials such as Al, where the equilibrium H concentration in bulk is extremely low.

8.3.4 Hydrogen Diffusion

H diffusion in metals plays an important role in H embrittlement. For example, it is observed experimentally that HELP occurs only when the thermal diffusion of H in the lattice is fast enough to follow the motion of dislocations [1]. In addition to their trapping of H, dislocations and grain boundaries are believed to provide short-circuit paths for accelerated diffusion. Dislocations and grain boundaries could act as fast paths for diffusing atoms whose mobility can be orders of magnitude higher than in bulk diffusion, and this phenomenon is often referred as *pipe diffusion*. Furthermore, H diffusion at the crack tip under stress is central to H-assisted cracking. Therefore, it is of great importance to determine H diffusion energy barriers in extended defects. To this end, we have employed QM/MM approaches to calculate H diffusion energy barriers along dislocations and grain boundaries in Fe, Al, and Mg [53, 76, 77]. Being important by themselves, these energy barriers could

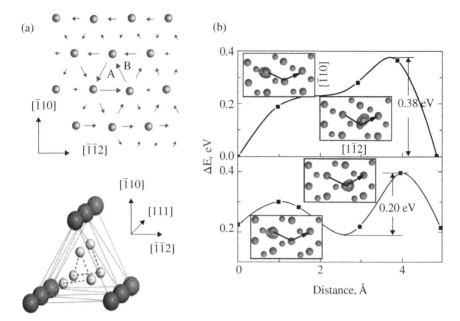

FIGURE 8.8 (a) H diffusion along the screw dislocation in Fe. Top panel: The differential displacement map of the screw dislocation. Bottom panel: The diffusion path represented by the red dashed curve passing through the tetrahedral sites along pipe A. (b) The energy profile for the H diffusion along partial core (top panel) and stacking fault ribbon (bottom panel). Inset: The schematic trajectory of the H atom along the diffusion path on the (111) plane. From Zhang et al. [53] and Zhao and Lu [76]. © American Physical Society.

also be incorporated in kinetic Monte Carlo or other models for sequential multiscale modeling of H embrittlement in metals.

We have carried out QM/MM calculation to study H diffusion in bulk and along dislocations in α-Fe. We find that the H diffusion energy barrier is 0.09 eV in the bulk Fe between two adjacent bulk tetrahedral sites. Along a screw dislocation, H could diffuse between adjacent tetrahedral sites with two different diffusion paths: (a) along the dislocation center in which all tetrahedra are equally distorted and (b) along a path adjacent to the dislocation center in which the distortions are the not same, as shown in Figure 8.8a. The simulation results show that the diffusion energy barrier is 0.035 eV along path A and 0.063 eV along path B [76]; both values are smaller than that in the bulk. Similar QM/MM calculations have been performed in Al, and we find that the energy barriers are 0.26, 0.38, and 0.20 eV for H diffusion along grain boundaries, partial dislocation cores, and the stacking fault ribbon, respectively [53]. Along the same crystallographic directions, the corresponding diffusion barriers in the crystalline bulk lattice are 0.40, 0.32, and 0.32 eV, respectively. Based on the diffusion energy barriers, one can estimate the H diffusivity

$$D = \nu l^2 \exp[-\Delta E / k_B T], \tag{8.22}$$

where ν is the attempt frequency, l is the hopping length, and ΔE is the diffusion energy barrier. At room temperature, there are two (one) orders of magnitude increase of diffusivity for H diffusion in Al (Fe) dislocations comparing to the bulk diffusivity. Therefore, dislocations in Al and Fe indeed can sever as fast pipes for H diffusion. The consequence of these results on H embrittlement is being pursued in our group.

However, we find that the basal dislocations of Mg are not easier paths for H diffusion [77]. In bulk Mg, the H diffusion consists of two types: (i) between the adjacent tetrahedral sites surrounded by octahedral sites and (ii) between the octahedral site and its neighboring tetrahedral site. The long-range diffusion requires a combination of the two types. The energy barriers of cases (i) and (ii) are 0.04 and 0.21 eV, respectively. The higher value 0.21 eV is the actual energy barrier for the long-range diffusion. These values are consistent with experimental results [98, 99]. Along the partial dislocation core of Mg, the QM/MM results show that the barriers slightly increase to 0.22 and 0.26 eV for the screw and edge dislocation, respectively. And in the stacking fault ribbon, the corresponding barriers become 0.25 and 0.47 eV. Therefore, the energy barrier for H diffusion in Mg dislocations is actually greater than that in the crystalline lattice, in contrast to the cases in Al and Fe. More investigations are desired to understand the ramifications of these results in H embrittlement of Mg. Finally, we have also examined H diffusion at the crack tip of Al and find that H can diffuse with relative ease on the crack surfaces with lower energy barriers than those in the bulk; the H atoms on the fracture surfaces can even recombine into H_2 molecules before redissociation at the fresh crack front surface to attack Al–Al bonds [85].

8.4 SUMMARY AND OUTLOOK

Based on the experimental observations and the simulation results, we can tentatively identify the following key processes leading to H embrittlement of metals: (i) Application of external stress produces local concentration of tensile stress in the vicinity of cracks, which attracts H since H prefers to stay in slightly enlarged interstitial sites. H impurities can also be adsorbed at the crack surfaces near the crack tip; the relative ease of H diffusion on the crack surfaces may provide a rapid means of transporting H to the fresh crack front. (ii) H can also segregate at vacancies inside the material, which may lead to microvoids with high local H concentrations; the microvoids could merge to form cracks under stress. (iii) The segregated H impurities at the crack tip can facilitate dislocation generation and deformation twining and enhance dislocation mobility, which will lead to extensive plastic deformation in front of the crack. The highly deformed region ahead of the crack tip becomes the *weakest link* where fracture takes place. (iv) The highly deformed and disordered region along with the localization of slip due to the inhibition of cross-slip allows the crack to propagate at lower stress levels, prior to general yielding away from the crack tip. These processes are not likely to be the only mechanism operating in H embrittlement [70], but are the most relevant processes in line with our multiscale modeling results on aluminum. The similar mechanism could be operative in other metals as well, but it should be further validated to be certain.

Multiscale modeling will continue to be important for understanding H embrittlement phenomena in metals. One area of particular importance and challenge is to model stress corrosion cracking under realistic conditions and in aqueous environment. The molecular structure of solid–liquid interface, pH of electrolyte, surface charge, field, and electrochemical potential are all relevant parameters that need to be included in the modeling. *Ab initio* methods for modeling the charged solid–liquid interface at the atomic level are starting to appear [93, 99–101]. An interesting future direction is to incorporate this type of *ab initio* modeling into the QCDFT framework so that the chemomechanical and electrochemical processes important to the stress corrosion cracking can be captured.

ACKNOWLEDGMENT

We acknowledge the collaborators who have contributed to our understanding of H embrittlement in metals and the development of the multiscale methodologies presented in this chapter. The related research was supported primarily by the NSF-PREM program, the DoE-SciDAC program, and the Office of Naval Research (ONR). We are particularly grateful to Kenny Lipkowitz for his encouragement and support.

REFERENCES

1. Myers SM, Baskes MI, Birnbaum HK, Corbett JW, DeLeo GG, Estreicher SK, Haller EE, Jena P, Johnson NM, Kirchheim R, Pearton SJ, Stavola MJ. Hydrogen interactions with defects in crystalline solids. *Rev. Mod. Phys.* 1992;64:559–617.

2. Oriani RA. Hydrogen embrittlement of steels. *Annu. Rev. Mater. Sci.* 1978;8:327–357.

3. Westlake DG. A generalized model for hydrogen embrittlement. *Trans. ASM* 1969;62:1000–1006.

4. Gahr S, Grossbeck ML, Birnbaum HK. Hydrogen embrittlement of Nb I—macroscopic behavior at low temperatures. *Acta Metall.* 1977;25:125–134.

5. Shi D, Robertson IM, Birnbaum HK. Hydrogen embrittlement of α-titanium: In situ TEM studies. *Acta Metall.* 1988;36:111–124.

6. Beachem CD. A new model for hydrogen-assisted cracking (hydrogen embrittlement). *Metall. Trans.* 1972;3:437–451.

7. Sofronis P, Robertson IM. TEM observations and micromechanical/continuum models for the effect of hydrogen on the mechanical behavior of metals. *Philos. Mag. A* 2002;82:3405–3413.

8. Eastman J, Matsumoto T, Narita N, Heubaum N, Birnbaum HK. Hydrogen effects in nickel embrittlement or enhanced ductility? In: Bernstein IM, Thompson AW, editors. *Hydrogen in Metals*. New York: AIME;1981. pp. 397–399.

9. Matsumoto T, Eastman J, Birnbaum HK. Direct observations of enhanced dislocation mobility due to hydrogen. *Scr. Metall.* 1981;15:1033–1037.

10. Oriani RA, Josephic PH. Equilibrium aspects of hydrogen-induced cracking of steels. *Acta Metall.* 1974;22:1065–1074.

11. Oriani RA, Josephic PH. Equilibrium and kinetic studies of the hydrogen-assisted cracking of steels. *Acta Metall.* 1977;25:979–988.

12. Steigerwald EA, Schaller FW, Troiano AR. The role of stress in hydrogen induced delayed failure. *Trans. Metall. Soc. AIME*, 1960;218:832–841.

13. Phillips R. *Crystals, Defects and Microstructures—Modeling Across Scales*. Cambridge, UK: Cambridge University Press; 2001.

14. Lu G, Kaxiras E. Overview of multiscale simulations of materials. In: Rieth M, Schommers W, editors. *Handbook of Theoretical and Computational Nanotechnology*. Stevenson Ranch: American Scientific; 2004. Chapter 22.

15. Kaxiras E, Yip S. Modeling and simulation of solids. *Curr. Opinion Solid State Mater. Sci.* 1998;3:523–525 and accompanying articles.

16. de la Rubia TD, Bulatov VV. Materials research by means of multiscale computer simulation. *MRS Bull.* 2001;26:169–175 and accompanying articles.

17. Yip S. Synergistic science. *Nat. Mater.* 2003;2:3–5.

18. Hohenberg P, Kohn W. Inhomogeneous electron gas. *Phys. Rev.* 1964;136:B864–B871.

19. Kohn W, Sham LJ. Self-consistent equations including exchange and correlation effects. *Phys. Rev.* 1965;140:A1133–A1138.

20. Abraham FF, Walkup R, Gao H, Duchaineau M, de la Rubia T, Seager M. Simulating materials failure by using up to one billion atoms and the world's fastest computer: Work-hardening. *Proc. Natl. Acad. Sci. U. S. A.* 2002;99:5777–5782.

21. Hughes TJR. *The Finite Element Method.* Englewood Cliffs, NJ: Prentice-Hall; 1987.

22. Peierls R. The size of a dislocation. *Proc. Phys. Soc. Lond.* 1940;52:34–37.

23. Nabarro FRN. Dislocations in a simple cubic lattice. *Proc. Phys. Soc. Lond.* 1947;59:256–272.

24. Joos B, Ren Q. Duesbery MS. Peierls–Nabarro model of dislocations in silicon with generalized stacking-fault restoring forces. *Phys. Rev. B* 1994;50:5890–5898.

25. Joos B, Duesbery MS. The peierls stress of dislocations: An analytic formula. *Phys. Rev. Lett.* 1997;78:266–269.

26. Juan Y, Kaxiras E. Generalized stacking fault energy surfaces and dislocation properties of silicon: A first-principles theoretical study. *Philos. Mag. A* 1996;74:1367–1384.

27. Hartford J, von Sydow B, Wahnstrom G, Lundqvist BI. Peierls barriers and stresses for edge dislocations in Pd and Al calculated from first principles. *Phys. Rev. B* 1998;58:2487–2496.

28. von Sydow B, Hartford J, Washnstrom G. Atomistic simulations and Peierls-Nabarro analysis of the Shockley partial dislocations in palladium. *Comput. Mater. Sci.* 1999;15:367–379.

29. Medvedeva NI, Mryasov ON, Gornostyrev YN, Novikov DL, Freeman AJ. First-principles total-energy calculations for planar shear and cleavage decohesion processes in B2-ordered NiAl and FeAl. *Phys. Rev. B* 1996;54:13506–13514.

30. Mryasov ON, Gornostyrev YN, Freeman AJ. Generalized stacking-fault energetics and dislocation properties: Compact versus spread unit-dislocation structures in TiAl and CuAu. *Phys. Rev. B* 1998;58:11927–11932.

31. Bulatov VV, Kaxiras E. Semidiscrete variational peierls framework for dislocation core properties. *Phys. Rev. Lett.* 1997;78:4221–4224.

32. Lu G, The Peierls-Nabarro model of dislocations: A venerable theory and its current development. Yip S, editor, *Handbook of Materials Modeling. Volume I: Methods and Models*, Netherlands: Springer; 2005.

33. Lu G, Kioussis N, Bulatov VV, Kaxiras E. Generalized-stacking-fault energy surface and dislocation properties of aluminum. *Phys. Rev. B* 2000;62:3099–3108.

34. Lu G, Zhang Q, Kioussis N, Kaxiras E. Hydrogen-enhanced local plasticity in aluminum: An *ab initio* study. *Phys. Rev. Lett.* 2001;87:095501.

35. Lu G, Kaxiras E. Can vacancies lubricate dislocation motion in aluminum? *Phys. Rev. Lett.* 2002;89:105501.

36. Lu G, Bulatov VV, Kioussis N. Dislocation constriction and cross-slip: An *ab initio* study. *Phys. Rev. B* 2002;66:144103.

37. Vitek V. Intrinsic stacking faults in body-centered cubic crystals. *Philos. Mag.* 1968;18: 773–786.

38. Eshelby JD. Edge dislocations in anisotropic materials. *Philos. Mag.* 1949;40:903–912.

39. Hirth JP, Lothe J. *Theory of Dislocations.* 2nd ed. New York: Wiley; 1992.

40. Bernstein N, Kermode JR, Csanyi G. Hybrid atomistic simulation methods for materials systems. *Rep. Prog. Phys.* 2009;72:026501.

41. Lin H, Truhlar DG. QM/MM: What have we learned, where are we, and where do we go from here? *Theor. Chem. Acc.* 2007;117:185–199.

42. Zhang X, Zhao Y, Lu G. Recent development in quantum mechanics/molecular mechanics modeling for materials. *Int. J. Multiscale Comput. Eng.* 2012;10:65–82.

43. Antes I, Thiel W. On the treatment of link atoms in hybrid methods. In: Gao J, Thompson MA, editors. *Hybrid Quantum Mechanical and Molecular Mechanical Methods Proceedings of ACS Symposium Series.* Washington, DC: ACS; 1998. 712:50–65.

44. Gao J, Truhlar DG. Quantum mechanical methods for enzyme kinetics. *Annu. Rev. Phys. Chem.* 2002;53:467–505.

45. Zhang X, Lu G, Curtin WA. Multiscale quantum/atomistic coupling using constrained density functional theory. *Phys. Rev. B* 2013;87:054113.

46. Huang P, Carter EA. Advances in correlated electronic structure methods for solids, surfaces, and nanostructures. *Annu. Rev. Phys. Chem.* 2008;59:261–290.

47. Nair AK, Warner DH, Hennig RG, Curtin WA. Coupling quantum and continuum scales to predict crack tip dislocation nucleation. *Scr. Mater.* 2010;63:1212–1215.

48. Leyson GP, Curtin WA, Hector LG Jr, Woodward CF. Quantitative prediction of solute strengthening in aluminium alloys. *Nat. Mater.* 2010;9:750–755.

49. Woodward C, Rao SI. Flexible *ab initio* boundary conditions: Simulating isolated dislocations in bcc Mo and Ta. *Phys. Rev. Lett.* 2002;88:216402.

50. Choly N, Lu G, Kaxiras EW. Multiscale simulations in simple metals: A density-functional-based methodology. *Phys. Rev. B* 2005;71:094101.

51. Zhang X, Lu G. Quantum mechanics/molecular mechanics methodology for metals based on orbital-free density functional theory. *Phys. Rev. B* 2007;76:245111.

52. Zhang X, Wang CY, Lu G. Electronic structure analysis of self-consistent embedding theory for quantum/molecular mechanics simulations. *Phys. Rev. B* 2008;78:235119.

53. Zhang X, Peng Q, Lu G. Self-consistent embedding quantum mechanics/molecular mechanics method with applications to metals. *Phys. Rev. B* 2010;82:134120.

54. Daw MS, Baskes MI. Embedded-atom method: Derivation and application to impurities, surfaces, and other defects in metals. *Phys. Rev. B* 1984;29:6443–6453.

55. Garcia Gonzalez P, Alvarellos JE, Chacon E. Nonlocal kinetic-energy-density functionals. *Phys. Rev. B* 1996;53:9509–9512.

56. Wang LW, Teter MP. Kinetic-energy functional of the electron-density. *Phys. Rev. B* 1992;45:13196–13220.

57. Wang YA, Govind N, Carter EA. Orbital-free kinetic-energy density functionals with a density-dependent kernel. *Phys. Rev. B* 1999;60:16350–16358.

58. Zhang X, Lu G, Curtin WA. Multiscale quantum/atomistic coupling using constrained density functional theory. *Phys. Rev. B* 2013;87:054113.

59. Liu Y, Lu G, Chen ZZ, Kioussis N. An improved QM/MM approach for metals. *Model. Simul. Mater. Sci. Eng.* 2007;15:275–284.

60. Tadmor EB, Ortiz M, Phillips R. Quasi-continuum analysis of defects in solids. *Philos. Mag. A* 1996;73:1529–1563.

61. Lu G, Tadmor EB, Kaxiras E. From electrons to finite elements: A concurrent multiscale approach for metals. *Phys. Rev. B* 2006;73:024108.

62. Peng Q, Zhang X, Hung L, Carter EA, Lu G. Quantum simulation of materials at micron scales and beyond. *Phys. Rev. B* 2008;78:054118.

63. Peng Q, Lu G. A comparative study of fracture in Al: Quantum mechanical vs. empirical atomistic description. *J. Mech. Phys. Solids* 2011;59:775–786.

64. Birnbaum HK. Mechanism of hydrogen related fracture of metals. In: Moody NR, Thompson AW, editors. *Hydrogen Effects on Material Behavior.* Warrendale: TMS; 1990. pp. 629–660.

65. Tabata T, Birnbaum HK. Direct observations of the effect of hydrogen on the behavior of dislocations in iron. *Scr. Metall.* 1983;17:947–950.

66. Tabata T, Birnbaum HK. Direct observations of hydrogen enhanced crack propagation in iron. *Scr. Metall.* 1984;18:231–236.

67. Robertson IM, Birnbaum HK. An HVEM study of hydrogen effects on the deformation and fracture of nickel. *Acta Metall.* 1986;34:353–366.

68. Bond GM, Robertson IM, Birnbaum HK. The influence of hydrogen on deformation and fracture processes in high-strength aluminum alloys. *Acta Metall.* 1987;35:2289–2296.

69. Bond GM, Robertson IM, Birnbaum HK. Effects of hydrogen on deformation and fracture processes in high-purity aluminum. *Acta Metall.* 1988;36:2193–2197.

70. Birnbaum HK, Sofronis P. Hydrogen-enhanced localized plasticity—a mechanism for hydrogen-related fracture. *Mater. Sci. Eng. A* 1994;176:191–202.

71. Ulmer DG, Altstetter CJ. Hydrogen-induced strain localization and failure of austenitic stainless-steels at high hydrogen concentrations. *Acta Metall. Mater.* 1991;39:1237–1248.

72. Tang X, Thompson AW. Hydrogen effects on slip character and ductility in Ni–Co alloys. *Mater. Sci. Eng. A* 1994;186:113–119.

73. Walston WS, Bernstein IM, Thompson AW. The effect of internal hydrogen on a single-crystal nickel-base superalloy. *Metall. Trans. A* 1992;23:1313–1322.

74. He J, Fukuyama S, Yokogawa K, Kimura A. Effect of hydrogen on deformation structure of inco-nel-718. *Mater. Trans. JIM* 1994;35:689–694.

75. McInteer WA, Thompson AW, Bernstein IM. The effect of hydrogen on the slip character of nickel. *Acta Metall.* 1980;28:887–894.

76. Zhao Y, Lu G. QM/MM study of dislocation-hydrogen/helium interactions in α-Fe. *Model. Simul. Mater. Sci. Eng.* 2011;19:065004.

77. Zhao Y, Lu G. H in Mg: Energetics of decohesion, slip, twinning and diffusion. unpublished.

78. Lu G, Orlikowski D, Park I, Politano O, Kaxiras E. Energetics of hydrogen impurities in aluminum and their effect on mechanical properties. *Phys. Rev. B* 2002;65:064102.

79. Rice JR. Dislocation nucleation from a crack tip: An analysis based on the Peierls concept. *J. Mech. Phys. Solids* 1992;40:239–271.

80. Daw MS, Baskes MI. Application of the embedded atom method to hydrogen embrittlement. In: Latanision RM, Jones R, editors. *Chemistry and Physics of Fracture.* Netherlands: Martinus Nijh-off Publishers; 1987. p 196–218.

81. Sofronis P, Birnbaum HK. Mechanics of the hydrogen-dislocation-impurity interactions: Part I-increasing shear modulus. *J. Mech. Phys. Solids* 1995;43:49–90.

82. Spitzig WA, Keh AS. The effect of orientation and temperature on the plastic flow properties of iron single crystals. *Acta Metall.* 1970;18:611–622.

83. Lynch SP. Metallographic and fractographic techniques for characterising and understanding hydrogen-assisted cracking of metals. In: Gangloff R, Somerday B, editors. *Gaseous Hydrogen Embrittlement of Materials in Energy Technologies.* Cambridge: Woodhead; 2012. pp. 274–346.

84. Farkas D, Duranduru M, Curtin WA, Ribbens C. Multiple-dislocation emission from the crack tip in the ductile fracture of Al. *Philos. Mag. A* 2001;81:1241–1255.

85. Sun Y, Peng Q, Lu G. Quantum mechanical modeling of hydrogen assisted cracking in aluminum. *Phys. Rev. B.* 2013;88:104109.

86. Apostol F, Mishin Y. Hydrogen effect on shearing and cleavage of Al: A first-principles study. *Phys. Rev. B* 2011;84:104103.

87. Gunnarsson O, Hjelmberg H, Lundqvist BI. Binding energies for different adsorption sites of hydrogen on simple metals. *Phys. Rev. Lett.* 1976;37:292–295.

88. Waghmare UV, Kaxiras E, Duesbery MS. Modeling Brittle and Ductile behavior of solids from first-principles calculations. *Phys. Status Solidi B* 2000;217:545–564.

89. Fukai Y, Okuma N. Formation of superabundant vacancies in Pd hydride under high hydrogen pressures. *Phys. Rev. Lett.* 1994;73:1640–1643.

90. Fukai Y. Superabundant vacancies formed in metal hydrogen alloys. *Phys. Scr.* 2003;T103:11–14.

91. Tateyama Y, Ohno T. Stability and clusterization of hydrogen-vacancy complexes in Fe: An *ab initio* study. *Phys. Rev. B* 2003;67:174105.

92. Birnbaum HK, Buckley C, Zeides F, Sirois E, Rozenak P, Spooner S, Lin JS. Hydrogen in aluminum. *J. Alloys Compd.* 1997;253–254:260–264.

93. Schnur S, Grob A. Properties of metal–water interfaces studied from first principles. *New J. Phys.* 2009;11:125003.

94. Lu G, Kaxiras E. Hydrogen embrittlement of aluminum: The crucial role of vacancies. *Phys. Rev. Lett.* 2005;94:155501.

95. Ismer L, Park MS, Janotti A, Van de Walle CG. Interactions between hydrogen impurities and vacancies in Mg and Al: A comparative analysis based on density functional theory. *Phys. Rev. B* 2009;80:184110.

96. Gunaydin H, Barabash SV, Houk KN, Ozolins V. First-principles theory of hydrogen diffusion in aluminum. *Phys. Rev. Lett.* 2008;101:075901.

97. Tsuru T, Latanision RM. Grain boundary transport of hydrogen in nickel. *Scr. Metall.* 1982;16:575–578.

98. Nishimura C, Komaki M, Amano M. Hydrogen permeation through magnesium. *J. Alloys Compd.* 1999;293–295:329–333.

99. Taylor CD, Sally A, Wasileski SA, Filhol JS, Neurock M. First principles reaction modeling of the electrochemical interface: Consideration and calculation of a tunable surface potential from atomic and electronic structure. *Phys. Rev. B* 2006;73:165402.

100. Filhol JS, Neurock M. Elucidation of the electrochemical activation of water over Pd by first principles. *Angew. Chem. Int. Ed.* 2006;45:402–406.

101. Rossmeisl J, Skulason E, Bjorketun ME, Tripkovic V, Norskov JK. Modeling the electrified solid–liquid interface. *Chem. Phys. Lett.* 2008;466:68–71.

INDEX

A

ab initio molecular dynamics, 41, 154
absorption, 14, 52, 81, 155
accelerated corrosion, 47, 212
accelerated molecular dynamics, 24
accuracy, 17, 19–20, 24, 37, 59, 66, 126, 128,
    158, 161, 164–8, 178, 195, 204, 229, 231
acid dissociation constant, 48, 127, 145
acidity, 154–5
activation barrier, 40, 90
activation energy, 23, 52, 69, 71, 82, 90, 94
active dissolution, 36, 48, 52
activity, 6, 44, 49, 52, 88, 93, 126, 149–50,
    155, 223
adatom, 10, 46, 57–9, 67, 69–71, 75, 80–82, 108
adduct, 141–2
adlayer, 46, 138–9
adsorption, 5, 8–9, 14, 27, 36, 38–9, 41–4,
    47–56, 59, 66, 81, 84, 86–93, 95,
    106–7, 125–6, 128, 130–136, 138–48,
    150–157, 166, 169–70, 173–4, 179–80,
    183, 191, 195–201, 203–14, 217, 220,
    238, 248
adsorption-induced phenomena, 142, 191, 205,
    211, 216–17
aerospace, 223

algorithms, 20–23, 27, 38, 59, 66, 102–5, 108,
    110, 113, 116, 122, 124, 186, 205, 228
alloys, 1–2, 4, 10, 12–13, 19, 25, 28, 51, 57, 64,
    100–102, 105, 108–11, 113–16, 121–5, 148,
    152, 154, 172, 187, 191–2, 216–18, 223,
    234, 247–9
amino acids, 127, 142, 150, 154
ammonium chloride, 47–8, 51–2
anion, 79, 84, 106–7, 110, 113–14, 139, 147,
    156, 197, 207–8
anisotropic, 246
anode, 99–100
anodic dissolution, 14, 47, 113–14, 123
anodic potential, 12
antibonding, 39, 42, 90, 126
antiferromagnetic, 199, 207
aqueous corrosion, 4, 10, 25, 151
assumptions, 22, 49, 73, 81, 112, 121, 161,
    186, 196, 232

B

band gap, 122, 166–7, 194, 198–9, 207
bare metal surface, 9, 47
basicity, 47, 200–201
basis set superposition error (BSSE), 195–6
benchmarking, 134, 148, 153, 159

Molecular Modeling of Corrosion Processes: Scientific Development and Engineering Applications, First Edition.
Edited by Christopher D. Taylor and Philippe Marcus.
© 2015 John Wiley & Sons, Inc. Published 2015 by John Wiley & Sons, Inc.

THE ELECTROCHEMICAL SOCIETY SERIES

Corrosion Handbook
Edited by Herbert H. Uhlig

Modern Electroplating, Third Edition
Edited by Frederick A. Lowenheim

Modern Electroplating, Fifth Edition
Edited by Mordechay Schlesinger and Milan Paunovic

The Electron Microprobe
Edited by T. D. McKinley, K. F. J. Heinrich, and D. B. Wittry

Chemical Physics of Ionic Solutions
Edited by B. E. Conway and R. G. Barradas

High-Temperature Materials and Technology
Edited by Ivor E. Campbell and Edwin M. Sherwood

Alkaline Storage Batteries
S. Uno Falk and Alvin J. Salkind

The Primary Battery (in Two Volumes)
Volume I *Edited by* George W. Heise and N. Corey Cahoon
Volume II *Edited by* N. Corey Cahoon and George W. Heise

Zinc-Silver Oxide Batteries
Edited by Arthur Fleischer and J. J. Lander

Lead-Acid Batteries
Hans Bode
Translated by R. J. Brodd and Karl V. Kordesch

Thin Films-Interdiffusion and Reactions
Edited by J. M. Poate, M. N. Tu, and J. W. Mayer

Lithium Battery Technology
Edited by H. V. Venkatasetty

Quality and Reliability Methods for Primary Batteries
P. Bro and S. C. Levy

Techniques for Characterization of Electrodes and Electrochemical Processes
Edited by Ravi Varma and J. R. Selman

Electrochemical Oxygen Technology
Kim Kinoshita

Synthetic Diamond: Emerging CVD Science and Technology
Edited by Karl E. Spear and John P. Dismukes

Corrosion of Stainless Steels
A. John Sedriks

Semiconductor Wafer Bonding: Science and Technology
Q.-Y. Tong and U. Göscle

Uhlig's Corrosion Handbook, Second Edition
Edited by R. Winston Revie

Atmospheric Corrosion
Christofer Leygraf and Thomas Graedel

Electrochemical Systems, Third Edition
John Newman and Karen E. Thomas-Alyea

Fundamentals of Electrochemistry, Second Edition
V. S. Bagotsky

Fundamentals of Electrochemical Deposition, Second Edition
Milan Paunovic and Mordechay Schlesinger

Electrochemical Impedance Spectroscopy
Mark E. Orazem and Bernard Tribollet

Fuel Cells: Problems and Solutions, Second Edition
Vladimir S. Bagotsky

Lithium Batteries: Advanced Technologies and Applications
Edited by Bruno Scrosati, K. M. Abraham, Walter van Schalkwijk, and Jusef Hassoun

Electrochemical Power Sources: Batteries, Fuel Cells, and Supercapacitors
Vladimir S. Bagotsky, Alexander M. Skundin, and Yurij M, Volfkovich

Molecular Modeling of Corrosion Processes: Scientific Development and Engineering Applications
Edited by Christopher D. Taylor and Philippe Marcus

Printed and bound by CPI Group (UK) Ltd, Croydon, CR0 4YY

16/04/2025

14658535-0001